DIVERSITY
AND UNITY

DIVERSITY AND UNITY

Relations Between Racial and Ethnic Groups

Martin Patchen

Purdue University

Nelson-Hall Publishers / Chicago

Project Editor: Steven M. Long
Typesetter: Precision Typographers
Manufacturer: Capital City Press
Cover Painting: *Ecos del Alma* by Oscar Luis Martinez

Library of Congress Cataloging-in-Publication Data

Patchen, Martin.
   Diversity and unity : relations between racial and ethnic groups
/ Martin Patchen.
     p.   cm.
   Includes bibliographical references (p.   ) and index.
   ISBN 0-8304-1441-X
   1. United States—Ethnic relations.   2. Ethnic relations.
   3. United States—Social conditions—1980–   I. Title.
E184.A1P344 1998   97-36611
305.8′00973—DC21   CIP

Manufactured in the United States of America

10 9 8 7 6 5 4 3 2 1

The paper used in this book meets the
minimum requirements of American
National Standard for Information
Sciences—Permanence of Paper for
Printed Library Materials, ANSI
Z39.48-1984.

to Nancy, with love

Contents

ix

Preface

Relationships among racial and ethnic groups—their attitudes and behaviors toward each other—is a subject that arouses strong passions. Those who belong to a particular group often harbor grievances against those in another group who, in turn, may feel mistreated by those in the first group. Moreover, there is often some justification for such feelings.

Understandably, many people want to take action that will bring greater equality among ethnic groups and reduce conflict among them. But before taking effective action, people need to understand the reasons for existing problems. Often, the explanations that are given are simplistic. For example, hostility against some minority groups is said by many to be the result of prejudice and of racism. While such statements may be true, they provide only a surface explanation. Why do particular ethnic groups display prejudice or racist behavior against particular other groups at specific times? Why is prejudice and racism less apparent in other intergroup relations?

Most texts on racial and ethnic relations have presented discussions on a series of particular ethnic groups—African Americans, Hispanic Americans, Chinese Americans, Native Americans, Jewish Americans, and so on. Such books are useful for understanding the position of each group in the larger society and its relations with other ethnic groups, especially with the dominant majority, but they contain only relatively brief discussions of the social and psychological factors that are common to (or that differ among) the many specific cases of intergroup relations. Thus, they are of limited usefulness for understanding intergroup relations as a general process.

A few texts have departed from the usual practice of focusing on a series of particular ethnic groups and have taken a more general and analytic approach. This text, while it gives considerable attention to specific racial and ethnic groups, is one of the relatively few that focuses on ethnic relations as a general phenomenon. It also aims to go beyond the treatments of other general texts in several ways.

Whereas the focus of many books is either on individuals' attitudes or on social structure as crucial determinants of ethnic relations, *Diversity and Unity* gives extensive attention both to the individual and to the larger society. I show how broad societal institutions and processes affect the types of interpersonal contacts that occur and the attitudes and behaviors of individuals. In tracing these linkages, I present a comprehensive model of relations between ethnic groups that spans the societal, interpersonal, and individual "levels of analysis." Relationships between specific groups, (e.g., between blacks and whites) are discussed in the context of more general explanations of intergroup relations and how they may apply to particular cases. When policy issues—such as the use of affirmative action programs—are discussed, it also is in the context of the general analyses of ethnic group relations that is presented in the book.

The broad policy issues that are considered are ones that are important for all Americans today and for people in many other nations as well. As ethnic diversity increases, we face the central question of whether different groups will assimilate into a single society or whether they will remain separate. Should ethnic diversity as a permanent feature of society be encouraged? To the extent that various ethnic groups do retain separate identities, what, if anything, will provide some measure of national unity? I do not attempt to answer the question of what kind of society is most desirable because that is a judgment that depends, at least in part, on personal values. I do try to clarify the processes that lead either to assimilation or to ethnic pluralism. I also try to show the problems of attaining assimilation, the problems that arise in a pluralistic society, and possible ways of dealing with each set of problems. Thus, I hope this book will help readers to move beyond simplistic slogans ("end racism," "celebrate diversity," and so on) to a more realistic consideration of the complex and sometimes difficult actions that may be necessary to make relations between ethnic groups more harmonious.

In writing this book, I have benefited from the comments of a number of people on drafts of one or more chapters. Those who have provided such help include Steven Cornell, Richard Heslin, Steve Hillis, Roberta Rehner Iversen, Liza Patchen-Short, Nancy Patchen, Robert Perrucci, and Thomas Pettigrew. I also benefited from the comments of Jennifer Eichstedt and of Benjamin Ringer who reviewed my draft manuscript for the publisher; the detailed comments of Dr. Eichstedt were especially useful. Much of the typing of the manuscript was done by Sherry Leuck. Additional typing on an earlier draft was done by Barbara Puetz and by Tara Booth. Sarah Elko also helped in preparing the manuscript, especially the references. I appreciate the help that all these persons have given to me.

Martin Patchen
West Lafayette, Indiana
July 1998

1) Problems In Ethnic Group Relations

Ethnic Conflict and Harmony

Relations among ethnic groups have assumed a central place in today's world. Distinctions based on other characteristics—such as occupation, wealth, and political affiliations—continue to be important. But in many parts of the world such other distinctions often are overshadowed by divisions of ethnicity—that is, of group identity based on the real or supposed sharing of ancestry. (By this definition, a racial group is one type of ethnic group.) Along with differences in ancestry, ethnic groups usually also differ in other characteristics, such as physical traits, language, religion, culture, and shared history.

Newspaper headlines and television screens bring to our attention many of the places around the world where hatreds and conflicting claims by different ethnic groups have led to overt conflict or to actual civil war. In former Yugoslavia, Serbs battled Croats for control of territory and killed, raped, and displaced tens of thousands of Muslims. In Sri Lanka, Tamils have carried on a guerrilla war against the dominant Sinhalese population. In Spain, Basque separatists have conducted terrorist operations to try to win an independent state, while militant Catholics and Protestants in Northern Ireland have waged campaigns of terror to win political goals. The list of ethnic groups in conflict can be easily extended: Chinese and Malays in Malaysia; Jews and Arabs in Israel; Hutus and Tutsis in Burundi: Armenians and Azerbajanis in the former Soviet Union; Kurds and Arabs in Iraq; and so on (Boucher et al. 1987; Gurr 1993).

1

In many cases, conflict between ethnic groups has resulted in, or threatens to result in, the division of formerly unified states into several separate states (e.g., in Yugoslavia, Nigeria, and Ethiopia). In other cases, ethnic conflict has led to the expulsion of a weaker ethnic group by its stronger neighbor (e.g., the expulsion of Asians from Uganda or the Muslims from parts of Bosnia) or to the slaughter of large numbers of one ethnic group by another (e.g., the mass killings of Armenians by the Turks, of Jews by Germans, and of Ibos by Hausa and Yoruba in Nigeria) (Horowitz 1985; Melson 1992).

While examples of ethnic conflict around the world abound, relations between ethnic groups are not always ones of conflict and division. In some countries, contact among different ethnic groups has resulted in cultural assimilation of a minority group into a larger population or in the physical amalgamation of two or more ethnic groups into one people (Shibutani and Kwan 1965). This happened in some western European countries, such as in the United Kingdom where Celts, Angles, Saxons, Normans, and Scandinavians amalgamated into a single English people. In Mexico, despite attempts by Spanish conquerors to maintain separation from Indians, cultural assimilation and physical interbreeding have produced a largely homogeneous mixed-race population.

Even where ethnic groups have remained somewhat separate in ancestry and culture, there are cases in which such pluralistic societies have been fairly harmonious—at least for long periods of time. Jews and Muslims lived in general harmony for many centuries in Spain before the fifteenth century. Christians and Muslims lived side by side in peace in Lebanon for many decades in the twentieth century. People of German, French, and Italian ancestry live in general harmony—albeit largely in separate areas—in Switzerland. However, even where there is relative harmony in pluralistic societies, serious conflicts may eventually erupt or the avoidance of conflict may depend on geographic separation (Abramson 1980; Kellas 1991).

Ethnic Relations in the United States

The United States is a nation which has been formed by a wide variety of ethnic groups from all parts of the world—the United Kingdom, Eastern and Western Europe, Scandinavia, Africa, Asia, Central and South America, the Caribbean, and many others. Immigration, which declined after the passage of restrictive legislation in 1924, has risen sharply in recent decades following the enactment of a less restrictive law in 1965. An influx of nearly five million immigrants between 1985 and 1990 brought the number of foreign-born people living in the United States to 19.8 million, 7.9 percent of the American population (up from 4.7 percent in 1970). Concentrations of foreign-born population are especially heavy in California and New York. Whereas earlier immigration to the United States was mainly from Europe, recent immigration has been largely from Mexico and other Latin American countries and from Asia (e.g., the Philippines, Korea, Vietnam, and China), thus further increasing the racial and cultural diversity of the American population (Bean 1989; Jones 1992).

Is ethnic diversity in the United States a problem? Does it threaten our unity as a nation? Should public policy aim at assimilating individuals of many diverse origins into one culturally, socially, and perhaps physically homogeneous people? Or is diversity a source of strength, something that should be celebrated and even encouraged?

Historically, the United States has not been immune from ethnic conflicts (Takaki 1993). The great issue, of course, has been the status of blacks. The struggle over slavery led to a great civil war. Emancipation was followed by economic, social, and political discrimination against blacks, segregation, race riots, and other manifestations of tension and conflict. In addition to the central problem of black-white relations, many other tensions and conflicts between ethnic groups in the United States have occurred. During the nineteenth century, a wave of anti-Chinese feeling on the West Coast led to Chinese being harassed, beaten, and sometimes killed. Irish immigrants in New England were denied jobs ("No Irish Need Apply," said many signs). Hostility to Jews led to discrimination against Jews in school admissions, jobs, and hotels.

"Nativist" groups in the nineteenth and early twentieth century campaigned against immigration and launched violent attacks on immigrants and Catholics. In more recent years, examples of ethnic conflicts include violence between blacks and Korean merchants in several cities, between white and Vietnamese fishermen, and a major racially-based riot in Los Angeles in 1992 (Marger 1994).

Dealing With Diversity

How should the United States deal with the great diversity of ethnic groups within its borders? The most common answer during the nineteenth and early twentieth centuries was a policy of assimilation—at least for whites. Faced with a continuing, sometimes massive inflow of immigrants, many prominent Americans saw a necessity to create a "melting pot" that would turn the newcomers into homogeneous Americans sharing a common language, common values, and common culture (e.g., see James 1907; Roosevelt 1923–26). In fact, cultural assimilation and social assimilation, including intermarriage, has proceeded rapidly among white Americans of diverse national and even religious origins (Alba and Golden 1986; Alba 1990).

A considerable degree of assimilation, including much intermarriage, has also occurred for some non-white groups, such as Mexican Americans and Japanese Americans (Levine and Rhodes 1981; Valdez 1983; Alba 1995). Some have argued that full assimilation is possible too for black Americans, at least in the long run. The frightening picture of warring ethnic groups in other parts of the world—such as the former Soviet Union, Sri Lanka, and Iraq—convinces some Americans today that diversity is dangerous and should be reduced through gradual assimilation.

But a policy of assimilation has its critics and diversity has its promoters. Those who have a particular ethnic heritage—be it African, Mexican, Chinese, Ukrainian, or other—often believe that there is great value in preserving their culture, their language, their group solidarity. They see demands for assimilation as an effort by a dominant Anglo-Saxon elite to force them to accept an alien culture and to lose their own identity and ethnic pride. Furthermore, especially among many African Americans, the goal of assimilation is viewed as an impossible one—at least in any foreseeable time span.

Diversity is seen by many as desirable, as a source of strength for the country. This view has roots in an earlier era. Images of a mosaic, or of an orchestra composed of diverse instruments playing in harmony, were opposed to the image of the melting pot (Kallen 1924).

More recently, the concept of "multiculturalism" has been advanced. In the view of its proponents, it is desirable to perpetuate and even strengthen diverse cultures within our society (*Culture Wars* 1994). They believe, for example, that use of a variety of languages should be encouraged in schools and other institutions (Sleeter and Grant 1988).

But some of the ideas of the advocates of multiculturalism have been criticized by others as having the potential for creating disunity within the United States (e.g., see Schlesinger 1992; Lind 1995). Those concerned with maintaining a single American culture have promoted state and local laws that make English the only official language in some areas and have opposed extensive bilingual programs in the schools. Advocates of multiculturalism sometimes have responded by labeling their critics as chauvinists or bigots.

The problems of ethnic relations in the United States and around the world raise important questions of fact and of policy. How can conflict between ethnic groups be reduced and harmony and cooperation be increased? If assimilation of separate groups into one organic whole is desired, how can this be accomplished? If we wish instead to maintain and even encourage diversity and pluralism, can this be done in a way that will preserve unity and promote equality?

These are the major questions to which this book is addressed. It will review the work of scholars concerned with ethnic group relations to see what is known on this subject and what answers to the central questions of ethnic relations are suggested by this body of work.

Research on Ethnic Group Relations

The subject of ethnic group relations has received considerable attention from scholars from many disciplines, including sociology, anthropology, political science, psychology, and history. Studies have been done from a variety of perspectives. These approaches may be grouped for purposes of this discussion into four types: 1) studies of total societies; 2) studies of smaller social units, especially of residential segregation in cities; 3) studies of interpersonal contacts between persons from different ethnic groups: and 4) studies of individual ethnic identifications and attitudes toward other ethnic groups (see Blalock and Wilkin 1979).

Study of Societies

Some scholars have presented broad descriptions of societies that contain two or more major ethnic groups (e.g., Weiner 1978; Schmid 1981). They have described such aspects of each society as the type of initial contact between ethnic groups (e.g., conquest or migration); the general nature of relations between the groups (e.g., paternalistic or competitive); the social class distribution of ethnic groups; the extent and nature of physical separation between ethnic groups (e.g., residential patterns or enforced segregation); and the policy aims of dominant and subordinate groups (e.g., pluralism or assimilation). Many scholars concerned with race relations in total societies have compared patterns of such relations in different societies and some have constructed typologies of ethnic relations (e.g., see Schermerhorn 1978; van den Berghe 1978). Most such studies have been conducted in countries of Africa, Southeast Asia, Latin America, and the Caribbean and less often in more developed countries.

Studies of this type—that is, of patterns of ethnic relations in total societies—have several important strengths. They provide an important context for understanding ethnic relations in any society by considering such relations against a background of the society's historical development and other aspects of the society (e.g., its economic and political institutions). This literature also encourages us to view ethnic relations in any given society with a comparative perspective—seeing how they are similar to and different from patterns in other societies.

Largely descriptive studies of total societies also have limitations. They usually do not contain detailed quantitative information about specific topics (e.g., about residential segregation) or about patterns of interaction between members of different ethnic groups. Nor do they provide systematic information about the identifications and attitudes of individuals.

5

Study of Cities

This body of research (e.g., see Tobin 1987; Massey and Denton 1993), conducted mostly by sociologists, has focused primarily on patterns of segregation in cities and metropolitan areas. (Other characteristics of cities, such as income differences among ethnic groups and crime rates, also have received some attention.)

Some early studies in this tradition, such as studies of Jewish ghettos in Europe and America and early studies of black-white segregation, were largely descriptive. But studies of segregation in recent decades—most of it done in the United States—have relied heavily on quantitative census data and sophisticated statistical techniques. Scholars working in this area have described differences in the extent of residential segregation between ethnic groups in cities (especially of blacks from whites, but also Hispanics and to a lesser extent Asians and European-descended groups from other ethnic groups). They have tried to explain differences in levels of segregation among cities by relating segregation to other city characteristics, such as the proportion of the population represented by various ethnic groups and the relative socio-economic status of different ethnic groups.

Such studies of cities give us considerable reliable information about the local circumstances under which people of different ethnic groups are more or less physically separated from each other. However, the literature in this tradition usually does not provide an extensive context of conditions in the larger society (e.g., of political and economic events and trends in the United States as a whole). Also, it usually does not provide information about interpersonal relations and about individual attitudes that might affect city characteristics such as segregation.

Study of Interpersonal Relations

A third line of research has examined systematically the interactions between members of different ethnic groups (for reviews, see Stephan 1987; Jackson 1993). Most of this type of work has been done by sociologists and by social psychologists who have used standard instruments (such as surveys and measures of friendship choices) to examine patterns of interaction in specific settings. How friendly or unfriendly are interactions between those in different ethnic groups? How often do they initiate interaction or, conversely, try to avoid contact?

One important question addressed by research on interpersonal relations is the following: what are the conditions under which contact between members of different ethnic groups leads to positive, or to negative, attitudes and behavior? Among the conditions that have been investigated are: the presence of common or conflicting goals; equality or inequality of status; similarity or dissimilarity of values; social norms supporting or opposing friendly interaction; and the numerical proportions of the groups.

Systematic studies of interaction between members of different ethnic groups (especially racial groups) have been conducted especially in schools, with some research also in interracial housing, work settings, and other situations. This body of research has provided us with much useful information about the types of interactions that occur across ethnic group lines and especially about the circumstances in the immediate situation which may affect the nature of interaction. This type of research does not give great attention to the larger societal factors outside the immediate interaction setting—that is, to the factors that shape the circumstances under which members of different groups come into contact.

Study of Individual Attitudes

While somewhat overlapping the work on interpersonal interactions, much research has focused on the subjective reactions of individuals to ethnic groups (e.g., Bar-Tal et al. 1989; Mackie and Hamilton 1993). These include ethnic identifications and loyalties, perceptions of other ethnic groups (including stereotypes), emotions (such as anger and fear), and other pertinent perceptions and values. This line of research has been carried out primarily by psychologists, although others (including some sociologists) also have done relevant work.

One of the main goals of work in this area has been to better understand the nature and structure of ethnic attitudes. For example, some researchers have investigated inconsistencies in the beliefs, feelings, and behavioral dispositions that whites have towards blacks. Another line of research has focused on the processes by which people categorize others according to particular characteristics (ethnic and other).

In addition to understanding the structure of ethnic attitudes, many researchers have tried to understand variations in such attitudes. Ethnic attitudes have been related to personality characteristics, to social characteristics (e.g., occupation and gender), to real or perceived conflicts of interest with other ethnic groups, and to personal and political values (e.g., see Rokeach 1960; Campbell 1971; Katz and Hass 1988).

This line of research has provided us with greater insights into people's ethnic identifications, and into the perceptions and feelings they have about other ethnic groups. It also has shed light on the reasons some people have relatively friendly and accepting attitudes toward members of other ethnic groups, while other people have negative beliefs and feelings toward those who are different from themselves. At the same time, the subjective reactions of individuals are only a part of the larger story of ethnic relations. The perceptions and feelings of individuals do not exist in isolation. Their attitudes toward other ethnic groups are shaped by the immediate social situations in which they are placed and ultimately by the larger society of which they are part.

Goals of this Book

The research summarized briefly in the preceding section provides important perspectives on ethnic group relations at several levels of analysis—the societal, the interpersonal, and the individual. One of the main goals of this book is to review and to try to integrate the work on ethnic group relations at each of these levels of analysis. We will consider how conflict or harmony between groups is affected by individual attitudes, by interpersonal contact between members of different groups, and by the overall structure of society (e.g., its economic and political institutions).

But to consider separately the body of work on ethnic relations at each level of analysis is not enough. An adequate understanding of ethnic group relations requires that factors at each level be related to each other. To fully understand the attitudes and behaviors of individuals, we must understand the types of interpersonal contact they have with those from other ethnic groups—in schools, at work, in neighborhoods, and so on. Furthermore, both individual attitudes and the amount and conditions of interpersonal contact are shaped by the societal context in which they occur.

Yet, work at the societal, or "macro," level usually has been done in virtual isolation from work at the individual, or "micro," level. Thus, Hubert Blalock and Paul Wilken comment, "There appear to have been very few efforts to link these macro-level and rather general discussions of pluralism with either the social psychological literature on prejudice or with the ecological literature on residential segregation (1979: 519).

In a similar vein, Thomas Pettigrew states during a review of theory and research on black-white relations:

> The contradictions between genuine progress in American race relations and continued racism, and even retrogression, operate at both the macro- and microlevels. Indeed, proposed structural explanations parallel proposed social psychological explanations. But attempts to combine levels in the same theoretical or empirical effort are rare. The necessary new conceptualization, however, must explain the diverse patterns of modern race relations at both levels. . . . The review's principal plea is an urgent call for integrative work across "theories" and across the micro- and macrolevels. (1985: 332, 344)

The need for more integrative work linking work at different levels of analysis still exists. Therefore, another goal of this book is to build more bridges to con-

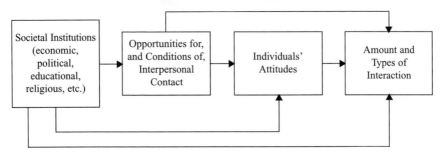

FIGURE 1.1

Overview of Determinants of Relations Between Ethnic Groups

nect the largely separate bodies of work on ethnic relations. We will outline an approach to understanding ethnic group relations that is inclusive of factors at the societal, the interpersonal, and the individual psychological level.

Figure 1.1 presents a broad overview of this perspective. It shows that the extent to which members of different ethnic groups interact with each other and the nature of these interactions (e.g., how friendly or hostile they are) are affected most directly by individuals' ethnic attitudes. These attitudes depend, in turn, on how much opportunity for contact there is between members of the different groups (in school, at work, in neighborhoods, and so on) and especially on the conditions under which such interpersonal contact occurs (such as the relative status and power of the groups and the extent to which they have common or competing goals). For example, greater opportunity for contact may tend to dispel negative stereotypes about members of another group but, if the groups are competing for the same rewards (such as jobs), then more opportunity for contact may result in more negative attitudes. In addition to affecting interaction indirectly, through their impact on individual attitudes, the opportunity for and conditions of contact may influence interaction between the groups directly. For example, if "Anglos" (non-Hispanic whites) occupy positions of authority in a work organization while Hispanics are subordinates, then the Anglos will tend to be dominant and the Hispanics submissive in their interactions.

Finally, figure 1.1 shows that all of the factors that affect interaction between ethnic groups may be influenced, directly or indirectly, by societal institutions (economic, political, educational, religious, and other). These institutions help to determine the opportunities and conditions of contact between members of different ethnic groups. For example, if the educational system assigns students to particular schools with an eye to achieving a mixture of ethnic groups, then the opportunity for students of different groups to have contact with each other in school is increased. Societal institutions—such as schools, churches and the media—may also influence individuals' ethnic attitudes directly, by providing information and by transmitting norms about "appropriate" attitudes. In addition, societal institutions may have a direct impact on the amount and types of interaction that occurs between those from different ethnic groups. For example, the work of a business firm may require that blacks and whites communicate often in order to coordinate their work.

The overview of connections between individual, interpersonal, and societal variables shown in figure 1.1 should help to give you a general idea of the perspective taken in this book. These ideas will be elaborated in much greater detail throughout much of the book. An integrated theoretical overview which links variables at different "levels" is presented especially in Chapter 8.

Such an integrated approach should be helpful not only for understanding ethnic group relations more fully but also for assessing relevant social policies. The latter part of the book considers especially two alternative policies for dealing with ethnic diversity. The first policy aims at assimilating various groups into a single society. The second policy aims at preserving and encouraging ethnic pluralism. Prospects for, and the problems of, each policy are discussed in the context of our overall analysis of ethnic group relations.

The main focus of this book is on relations between racial and ethnic groups in the United States. However, to understand ethnic group relations in any one country, it is useful to view relationships there in the context of the broader range of ethnic relations around the world. Cross-national similarities and differences highlighted by comparative studies (e.g., Horowitz 1985; Gurr 1993) can help to illuminate interactions between particular ethnic groups in any one country. Thus, while primary attention will be given to events in the United States, we will try to make use of generalizations and insights derived from studies of other societies. Moreover, most of the ideas about ethnic group relations discussed in this book are sufficiently general that they are relevant to these types of relations in any country.

Terminology

At this point to indicate how the terms "ethnic group" and "race" are used in this book and to note some of the terms used for specific groups would be useful. While the term "ethnic group" has been defined in a number of different ways by different scholars, the sharing of common ancestry is essential to its meaning (Burkey 1978:5). Thus, "ethnic group" will be used here to refer to any categorization of people based on real or supposed common ancestry.

In this usage, a racial group (such as blacks or whites) is a particular type of ethnic group based on real or perceived common racial ancestry. Another type of ethnic group is the "national ethnic" group based on common national ancestry, such as Polish Americans or Greek Americans. Sometimes the term "racial/ethnic groups" is used to refer to a variety of racial and national ethnic groups.

"Race" is a term that has no clear scientific meaning. Physical characteristics used to define race are arbitrary and interbreeding among populations has further reduced the scientific basis for clearly defining separate racial groups (Shanklin 1994). However, there are some physical differences (skin color, shape of nose, hair type, and so on) that are often seen by members of a given society as a basis for classifying people by "race." Thus, despite its scientific ambiguity, race has an important social reality. The term "race" is used in this book to refer to those social classifications commonly recognized as racial (e.g., black, white, Indian), both by those included in those categories and by others.

Specific ethnic groups in the United States often are referred to by more than one term. Where such terms seem equally appropriate, we often will alternate between them—sometimes for variety, sometimes because one term may fit a given context of discussion better than another.

"Blacks" and "African Americans" are both widely-used terms, by those in this racial group and by other Americans. We use these terms interchangeably, though African American is more likely to be used here when the context is the ties of various American ethnic groups to other parts of the world.

The terms "Native American" and "Indian" are both used in this book, usually without any intended difference in meaning. However, "Native American" conveys more the historical priority of this group in America. Indian is used here also because of its wide usage in American society, both within and outside the in-group, and to provide some variety in terminology.

"Hispanic" and "Latino" are both used in this book to refer primarily to people whose origins are in the Western Hemisphere south of the United States. "Hispanic" is the term most often used by the U.S. Census and often will be used here when citing statistical comparisons among groups. "Latino" is more the "in-group" term used by those with origins south of the U.S. border, especially by those in the western United States. Neither "Hispanic" nor "Latino" is a racial category, since the category includes persons who are white, black, Indian, or some mixture of these races.

The term "white" usually is used in this book to refer to those whose origins are in Europe or the Mediterranean region. However, we sometimes use "European American" to emphasize the geographical origins of most white Americans. Also, we use the term "Anglo" at times to refer to non-Hispanic whites. Finally, the term "Asian Americans" is used to refer to persons whose origins are in Asia or the Pacific.

Note that none of the broad ethnic groups referred to in this section (whites, African Americans, Hispanics, Asian Americans, and Native Americans) are homogeneous. Each is composed of subgroups that share much in common but that also differ in many ways. Asian Americans include people originating from

11

many different nations (China, Korea, the Philippines, India, and so on), and having different physical and cultural characteristics. "Hispanics" also include people deriving from a wide variety of countries (Mexico, Puerto Rico, Cuba, and so on), with different cultures, physical traits, and identities. "Native Americans" include people from a large number of different tribes, which had different languages and different ways of life. White Americans derive from many different European and Mediterranean countries (such as Italy, Poland, Germany, and Ireland) and African Americans include some recent immigrants from the Caribbean and from different parts of Africa as well as a much larger number whose black ancestors came from Africa centuries ago.

While keeping in mind these differences within each ethnic group is important (and we sometimes will note such differences), speaking in terms of the broader ethnic groups is useful for two main reasons. One is that today many American institutions (such as government agencies, universities, business firms) categorize and sometimes treat people in terms of these broad ethnic categories. The second reason is that, given current conditions in America, people who once thought of themselves solely in terms of a relatively narrow ethnic group (e.g., as Salvadorans, Koreans, or Navahos) now tend to identify themselves also as part of some broader grouping (e.g., Hispanics, Asian Americans, or Native Americans).

Plan of Book

The plan of the rest of the book is as follows:

Chapter 2 addresses the subject of the individual's psychological attachment to a particular ethnic group to which he or she belongs. Under what conditions does a person identify strongly with an ethnic group? Is strong identification with a particular group inconsistent with attachment to the broader society? Does it result in hostility toward other ethnic groups?

Chapter 3 considers people's attitudes toward other ethnic groups. How can "prejudice" be explained? What types of perceptions of and feelings toward members of other groups cause a person to seek them out or to avoid them? To try to help them or to harm them?

Chapter 4 discusses the effects of contact between members of different ethnic groups on their attitudes and behavior. What are the conditions under which interpersonal contact produces positive changes in images of, feelings toward, and behaviors toward another ethnic group? Under what conditions does interpersonal contact across ethnic group lines produce no positive effects or even greater hostility?

Chapters 5, 6, and 7 focus on aspects of the broader society. How are the relations between ethnic groups affected by demographic factors, such as the relative size of groups? By economic factors, including inequality and competition? By the political and legal institutions of society? By other institutions, including religion and the media? In these chapters, the position of major ethnic groups in American society with respect to such outcomes as occupation, income, and education is also considered. Reasons for differences among groups in socioeconomic status, educational attainment, and other outcomes are discussed.

Chapter 8 ties together the discussions in the previous chapters with a comprehensive model of ethnic group relations. Links are drawn between characteristics of the larger society; the frequency of and conditions under which members of different ethnic groups come into contact (at work, school, neighborhood, and so on); perceptions of and feelings toward members of another group; and finally, behaviors toward those from the other group. Four different patterns of ethnic group relations are described and examples of each type of relationship are discussed.

Chapters 9 and 10 discuss some important policy issues in the context of the evidence and analyses presented in previous chapters. Alternative approaches to dealing with a multiethnic society are considered. Chapter 9 examines the possibilities and problems of trying to assimilate various ethnic groups into a single society. Chapter 10 focuses on the experiences and problems of encouraging ethnic pluralism. In the course of discussing these broad societal directions, a variety of specific policy issues, such as economic inequality, housing segregation, and affirmative action programs are considered. The book concludes by briefly discussing some policy approaches that attempt to find a middle ground between ethnic assimilation and ethnic pluralism.

This book cannot provide complete answers on policy issues, and certainly not answers that satisfy everyone. But it does aim to link consideration of such issues more closely to an understanding of the determinants of ethnic relations. By understanding better the roots of both ethnic conflict and ethnic harmony, the effects that any particular policies will have on ethnic relations can be judged better.

Summary

Relations among ethnic groups are important in countries around the world. Conflict between ethnic groups often occurs, but examples of harmony also can be found. Research on ethnic group relations has examined how conflict and cooperation are affected by individual attitudes; by contact between those from different groups; and by the institutions of the larger society. However, research on the "microlevel" of individual attitudes and behavior has been rarely connected systematically to "macroresearch" concerning the effect of societal institutions on ethnic relations. This book will summarize and attempt to integrate much of the research at each level of analysis as well as link research at the individual, interpersonal, and societal levels.

The book is also concerned with some important policy issues. The book will ask what implications the evidence and theoretical analysis presented have for reducing conflict and increasing cooperation between ethnic groups. In addition, it will explore their implications for alternative ways of dealing with ethnic diversity within society.

2) "We" and "They": Group Boundaries and Loyalties

In any society people see certain groupings as important. A variety of differences among people—for example, in age, gender, income, occupation, geographical location, or ethnicity—may be salient. People tend to see those in their own group or category as "we" or insiders, and others as "they" or outsiders. They tend to identify with and feel loyalty to their own group and often treat those in other groups less favorably (LeVine and Campbell 1972; Hogg and Abrams 1988; Rose 1990).

When does ethnicity, rather than other characteristics, become an important way in which people are categorized? What determines the strengths of attachment that people feel towards their own ethnic group? Does strong attachment and loyalty to one's ethnic group lead to hostility towards other ethnic groups and to a weak attachment to the larger society? These are some of the major issues that will be discussed in this chapter.

Drawing Boundaries Between Groups

When people draw sharp lines between themselves (their in-group) and others, hostility and conflict between "insiders" and "outsiders" become more probable. Weakening or eliminating the boundary lines that people see between their own group and some other group may be a key to reducing intergroup conflict (Hogg and Abrams 1988).

On what basis do people draw lines between themselves and others? Who do they see as being members of their in-group ("we") and who do they see as belonging to a different group ("they"). Psychologists interested in perceptions have studied the factors that lead people to see individual elements as part of a unified whole (Wertheimer 1950). The general principles that they discovered have been applied to perceptions of social groups (Zander, Stotland, and Wolfe 1960; LeVine and Campbell 1972). LeVine and Campbell suggest that we are more apt to see individuals as part of a group the more that these individuals:

1. are similar to each other.
2. share a common fate.
3. are in close proximity to one another.
4. form a bounded figure in space—that is, the area they occupy has a clear spatial boundary with other areas.
5. have a high level of communication and interaction among themselves but a low level of communication and interaction with others (1972).

How do these principles help us to understand the ways in which lines between ethnic groups are drawn?

Similarities and Differences

Among the important bases on which people categorize others are the perceived similarities and differences among them. Similarities and differences are found, of course, with respect to many characteristics—age, gender, occupation, and so on. Two kinds of characteristics—physical differences and cultural differences—have been used widely to categorize people into ethnic groups (Shibutani and Kwan 1965; Marger 1994). Physical differences such as those in skin color, hair type, and facial features have often been used as the main criteria for classifying people.

However, the number and nature of physical categories that are socially recognized is arbitrary and varies among different societies. Thus, while Americans generally divide people into "black" and "white," South Africans recognize three major racial categories (black, white, and a racially mixed or "coloured" group) and Brazilians distinguish over a dozen mixtures of black and white ancestry (Banton 1967). In some societies, cultural similarities and differences are more important than physical characteristics as the basis for distinguishing ethnic groups. For example, in Mexico many people believe that "an Indian is someone who lives like an Indian"; in the African city of Timbuktu, the various ethnic groups have been distinguished by differences in their clothes and haircuts (Shibutani and Kwan 1965). Cultural elements that often are used by people to categorize others include language, religion, clothing, foods, and names (Skardel

1974). Both similarities in physical characteristics and in culture generally are seen as stemming from common ancestry and common history.

Range of Variation

What is seen as similar and what is seen as different depends on the range of variation among the objects or people being observed. If a person is shown a set of cards of different shades of blue, she is likely to distinguish among these colors. But if she is shown a set of cards with colors across the entire rainbow and asked to divide them into groups, then the cards of different shades of blue will be grouped together and distinguished from, say, those that are red and yellow (Sherif and Hovland 1961).

Similarly, the way in which people divide themselves and others into ethnic groups depends in part on the range in variation among those with whom they have contact. Horowitz comments: "What looks like a major characterological or behavioral deviation in a parochial environment with a restricted range of difference may begin to look trivial when the range is expanded" (1975: 122–123). For example, people from different parts of China who migrated to countries such as the Philippines and Indonesia or to the United States, found that regional differences in ethnicity, language, and customs that had loomed large back in China were less important than the similarities they shared in a new setting that had a much wider ethnic variation (Garth 1974). The same was true for people who emigrated from different sections of Italy to the United States in the nineteenth century. Differences of dialect and customs that had divided them in Italy seemed less important than their similarities within a much broader range of ethnic variation in America. In more recent times, white Americans deriving from many different European countries are beginning to see their differences in origin as relatively minor in the context of an America that is increasing in racial diversity (Alba 1990). The wider the range of variation in ethnic group characteristics, the broader are the perceived boundaries of each group.

Accentuation of Differences

People do not simply notice similarities and differences and group people accordingly. Once they categorize people according to some criterion, they tend to accentuate the similarities among those in the same category and the differences among those in different categories. They are especially likely to overestimate similarities within, and differences among, people on characteristics that they believe are associated with category membership (Hogg and Abrams 1988). For example, if people are categorized as Hispanic and Anglo, and if Anglos are believed to do better in school than Hispanics, then all Hispanic students will tend to be seen as not capable academically. This tendency to see more similarities among those in each category and to see greater differences among those in different categories than actually exist is one of the important bases of ethnic stereotyping.

Emphasis on Ethnicity versus Other Criteria

There are a variety of possible criteria on which people may be seen as similar or different—education, occupation, political affiliations, physical traits, and so on. When do people pay most attention to similarities related to ethnicity, to the physical and cultural differences among people?

First, the type of similarities and differences between people that become most salient tend to be those that most affect their interactions. When cultural differences affect the transactions between people, distinctions based on such differences become important (Shibutani and Kwan 1965). For example, if those of different ancestry speak different languages or have different norms about bargaining, these differences may make commercial transactions between them more difficult. If this distinction is true, business owners, merchants, and customers may tend to categorize others in terms of their similarities or differences with respect to ancestry and culture.

Given that people are concerned about particular behaviors of others, such as their tendencies to bargain about price when concluding transactions, some general cognitive tendencies discussed by psychologists become relevant. They assert that the social categorization which will be most salient to people is that which best fits the information available to people in the sense that it accounts best for similarities within categories and differences between categories (Turner et al. 1987; Hogg and Abrams 1988). For example, suppose that traders were interested in accounting for peoples' tendencies to bargain hard and thought that similarities and differences in such dispositions could be predicted better from their ethnicity than from other characteristics, such as income. They would then tend to categorize people by ethnicity (Hogg and Abrams 1988). If age, gender, or religion was thought to be a better predictor of behavior in business transactions, this latter characteristic would be the one on which people were most often categorized.

Sometimes, particular individuals within a society try to persuade people that ethnicity is important. They may stress similarities that exist between members of their own ethnic group and differences, real or imagined, between members of their group and other ethnic groups. Leaders or elite members of a particular ethnic group may do this because they have a vested interest in seeing a sharp division and even conflict between ethnic groups. Such divisions may protect and enhance their leadership positions and may lead to their gaining advantages (such as land or positions) as a result of the interethnic competition. Calling attention to this tendency among elites around the world, Paul Brass writes:

"In the movement to create greater internal cohesion and to press more effectively ethnic demands against rival groups, ethnic and nationalist elites increasingly stress the variety of ways in which members of the group are similar to each other and collectively different from others" (1991: 21).

Sometimes there is an effort among some ethnic group members to create more differences from other ethnic groups than have existed previously. For example, the adoption of the religion of Islam, rather than Christianity, among blacks belonging to the Nation of Islam in the United States may be motivated in part by a wish to become more separate from white Americans.

Common Fate

What happens to people in their lives—for example, what occupations they enter, how successful they are in their careers, where they live, how they are treated by government authorities—may be affected by a variety of their characteristics. These include gender, the social class of their families, their abilities, and their ancestries. When the life chances and outcomes of a group of people are similar because of their common ancestry, they are likely to see those who share this similar fate as being an in-group and those who do not share this fate as an out-group (Kriesberg 1982; Alba 1990). For example, Jews in Europe at the time of Nazi domination knew that they all shared a common fate of oppression and thus felt very separate from their countrymen.

In addition to sharing a common fate with respect to their overall life situation, people may also share a common fate in more limited situations. Members of athletic teams share in victory or defeat. Members of a work group may share the rewards of business success (such as bonuses) or of business failure (such as loss of job). Students who cooperate on a class project may all receive a common grade. In these types of situations, individuals are likely to see those who will receive the same outcomes as being part of their in-group and those who do not as being outsiders.

Which people will be perceived as sharing a common fate with oneself will depend to a considerable extent on: 1) the treatment of various people by others, especially those with authority or power; and 2) the extent of common interests and of conflicting interests that various people have.

Treatment by Others

The way in which people define who is a member of their own group and who is an outsider is affected strongly by how others behave towards them and towards others (Shibutani and Kwan 1965; LeVine and Campbell 1972). What kinds of people are treated by the government, by employers, by

19

school officials, by rental agents, by potential friends and lovers, in a way similar to how they themselves are treated and who is treated differently?

If important others allocate rewards and penalties, are friendly or unfriendly, or act arrogantly or submissively on the basis of people's education or income, then people in that society are likely to classify themselves and others primarily on the basis of these criteria. But if a person's ancestry is an important determinant of the way he is treated by others, then ethnicity will be a major way in which people group themselves. Moreover, the way in which they define their own ethnic group and other ethnic groups will be affected strongly by the specific lines along which differences in treatment by others occur. For example, if the American government treats all Native Americans in a similar way, regardless of their tribal affiliation, then Native Americans will tend to categorize themselves as Native American rather than as members of particular tribes. LeVine and Campbell state that ". . . if an aggregate of people is treated as a unit long enough by those who control its strategic resources, it will become a unit based on common interest and common fate" (1972: 103).

In many instances, people whose previous ethnic identity had been more narrow expanded their definition of their group boundaries as a result of the behavior of others. Members of tribes living in areas of Africa that came under colonial rule redefined themselves as Nigerians or Ghanaians or members of some other nationality as a result of being treated the same by colonial rulers (Horowitz 1985). Similarly, many Americans deriving from a variety of Latin American countries are coming to define themselves as Latino Americans (Padilla 1985) and people from a number of separate Asian countries are coming to define themselves as Asian Americans in part because they are treated similarly by the government and by other Americans (Espiritu 1992). For example, the efforts of people from many disparate Asian nations to combine into a single Asian American ethnic group owes much of its impetus to policies of the federal government (such as affirmative action programs), and sometimes of state governments, that often treat them in a common way. Writing about "pan-ethnicity" among Asian Americans, Espiritu states: ". . . ethnic resurgences are strongest when political systems structure political access along ethnic lines and adopt policies that emphasize ethnic differences" (1992: 10).

During World War II, when the United States was at war with Japan, the American government treated Japanese Americans very differently from others of Asian descent (many Japanese were held in detention camps). During this period, Chinese Americans were anxious to distinguish themselves from Japanese Americans in order to avoid the hostility directed at the latter group (Lyman 1974).

Common Interests

People are likely to perceive that they share a common fate not only when they are treated similarly by others but also when they have common interests and goals. Who is defined as a member of one's in-group and who is seen as an outsider depends in part on such perceptions of common interest. Closely related to common interest is interdependence in reaching common goals. Boundary lines between groups tend to blur when their members are dependent on each other for reaching common goals (Tajfel 1981).

Members of an athletic team, for example, often have a strong sense of group unity, and of separateness from others, because they share a common goal of winning, need to cooperate towards that end, and will share in the satisfactions and rewards of victory. People may have common interests on a variety of bases—for example, with those in the same occupations, those in the same workplace, those living in the same geographical area, and those of the same ethnicity. The extent to which ethnic boundaries, rather than other social groupings, are important to people will depend in part on the extent to which they believe they have common interests with those of similar ethnicity, as compared to shared interests with those in other possible groupings (Shibutani and Kwan 1965; Kriesberg 1982). For example, when employers brought in new black workers in order to break a strike in southern Illinois in 1917, white workers perceived common interests only with other whites, and conflicting interests with blacks, thus leading to race riots (Rudwick 1964). On the other hand, when white and black workers have seen a common interest in cooperating in strikes against employers, the dividing line between workers and management has been seen as more important than that between races (Blalock 1967).

Common material interests have played an important role in the coming together of Native American tribes differing in language, culture, and region; in the growth of an Asian American identity among those with roots in widely differing Asian nations (often at war with each other in the past); and in the development of a Hispanic or Latino identity among Americans whose forebears thought of themselves only as Mexicans, Puerto Ricans, Cubans, and so on. Espiritu has related the development of Asian American "panethnicity" to the desire of Asian Americans to advance their common interests with respect to government policies and programs in such areas as education and immigration. She asserts:

> If Asian American subgroups can come together, then the united Pan-Asian group will have more members and thus more political power than would any Asian subgroup operating alone. For these reasons, politically minded Asian Americans find it expedient—and at times necessary—to aggregate Asian American subgroups when seeking to recast their history of disenfranchisement and obtain political recognition. (1992: 53–54)

The same motivation of combining in order to win social and political advantage has been at work also in the case of Latino groups, Native American groups, and even among the dominant white majority. Alba argues that Americans from a variety of European backgrounds (Irish, Polish, German, Italian, and so on) are coming increasingly to think of themselves as European Americans in large part because they perceive members of all these groups as having common interests in resisting the demands of other groups, such as African Americans. Commenting on the importance of shared interests in forming ethnic boundaries, Roosens asserts that ethnic groups may be considered "pressure groups with a noble face" (1989: 14).

Communication Patterns

Communication among members of a category is a prerequisite for their having a sense of common identity (Kriesberg 1982). Shibutani and Kwan comment: "The lines along which human beings identify depend not so much on genetic continuity as upon avenues of communication—conditions that isolate from without and facilitate intense communication within" (1965: 216). Those who communicate frequently with each other, and little with others, have shared experiences and develop common perspectives and a common culture, different from that of outsiders. For example, the Jews of nineteenth-century Poland, isolated in ghettos and small villages, communicated primarily with each other and developed a culture separate from that of their Christian neighbors (Roskies and Roskies 1975).

Communication patterns may be centered within a variety of types of social groupings. These patterns may be occupational—where lawyers, physicians, or steelworkers work with and socialize with each other. Or interactions may occur in schools, in churches, in neighborhoods, in political organizations, in sports teams, and in other settings. Such groupings in which communication is centered may or may not be divided along lines of race or ethnicity.

Patterns of residence and of occupation affect the extent to which communication tends to occur within particular ethnic groups. When residential segregation is high, neighbors, friends, and acquaintances usually will be people of the same ethnic background. Similarly, when those in a given occupation generally come from the same ethnic group, communication will tend to be between those of the same ethnic background. For example, the concentration of those of Mexican descent in Californian agricultural jobs means that communication in such settings often will be focused among members of this ethnic group.

The extent to which communication takes place within a given category of people depends also on the size and homogeneity of that category. The larger the unit, the more likely it is that members will interact only (or primarily) within their own unit (Sawyer 1967). For example, the larger the African American population of a city, the more likely that a black person will meet other black people and that her needs for recreation, entertainment, professional services, and so on, can be met while staying in the black community. Homogeneity among those within a given category also promotes within-category communication, while heterogeneity may encourage outside communication. For example, the extent to which Chinese Americans in New York City communicate to each other may be affected by whether they are born in the U.S. or are recent immigrants and by their educational and occupational levels.

The more that lines of communication between members of different ethnic groups are expanded, the less likely people are to make sharp divisions between "we" and "they." Perceptually, a person with whom one is communicating is seen as less of an outsider. Moreover, more communication tends to contribute to some of the other main determinants of ethnic groupings—perceptions of similarity and of common interests.

Proximity and Spatial Boundaries

While the lines that people draw between ethnic groups depend mainly on the factors just discussed (perceptions of similarity, common fate, and communication patterns), proximity and spatial boundaries may serve to reinforce and strengthen these boundaries. People often observe that members of a given ethnic or racial group live close to one another. Such spatial groupings may be the result of historical patterns of settlement, people's preferences for living with their "own kind," or—especially in the case of groupings by race—discrimination against and purposeful segregation of an ethnic minority (Massey and Denton 1993).

In many countries of the world, members of particular ethnic groups are concentrated in a clear-cut geographical region. For example, Ibos in Nigeria live mostly in one area while Yorubas and Hausa people are found primarily in other areas. In Canada, those of French descent are concentrated in the province of Quebec. Being from the same region may, in fact, be one of the factors that helps people to define who belongs to a particular group (Skardel 1974).

In the United States, there are no large geographical areas that are populated entirely or even primarily by members of a particular group. But there are smaller areas, such as neighborhoods in cities, which do have a distinct ethnic identity—African American areas, the Mexican American "barrio," "Greektown," "Chinatown," "Little Italy," even some suburban areas, such as Skokie, Illinois, a suburb of Chicago with a large Jewish population.

Such neighborhoods often have fairly well defined boundaries—for example, a major road, a railroad track, a park—that people recognize as setting them apart from other neighborhoods (Hunter 1974). Such spatial boundaries help to reinforce the perceptual boundaries along which people divide people of different ethnic groups.

23

Changes in Ethnic Boundaries

Many people have grown up knowing a set of fixed ethnic categories—Irish, Polish, Italian, Jewish, Puerto Rican, and so on—and looking on such categories as unchangeable features of society, perhaps established by God or nature. While ethnic groupings in a given location indeed may be the same for a long period of time, such divisions are not necessarily fixed.

First, ethnic group boundaries are not always clear. Anthropologists have found that the assumed boundaries between ethnic groups often turn out on closer inspection to be rather fuzzy. LeVine and Campbell note: "Peoples may disagree about the ethnic names themselves (that is, whether they are applicable in the region), about the inclusiveness of the groupings they designate, and about the criteria by which inclusion and exclusion are determined" (1972: 90). For example, the Lue people of northern Thailand have disagreed among themselves about which criteria—such as dialect and particular cultural traits—distinguish them from neighboring ethnic groups (Moerman 1968).

Even when fairly clear-cut perceptual boundaries between ethnic groups exist, such boundaries may change over time. Groups may become larger in one of two ways: they may unite to form a larger group that is different from its component parts (amalgamation); or one group may lose its identity by being absorbed into another group, which retains its former identity (incorporation). For example, each of these types of group expansion has occurred in India as previously separate castes have combined. Similarly, Roosens (1989) describes the process by which previously separate Native American groups in Canada have become amalgamated into a new Huron ethnic group.

Ethnic groups may become smaller when one group divides into separate parts (division). In such a case, the original group no longer exists. This has occurred for some Indian castes (Horowitz 1975). Alternatively, a new group may come into existence without the original group or groups losing their identity (proliferation). An example of the latter is the creation of a "coloured" group in South Africa through the biological mixing of blacks and whites.

In the United States as well, ethnic boundaries have been changing in recent decades, as noted earlier. There has been a movement towards amalgamation of ethnic groups along broad lines of race and region of origin. An Asian American ethnic identity has been emerging, which groups together people with ancestry from a variety of Asian countries, including China, Japan, Korea, India, and the Philippines (Espiritu 1992). Previously separate groups deriving from many different Latin American nations—including Mexico, Puerto Rico, and Cuba—are increasingly identifying and being identified as Hispanic or Latino Americans (Padilla 1985). Those from a variety of tribes are combining as Native Americans (Nagel 1982; Cornell 1988). And Americans whose ancestors came from many different European countries—Great Britain, Germany, Ireland, Italy, Poland, and so on—are increasingly seeing themselves as one ethnic group: European Americans (Alba 1990).

The extent to which previously separate national groups are amalgamating should not be exaggerated. For example, a survey conducted in 1989 and 1990 (de la Garza et al. 1992) found that most Americans of Mexican, Puerto Rican, and Cuban origin preferred to identify themselves by a term based on their specific national origin. However, among American-born persons, the proportion preferring a "pan-ethnic" identification—for example, Hispanic or Latino—was substantial (one-fifth to one-fourth) and larger than among the foreign born, suggesting a trend toward a broader Latino ethnicity in the American context.

When ethnic amalgamations occur or are in progress, people do not discard their old identities. Rather, they add an additional identity at a "higher" level—for example, Asian American as well as Korean American, European American as well as Polish American. Which of these identities is more relevant may depend on the context. In a meeting with other Asian Americans, a person's Korean identity may be more salient. In a meeting with mainly white Americans, his Asian American identity may be more salient. Despite the fact that a "lower-level" identity remains, its overall importance in each individual's life may be less than it was before the broader identity was developed as well.

Changes in perceived ethnic boundaries are affected by the same factors discussed earlier in this chapter as determinants of such boundaries. Thus, the perceived boundary between two ethnic groups is likely to weaken as members of the groups: 1) become more similar—for example, in language and culture (or seem more similar when there are wider variations among groups); 2) share a more common fate than previously, because they are treated more alike by others and/or because they now share more common interests; 3) have become spatially closer—for example, in where they live; and 4) interaction and communication between the groups has increased. Conversely, the perceived boundary between groups is likely to strengthen as each of these conditions changes in an opposite direction—that is, toward less cultural similarity, less common fate, less proximity, and less intergroup communication.

Political Boundaries

Changes in perceived group boundaries may be affected also by an additional factor: the political boundaries of a society. As the political boundaries of an area expand, ethnic boundaries tend to expand as well; as political boundaries contract, so do ethnic boundaries.

However, ethnic boundaries do not necessarily correspond to political boundaries. Horowitz asserts: "As the importance of a given political unit increases, so does the importance of the highest available level of identification immediately *beneath* the level of that unit, for that is the level at which judgments of likeness are made and contrasts take hold" (1975: 137).

For example, at the time just before its independence from Britain, India was a single political unit and the principal ethnic division was that between Hindus and Muslims. However, after the partition of the country into two separate political units, one mostly Hindu (India) and the other heavily Muslim (Pakistan), the Hindu-Muslim distinction became less important within each country. Ethnic distinctions more relevant to competition within each country—such as provincial loyalties in Pakistan—became more important.

The relevance of political borders to ethnic identity may be seen also in the question "Can Europe save Belgium?" posed by some observers at a time of great tension between the Flemish and Walloon ethnic groups in Belgium (Hunt and Walker 1974). In a more united Europe, the various nationalities making up Europe (French, German, Belgian, and so on) would interact much more than before. Therefore, differences of nationality would be highlighted and ethnic differences within each country (such as those between Flemish and Walloons in Belgium) might tend to be overshadowed. In general, then, political boundaries may define the arena within which particular ethnic differences and ethnic competitions are most salient.

Strength of Ethnic Identification

People's social identity—that is, their concepts of who they are and of how they relate to others—is determined mainly by their perception of what groups they belong to (Hogg and Abrams 1988: 20). Ethnic identity has been defined as the ethnic component of social identity, "that part of an individual's self-concept which derives from his knowledge of his membership in a social group (or groups) together with the value and emotional significance attached to that membership" (Tajfel 1981: 255). Such self-identification usually is accompanied by feelings of belonging and commitment (Singh 1977).

Individuals differ in how *salient* their ethnic identities are to them—that is, how relevant they see these identities to be to their actions and relationships. People vary also in how important or *central* their ethnic identities are to them—for example, how much they value their ethnic identity. Salience and centrality of identities are usually related (Stryker and Serpe 1994) and we may consider them together as representing the *strength* of identification (Abrams 1992).

How can we explain the strength of identification that many people have with their ethnic group? Why is their membership in a particular ethnic group important to them? Why do some people value their ethnic identity less, downplaying it in comparison to other social identities?

The strength of any social identity depends both on the extent to which the reactions of other people depend on this identity and on the extent to which the individual finds this identity to bring satisfaction.

Reactions of Others

People's concepts of themselves are formed to a large extent by how others see and label them (Rosenberg 1979). Thus, gender is important to the identity of most people because how others act towards a person from the time he or she is a baby usually depends in part on the person's gender. So too with ethnic identity. Being African American or a Jew or a Native American becomes important to a person if the way others react to him or her—with hostility or friendliness, deference or arrogance, and so on—is often affected by ethnic background.

For this reason, a given identity tends to become more salient to the person when that identity is distinctive (Michener et al. 1990). People tend to treat those who are distinctive from most others—whether it be in height, or gender or ethnicity—differently than those who are more typical in their social milieu.

When asked to describe themselves, American sixth graders tended to select features that distinguish themselves from others. They mentioned their age, birthplace, gender, hair color, or weight more often when that characteristic placed them in the minority in their class or when they differed considerably on it from their class average (McGuire and Padawer-Singer 1976). When members of minority religious groups in India (Muslims and Sikhs) and in the United States (Catholics and Jews) were asked to describe themselves, they mentioned their religious identity three times as often as members of majority religious groups (Kuhn and McPartland 1954; Driver 1969). And white and Hispanic grade school and high school students were more likely to describe themselves in terms of their ethnic identity as they became more distinctive (their ethnic group a smaller percentage) in their schools (McGuire et al. 1978).

Of course, a person may be distinctive in one situation and not distinctive in another situation; for example, a person may be the only black in a classroom but be in the racial majority in his neighborhood. Thus, the salience of a given identity may vary in different situations. However, some characteristics, such as gender and race, are salient for most people in a wide range of situations and so become a permanently salient part of the self-concept (McGuire et al. 1978).

Rewards Attached to Identity

The extent to which people see themselves in terms of a given identity depends also on how rewarding it is to them to think of themselves, and to present themselves to others, in terms of that identity. Three types of rewards appear especially important: social, self-esteem, and material.

Social Rewards

The strength of a given identity depends on the extent to which it is anchored in important and satisfying social relationships (Stryker 1981; Stryker and Serpe 1994). The more numerous and the more significant are the social relationships that depend on enacting a given identity, the more important to the person that identity will be.

Like other important social identities, a person's ethnic identity often is anchored in a network of important social relationships. To have a certain ethnic identity (as a Greek, a Chinese, a Cuban, and so on) means to many people that they are part of a community. When needed, an ethnic community (like other types of communities) may provide friendships, emotional support, and help to its members. In these and other ways, it may provide the warmth and security of an extended family. As Donald Horowitz has written: "among the most important needs met by ethnicity is the need for familiarity and community, for family-like ties, for emotional support and reciprocal help, and for mediation and dispute resolution—for all the needs served by kinships, but now on a larger canvas" (1985: 81). Bonds to a group are especially valued by those who have strong needs for affiliation and affection (Nixon 1979).

The best predictor of people's identifications with any community is their social relations with people in the community (Hunter 1974; Abrams 1992). Consistent with these more general findings, ethnic identity is most likely to be strong among those whose close social relations center around persons (family and friends) of the same ethnicity. These people are likely to share traditions, values, and a lifestyle that are intertwined with their common ethnic background. Thus, to embrace one's ethnicity is to embrace one's community. To reject one's ethnicity would be to reject one's social world.

A number of studies confirm this linkage. Americans of European origin are more likely to have a strong ethnic identification (such as Italian, Irish, or German in ancestry) if they live close to relatives (Alba 1990). Similarly, a study of ethnic loyalty among Mexican-Americans living in California found that the strongest predictor of such loyalty was residence in the barrio (i.e., a predominantly Mexican-American neighborhood) (Keefe and Padilla 1987). And Jewish identification among American men of Jewish origin increases as ties to parents and to Jewish friends are stronger (Dashevsky and Shapiro 1974). Social relations centered in an ethnic group and ethnic identity are closely related.

Self-esteem Rewards

Like other social categories, ethnic groups are compared and often ranked in terms of their characteristics (Billig 1976; Horowitz 1985).

How do groups compare in intelligence and skills? In ethical standards? In courage? In wealth? And so on. People derive much of their self-esteem from their group memberships. As a member of a particular ethnic group a person can share in the generally high prestige or in the generally low prestige of her own group. The individual can experience a sense of personal power from belonging to a powerful group or a sense of impotence as a member of a weak group.

People generally try to compare their own group to other groups in ways that will make their own group appear superior. They focus on criteria on which their own group is better than other groups and tend to ignore criteria on which their own group is inferior (Hogg and Abrams 1988). For example, students at a school whose teams have won championships in football and basketball may compare themselves to students in other schools on the basis of athletic success, while those attending schools whose students have excelled academically may compare themselves to others on this basis.

The feelings of self-esteem that people derive from membership in an ethnic group appear to be an important determinant of their identifying with that group. During the era of racial segregation especially, white workers received a "sort of public and psychological wage" in the form of better per-sonal treatment and greater deference than non-whites (Feagin and Vera 1995: 15). Such "preferred" treatment would tend to raise individuals' self-esteem. Getting these psychological benefits from "whiteness" was one reason that European immigrants, who initially defined themselves by their national ori-gins, began to view themselves as "white Americans" (Roediger 1991; Allen 1994).

European Americans' identification with ethnic groups in their ancestry is also related to their feelings of self-esteem. Alba (1990) found that white Amer-icans who gave great importance to their ethnic identity (i.e., ancestry from a specific country such as Ireland or Germany) frequently gave as a reason their pride in their background and their admiration for the traits of their ethnic group. White Americans who "like being ethnic" also often describe the way in which their ethnic identity makes them feel "unique and special" (Waters 1990). If members of a minority group feel secure in its characteristics, they are likely to "assert their peculiarity" and show strong in-group identification (Liebkind 1992). Feelings of pride may be linked also to the prestige that is attached to membership in a group. Studies of group cohesiveness indicate that the prestige of a group is one of the sources of attraction for its members (Ridgeway 1983). For groups such as the Daughters of the American Revolution, the prestige of their ancestry may contribute to a strong ethnic identity (as early Americans, usu-ally of British descent).

People may also identify with a group because such an identity makes them feel more powerful. Thus, as a result of his studies on the roots of nationalism,

Daniel Katz concludes that one source of identification with a nation is the desire to be part of a powerful group (1965).

In general, we would expect that the more that individuals can derive feelings of self-esteem from membership in an ethnic group, the more likely they are to identify with that group. Liebkind asserts that a child who is ashamed of the characteristics attributed to his ethnic group will try to avoid identification with this ethnic group; on the other hand, pride in the perceived group traits will lead the young person to make ethnicity an important part of his or her identity (Liebkind 1992). If a member of a group that is accorded low status in society is able to leave that group, he may "disidentify" with the group (Abrams 1992).

While some members of a low-status ethnic group may try to escape identification with this group—for example, by changing their names—others may not be able or may not want to do so. They may not be able to escape being labeled by others as a member of a given ethnic group (perhaps because of physical characteristics) and/or their emotional ties of belonging to this group may be strong.

However, members of low-status groups may escape the possible feelings of low self-esteem deriving from their group membership by rejecting the status judgments of the larger society and substituting new criteria and new judgments (Hogg and Abrams 1988). Thus, many African Americans have emphasized the positive qualities and achievements of their race as a basis for black pride (Rose 1990). They have taken pride too in the history and culture of black people. Some African Americans say that they do not want to "act white." Rather, they assert their own special identity and want to build and preserve a viable black community. Describing the development of what he calls "black consciousness," Peter Rose states:

> ... those who had been seen, and often saw themselves, as second class, increasingly rejected such imagery and took pride in who they were. That pride undoubtedly gave impetus to the outward expressions of a collective sense of selfhood, which I previously called "peoplehood."As members of a group, an ethnic group, not just a social category, more and more Black Americans sought to use their collective strength to get their piece of the action. (1990: 174)

Similar efforts to focus on the positive aspects of one's own group's history, culture, and achievements have been made by other low-status minorities such as Mexican Americans (Trejo 1979), as well as by some white ethnic groups (Rose 1990). The more favorable comparisons of one's own ethnic group with other groups that result from such perspectives not only raise the self-esteem of group members but also increase the value of this identity to group members.

Practical Rewards

Asserting a particular identity also is more likely to occur if that action leads to extrinsic rewards (Michener et al. 1990). Studies of the attachment of individuals to all kinds of groups show that such attachments are stronger as the group provides practical rewards to its members and/or provides defense against the environment (Nixon 1979; Ridgeway 1983). For example, among employees of the Tennessee Valley Authority (TVA), the more that individuals expected promotion, the more likely they were to have a self-image as a member of TVA (Patchen 1970). Similarly, the commitment of people to a communal living group may be based in part on their financial interest in the success of the group (Kanter 1972).

As with other types of groups, the strength of identification and attachment to ethnic groups is strengthened if such an identity brings external rewards (Lawler 1992). In many nations, whether or not one gets a government job, gets admitted to a university, or gets a government loan or contract depends on one's ethnic identity (Horowitz 1985).

In the United States, especially prior to the middle of the twentieth century, being white often has brought important advantages with respect to jobs, promotions, housing, loans, and other rewards. The privileges associated with being white was one of the factors that encouraged European immigrants to view themselves as "white Americans" (Feagin and Vera 1995). In more recent times, while "whiteness" often continues to bring advantages, those belonging to other racial and ethnic groups sometimes get preferences, under various affirmative action and "diversity" programs.

A number of social analysts have concluded that politics is increasingly replacing the market as an instrument for the distribution of economic benefits (Glazer and Moynihan 1970; Nielson 1985). The political system responds to the demands and influence of groups, including ethnic groups. Thus, assertions of one's ethnic identity and solidarity with an ethnic group is one way for many individuals to try to increase their material benefits in society.

Conversely, when a particular ethnic identity does not help (and may hinder) the attainment of external rewards, individuals may downplay that ethnic identity (Brass 1991). For example, before World War I assimilation into Hungarian society was helpful to Jews for success in commerce and industry in Hungary. Describing that historical period, Barany observes, "Among the assimilated Jews, many knew no Yiddish at all and took no particular pride in developing a special Jewish ethos" (1974).

In general, then, we may expect that the more that ethnic identification brings extrinsic rewards to people, the more likely they are to emphasize this identity; conversely, the more such identity is an impediment to such rewards, the less likely they are to emphasize it. Similarly, the higher the overall rewards of a given ethnic identity—based on social ties and self-esteem, as well as on extrinsic rewards—the stronger the ethnic identity is likely to be.

Ethnic Identity and Hostility Towards Other Groups

Is there a relationship between the amount of attachment that people have to their own particular ethnic group and their feelings and behavior toward members of other ethnic groups? In particular, does strong ethnic identity tend to be linked to hostility towards those in other ethnic groups?

There are some reasons to suspect that solidarity within a group may be associated with antagonism towards other groups. First, the very definition of a group may be framed in terms of its differences from, or opposition to, one or more other groups (Kriesberg 1982). Southern Baptists are defined in terms of their differences from other Baptists, "Right to Life" groups are formed in opposition to "Freedom of Choice" groups, and vice versa, with regard to the abortion issue. Ethnic groups may also be defined in part in terms of their differences from other ethnic groups. Thus, the category "coloureds" in South Africa is meaningful primarily in terms of their differences from whites and from blacks. Many white ethnics in the United States have formed or strengthened their ethnic identity (e.g., as European Americans) in reaction to the ethnic assertiveness and sometimes perceived threat from non-European Americans, especially African Americans (Alba 1990; Rose 1990).

There are, of course, sometimes realistic conflicts between members of ethnic groups. Ethnic groups may compete for resources, for power, and for status within any society (Horowitz 1985; Abrams 1992). When groups are engaged in conflict, whether it is groups of boys at a summer camp (Sherif et al. 1961) or nations engaged in war (Cashman 1993), cohesion within each group tends to increase. Members of an ethnic group that is the target of external enemies or engaged in conflict with another ethnic group will also tend to draw closer to others in their group. For example, studies of Hispanic students found that when interethnic tensions in schools were greater, the students' ethnic identity became more salient to them (Rotheram-Borus 1993).

If hostility between ethnic groups may lead people to identify more strongly with their own ethnic groups, may the reverse also be true? Does strong commitment to an ethnic group lead people to be hostile towards those in other ethnic groups?

A sociologist of an earlier era, William Graham Sumner, saw peace within social groups as necessary in order for these groups to function effectively in their warfare with other groups. But he also described as universal the phenomenon of "ethnocentrism," a view that places one's own group at "the center of everything" and denigrates other groups: "Each group nourishes its own pride and vanity, boasts itself superior, exalts its own divinities, and looks with contempt on outsiders. Each group thinks its own folkways the only right ones and if it observes that other groups have folkways, these excite its scorn" (1906: 13).

A more modern and much more fully developed analysis of the relation between group identity and reactions to outsiders—but one that echoes Sumner's ideas—is that of social identity theory (Hogg and Abrams 1988). Social identity theorists point out that individuals base their self-esteem in large part on their identities as members of the social groups to which they belong. If a group to which a person belongs is superior in some way to other groups, then the person's own self-esteem is likely to be raised. For example, sports fans whose team has just won a championship and who shout "We're number one!" have a surge of heightened self-esteem (though only temporarily).

People are motivated to focus on and to accentuate differences among groups that are important to them and on which they believe their own group rates favorably. For example, Jews may focus on group differences in scholarly accomplishment, French on differences in cultural sophistication, and Japanese on differences in cooperation.

A number of experimental studies (Tajfel 1982) have supported the view of social identity theorists that simply dividing people into groups—even on the basis of arbitrary or random criteria and even when the groups are not in conflict—will result in more favorable evaluations of in-group than of out-group members. Such results are consistent with the proposition that people see their own groups as superior in order to raise their own self-esteem. Moreover, some studies found that just being divided into groups led people to discriminate in favor of their own group members when allocating rewards. Such discriminatory actions may reflect a greater sense of solidarity with those "like oneself" (at least in group membership) and perhaps adherence to a general norm that dictates helping those in one's own group.

The competition among ethnic groups for "group worth" and the role of invidious comparisons in ethnic conflict has been emphasized by Donald Horowitz (1985) in his study of third-world (primarily African and Asian) ethnic groups. Horowitz finds that ethnic groups that are more backward (in the sense of having less mastery of modern skills) feel humiliated and resentful at unfavorable comparisons with more advanced neighboring ethnic groups. For example, Malays comparing themselves to more successful Chinese in Malaysia and Hausa comparing themselves to more advanced Ibos in Nigeria have resented their inferior position. Horowitz writes: ". . . the unflattering images of group characteristics generated by the comparison give rise to powerful efforts to use the political system for the confirmation of group worth" (1985: 167). The resentments of "backward" ethnic groups also lead them sometimes to initiate violence against more advanced groups.

The tendency for solidarity with one's own in-group to be associated with some antagonism towards out-groups is supported also by other types of evidence. One relevant body of research has investigated the social distance at which various out-groups are held by a given in-group. For example, members of a given group are asked how willing they would be to accept various other racial, ethnic, or religious groups as neighbors, as co-workers, as members of their social club, as marriage partners for family members, and so on. This research has found that all in-groups hold most out-groups at some degree of social distance (LeVine and Campbell 1972: 15–16).

Some studies on group cohesiveness (the extent to which members have strong ties to their groups) have found that more cohesive groups express more hostility towards outsiders (Stein 1976). And some ethnic organizations and movements include hostility towards other ethnic groups as part of their philosophy and message. This hostility is found, for example, in "white-only" groups, such as the Ku Klux Klan, and also in some black nationalist groups in the United States, which have proclaimed blacks' superiority to whites, rejected association with white liberals, and endorsed separation of blacks from whites (Cross 1971; Rose 1990). Similarly, one study found that Hispanic students who identified more strongly with their ethnic in-group were more separatist in their orientation towards out-groups (Rotheram-Borus 1993).

However, although identification with one's own ethnic group is sometimes linked to hostility towards and/or avoidance of other groups, this does *not* appear to be a uniform or inevitable linkage. First, an association between in-group solidarity and negative reactions toward out-groups is not always found. Greater cohesiveness within in-groups is not necessarily associated with greater discrimination towards out-groups (Dion 1973). Also, experimental studies indicate that the relationship between identification with an in-group and bias towards out-groups is not consistent (Abrams 1992).

Moreover, when bias occurs it is not necessarily in the form of a negative reaction towards an out-group. Those with strong attachments to their own group typically are more positive than others towards members of their own group but are not necessarily more negative towards members of other groups (Dion 1973; Brewer 1979; Hinkle and Brown 1990; Abrams 1992). Thus, they may give preference to members of their in-group but such preference does not necessarily mean antagonism to outsiders.

When those who identify strongly with a given group express hostility and/or negative evaluations of an out-group, this tends to occur under a particular set of circumstances. One circumstance that may contribute to derogation of an out-group is that this out-group is a salient comparison group—that is, members of one's own group judge their own worth on one or more characteristics that are important to them (such as intelligence, skill, or honesty) by comparing their position to that of members of the particular out-group (Hogg and Abrams 1988).

If a particular out-group serves as such an important comparison group, social identity theory predicts that members of the in-group will tend to derogate the comparison out-group in order to enhance their own self-esteem. This tendency to derogate an out-group is likely to be especially strong when the out-group occupies an unstable position in a group status hierarchy and especially if its status position is viewed as illegitimate (Hinkle and Brown 1990). For example, Catholics in Northern Ireland (Ulster) may compare themselves to Ulster Protestants with respect to a variety of characteristics (abilities, morality, cultural contribution, and so on). They may tend to derogate the Protestants, especially since the higher position of the Protestants with respect to economic and political status is under attack as illegitimate. In addition, bias against an out-group may be enhanced among members of groups that emphasize group rather than individual achievement and whose ideologies emphasize comparisons to other groups (Hinkle and Brown 1990). For members of such groups, superiority of their own group to out-groups is of great importance.

Probably the most important circumstance that affects the possible linkage between in-group affiliation and out-group enmity is whether there is conflict of interest between the groups. For a person who identifies strongly with a particular group, another group that is attempting to thwart or hurt the in-group must evoke hostility. The role of conflict of interest is shown in a study of aggressive reactions toward ultraorthodox Jews by other Israeli Jews who are secular or more mainstream in their religious views (Struch and Schwartz 1989). In general, the strength of identification that respondents indicated with their own religious group was *not* related to their expressed willingness to take aggressive actions against the ultraorthodox (such as organizing boycotts against their stores in one's neighborhood). However, among those highly identified with their religious in-group, perceived conflict with the out-group (the ultraorthodox) had a stronger effect on aggression than it did among those with less in-group identification. Also, among those highly identified with their in-group, perceptions of value dissimilarity with the out-group and of firm boundaries between in-group and out-group had a much stronger effect on aggression. Thus, identification with a particular group makes the individual react negatively to outsiders when he sees those outsiders as being very separate from, having different values from, and threatening the welfare of his own group.

Ethnic Identity and Other Identities

A person's identity as a member of a given ethnic group is, of course, only one of many identities that he or she is likely to have. The person is also male or female, usually has a certain occupation, a religious affiliation, lives (or was raised in) a certain region, may have a political affiliation, may belong to civic or fraternal organizations, a sports team, and so on. Thus, the person has multiple identities and allegiances (Liebkind 1992).

Some identities may be broader than others and may subsume the narrower ones. For example, a person may be a resident of Peoria and a resident of Illinois. She may be a Baptist and a Christian. He may be a Chinese American and an Asian American.

The salience of a given identity will vary as a person enters different situations. In part this results from different characteristics of the person being distinctive in different situations (Michener et al. 1990). For example, a black woman is likely to be more aware of her race than her sex in a gender-mixed but primarily white social event. Her gender is likely to more salient to her in a racially-mixed but primarily male gathering.

While the salience of particular identities varies across situations, a particular identity will have an overall importance, relative to other identities, for each individual. Thus, we may consider what affects the relative importance of ethnic identity as compared to people's other social identities. This question has been discussed by social scientists and by others especially with respect to the relative importance of ethnicity and social class.

Many analysts have seen social class (especially occupation and income) as the crucial determinant of each individual's position, life chances, and interests in society. Some have expected that allegiances based on social class would be stronger than those based on ethnicity (McCall 1990). For example, some have expected—or at least hoped—that working class solidarity would overcome racial and ethnic divisions among poor people. In some instances, class identity indeed has appeared to be more salient than ethnic identities. For example, black and white workers in industrial unions, such as the United Auto Workers, have shown strong solidarity during strikes against employers (Blalock 1967). But in other settings, ethnic allegiances have overshadowed common economic interests. Thus, much of the ethnic mobilization among white ethnic groups has come at the expense of possible class-based solidarity across racial lines (Rose 1990).

Relative Strength of Different Identities

What will determine the relative strength of identification based on ethnicity as compared to that based on other social characteristics? To try to answer this question, it is useful to consider the factors discussed earlier with respect to the strength of an ethnic identity: the reactions of other people and the rewards and costs that result from a given social identity.

Reactions to an individual by other people—co-workers, neighbors, police, store clerks, and so on—usually will be influenced to some extent by that individual's ethnicity. However, in some settings people may react to a person primarily in terms of characteristics other than her ethnicity. For example, people in a work organization may base their behavior towards a Latino American primarily on her occupation and status in the organization. In other situations—for example, where a Mexican American family has just moved into a previously all Anglo neighborhood—others' reactions may be based primarily on the individual's ethnicity. Where people do not know each other—as is generally true in public areas such as shopping malls, buses, parks, and theaters—they are likely to base their behavior towards others on visible characteristics, such as age, gender, and (often) ethnicity. The wider the range of situations in which people react to an individual primarily in terms of his ethnicity, and the larger the number of people who do so (especially important others such as bosses and co-workers), the stronger his ethnic identity is likely to be. Conversely, the more that, overall, people's behavior towards him is affected by his other characteristics (occupation, religion, age, gender, interests, and so on) the stronger these other identities are likely to be.

The relative strength of ethnic and other identifications also may be expected to vary with the rewards that each type of identity brings to individuals. We have noted that, since social ties of family and friendship are likely to involve people of like ethnicity, ethnic identity usually brings great social rewards. To the extent that the social rewards of particular individuals are linked also to other, non-ethnic identities—as members of a particular occupation, church, sports team, and so on—those other identities are likely to increase in importance.

Whether greater rewards of self-esteem are derived from ethnic identity or from other social identities will depend on the specific identities involved. If a member of a low-status ethnic group has another social identity of high prestige—for example, an African American who is a physician—one might expect that (other things equal) she would tend to emphasize the high-prestige identity that would bring high self-esteem (as well as high social esteem).

Finally, the relative importance of given social identities may be affected by the extent to which each identity results in practical rewards or costs. Will individuals improve their life chances by emphasizing their ethnicity, their social class, their gender, their religion, or some other characteristic? A related question is the basis on which organized political action can be effective. Some analysts have suggested that ethnicity may offer a wider base for political mobilization than does social class (Olzak 1983; Nielsen 1985). Thus, Latinos, or Asian Americans, or African Americans may mobilize politically on the basis of ethnicity because they believe that is the most effective way to achieve their goals. On the other hand, some have argued that broader class-based movements that push for government programs based on income, rather than race or ethnicity, are politically more effective. Those minority group persons who accept the latter perspective are more likely to emphasize class identities.

American Identity and Ethnic Identity

So far in this chapter we have discussed the identities that people may have as members of ethnic and other groups within a country, especially within the United States. But what about identity as a citizen of the nation as a whole—for example, as an American? Is America simply "a collection of ethnic groups" as some have claimed; if there is a national identity, "a sense of peoplehood," what is the basis of this national identity (Gleason 1980)? And how is identity as an American related to identity as a member of a particular ethnic group within the nation?

The sense that there is a meaningful collectivity of people called Americans that is distinguished from other people is based on the same kinds of perceptions on which perceptions of other social groupings are based—especially on similarity, common interests, interdependence, and common fate. National identity may be based first on the perception of shared history, culture, and values (D. Katz 1965; A. Smith 1979). Many Americans share some historical memories or experiences—of the struggle to establish an independent nation, the common immigrant experience, of wars against external enemies, of economic and political crises. There is much shared culture—including a common language (English), music, sports, movies, and television. Most important, perhaps, the great majority of Americans share some important common values and beliefs—for example, in democracy, in political and religious freedom, and in the opportunity for all individuals to achieve material success (Eitzen 1985).

National identity also may derive from the perception of a shared fate, based on shared interests and shared treatment by others. Like citizens of other nations, Americans share a common fate in some important ways. For example, Americans from all ethnic groups tend to suffer if the United States does poorly in international trade competition and if the American economy does poorly; if the United States does well economically members of all groups tend to benefit (though not necessarily equally). Moreover, people of differing ethnic and other backgrounds depend on each other to perform important work that others rely on. They form a network of economic interdependence. Americans also share a common fate at times of conflict with other nations or foreign groups—for example, in times of war, of oil boycotts by foreign producers, and of terrorism directed at U.S. cities. These perceptions—that Americans share some important cultural similarities, are interdependent, and have a common fate in some important ways—combine with the perceptions of proximity and national boundaries to create a sense among many Americans of being part of a national entity, a national "we."

How is a person's identity as a member of a particular ethnic group related to his identity as a member of the larger community of Americans? Are these identities compatible? Or does strong ethnic identity tend to reduce attachment to the larger society?

Some have conceptualized ethnic identity as ranging along a continuum from strong ethnic ties at one extreme to strong ties to the larger society at the other extreme (Phinney 1990). The assumption underlying this model is that a stronger identity with either the narrower ethnic group or with the broader society is accompanied by a weakening of the other identity. An alternative view is that the strength of the narrower and the broader identities may be independent (1990). Members of minority groups may have either strong or weak identifications with both their own and with the mainstream cultures. According to this view, a strong ethnic identity does not necessarily mean a weak attachment to the larger society (Berry et al. 1986).

A number of studies support the latter view—that ethnic identity and national identity tend to be independent; a person can be high on both types of identity, low on both, or high on one and low on the other. For example, among adolescent girls of Indian background living in England, the strength of identification as an Indian was unrelated to identification as British (Hutnik 1986). Similarly, among Armenian Americans (Der-Karabetian 1980), Jewish Americans (Zak 1973), and Chinese Americans (Ting-Toomey 1981), the strength of people's American identity did not depend on the extent to which they identified with their narrower ethnic group.

Some people—especially some members of racial minorities—do not see their ethnic identity as consistent with an identification with the broader American society. They see the nation as essentially a white European society that has a history, a culture, and a set of values that are different from their own. The heroes of that society—men such as George Washington and Thomas Jefferson—have been viewed by some as slaveholders and oppressors. Asserting their own ethnic pride is seen as requiring a rejection of the history and culture of the white Europeans who have dominated American life. The interests of their own ethnic group are viewed as basically different from those of the dominant group in American society. While economic interdependence is recognized, it is seen as desirable to minimize dependence on the dominant group as much as possible.

For many other Americans, however—both members of racial minorities and others—a strong ethnic identity, and a strong attachment to the larger society are seen as entirely compatible. Many people from diverse ethnic backgrounds—Irish, Mexican, Jewish, Polish, African, and so on—see the fundamental values of the United States, such as freedom and opportunity for all, as corresponding with important values from their own heritage. They greatly value the economic and other rewards that have been earned by members of their own group as American citizens. They are often intensely proud of being American. For example, writing about Mexican Americans, whom he describes as generally having a strong attachment to their ethnic heritage, Lawrence Fuchs states: "Mexican settlers and especially their children discovered that their participation in American political life, as for other immigrant-ethnic groups, encouraged them to espouse a wider patriotism as advocates of opportunity and civil rights for all" (1990: 270–271).

39

The extent to which identity as a member of a particular ethnic group and identity as a member of the broader society (e.g, as an American) are seen by people as compatible may be expected to depend in part on whether they believe that the values and culture of the two units are compatible. It may depend also on whether they believe that the interests and goals that they share with members of their ethnic group are compatible with or opposed to the interests and goals of those in the broader society. For example, if African Americans see their own subculture as compatible with the broader American culture, and especially if they see most other Americans as sharing their goal of equality, they are able to identify both as blacks and as Americans. But if they see most non-black Americans as trying to keep black people down, then it is more difficult for them to identify both as blacks and as Americans.

Most Americans have both an identity as an American and an identity as a member of a particular ethnic group. The strength and salience of each of these identities will depend on factors we have considered earlier as affecting the strength of identifications: reactions of others and the rewards for particular identities.

Since Americans usually interact primarily or entirely with other Americans, national identity usually is not salient in everyday relationships; since almost everyone is American, people do not differentiate their reactions to others on this basis. People do vary in ethnic background and, of course, the reactions of others to a person are often affected by that person's ethnicity. This fact would tend to make ethnic identity more salient than national identity in most situations.

The relative importance of ethnic, national, state, or city identities will depend also on the rewards (and costs) associated with each identity. For the person whose social ties are limited to a particular ethnic group, social rewards are tied to this ethnic identity. For someone whose circle of friends is drawn from a broader community, social rewards may be tied more to identity as a member of the broader society. Self-esteem may come from pride in the culture and heritage of one's ethnic group. It may come also from pride in the strength or ideals of one's broader community or nation. Most crucial perhaps is whether practical advantages are seen as deriving more from the narrower ethnic community or from the larger society (Lawler 1992). Does one owe her chances and material advantages in life to membership in an ethnic community or primarily to her identity as an American or, say, as a Californian or as a citizen of San Francisco? Are educational opportunities, jobs, and eligibility for government programs based primarily on ethnic ties and ethnic identity or primarily on citizenship in the broader society? If it is one's ethnicity that is more crucial, ethnic identity will tend to be stronger. If citizenship in the larger society is more important, a broader identity will tend to be stronger.

Summary

People's attitudes and behaviors toward others often depend on whether the others are perceived as outsiders or as part of their own in-group. People are more likely to see particular others as part of their own in-group the more that these others: 1) are physically close; 2) are similar to themselves, especially in ways that facilitate smooth social relations; 3) share a common fate, as determined especially by common interests and by similar treatment by third parties; and 4) communicate often with them. The boundaries of ethnic groups, including those in the United States, change over time as these factors vary.

The strength of individuals' identifications with their ethnic groups depends in part on how much others' actions toward them are influenced by their ethnicity. In general, the more distinctive a person's ethnicity is, the more likely she is to be treated according to this ethnicity and to have a strong ethnic identification.

How strongly people identify with their ethnic group also depends on how rewarding this identification is. The more that people's social relationships (to friends, family, co-workers, and so on) are with those of the same ethnic group, the more their social rewards are tied to their ethnicity and the stronger their ethnic identification is likely to be. Ethnic identification also is likely to be stronger as a person derives more feelings of pride or power from his ethnic identity. In addition, the more that practical rewards (such as job opportunities) are enhanced by one's ethnicity, the stronger ethnic identification is likely to be.

Feelings of strong solidarity with their own ethnic groups may lead people to show hostility or bias towards members of other ethnic groups. Because people tend to compare their own group to other groups, they may derogate the characteristics of those in other groups in order to raise their own self-esteem. However, available evidence indicates that, while people do tend to act more positively towards members of their in-group than towards others, this is not necessarily accompanied by hostility towards members of out-groups. Hostility towards out-groups appears to occur when comparisons among groups are highly salient and especially when people see a conflict of interest between their own group and another group.

Identification with an ethnic group and identification with a nation, such as the United States, are generally not alternatives; rather, they are independent. A person may have a strong identification with only one or with both. The relative strength of people's identification with an ethnic group, with some other social group, and with the broader community depends on: 1) how much each identity affects the way they are treated by others; and 2) the relative rewards—with respect to social relationships, self-image, and practical advantages—that each affiliation brings.

3) Individual Attitudes and Behavior Toward Other Groups

In chapter 2, we focused on the attachments that people feel to their *own* ethnic groups. In this chapter we consider people's attitudes and behaviors toward members of *other* ethnic groups.

Why do many people try to avoid, or act in a hostile way towards, those who belong to an ethnic group other than their own? Why do many people oppose programs aimed at reducing inequality between ethnic groups? Why are some people friendly towards and supportive of those in different ethnic groups?

This chapter addresses these questions at a psychological level. It considers the kinds of subjective reactions (images, beliefs, feelings, and so on) that people have regarding people in other ethnic groups and how such reactions affect their behavior.

Attitudes Toward Other Groups

A person's behavior towards members of another ethnic group may depend on his attitude towards that group. An attitude towards some group may be defined as a positive or negative evaluation of that group (Zanna and Rempel 1988). For example, a Mexican American who has a positive attitude towards Anglos (non-Hispanic whites) would see them as having good traits (such as being friendly) and would like them; a Mexican American who had a negative attitude towards

Anglos would see them as having undesirable traits (such as acting superior) and would dislike them.

Some attitudes are based primarily on personal experience or on facts. For example, a black student may have a negative attitude towards his white classmates because most of them have harassed him. In other cases, attitudes toward a given ethnic group are based on what people have heard about this group, from family, friends, or the media—much of which may be inaccurate or distorted.

When a person has a negative attitude towards some group that is erroneous or overgeneralized and that is also inflexible even when contrary evidence is presented, we may say that he is prejudiced. Those who have racist attitudes not only view a target group as inferior but see their behavioral traits as inherited (e.g., see Marger 1994: 26–27, 74–75).

Two components of ethnic attitudes are central: 1) images and perceptions of another ethnic group; and 2) affect (feelings) towards members of the group (Eagly and Chaiken 1993).

Images

Members of one ethnic group tend to agree on the traits that they attribute to most members of another ethnic group. Several studies show, for example, that black people have been seen by many white Americans as having less ambition than whites and as more violent than whites (e.g., Schuman et al. 1985; Sniderman and Piazza 1993). Many majority-group Americans have had a similar "lower-class" image of Mexican Americans, seeing drunkenness, criminality, shiftlessness, and dirtiness as characteristic of Mexicans (Simmons 1971; T. W. Smith 1990). Many non-Jewish Americans have seen Jews as tending to be unscrupulous, clannish, too powerful, and having mixed loyalties (Martire and Clark 1982).

People's perceptions that members of a given ethnic group have a certain characteristic, or that they are more likely than members of other ethnic groups to have this trait, sometimes are based on fact. For example, African Americans are more likely than others to be arrested for crimes (Conklin 1995) and Jews are, on average, more wealthy than other Americans (Marger 1994). Such differences between ethnic groups may derive from differences in their histories, in their cultures, and in their position within society (e.g., greater poverty among blacks).

But while perceptions of ethnic groups may be based partly on fact, there is a widespread tendency for people to overgeneralize the actual differences that exist. A number of studies have found that, once having categorized people into groups, people tend to overestimate the *similarities* among members of the same group and the *differences* between members of separate groups (Messick and Mackie 1989).

In addition, people often are resistant to changing their images of another group when presented with new information that contradicts this image (Snyder

1981; Weber and Crocker 1983). For example, if a person who sees Mexican Americans as unintelligent meets several successful Mexican American scientists, he may downplay their accomplishments or see them as not typical, thereby maintaining his previous viewpoint. A perception of a group that is overgeneralized and rigid has been called a "stereotype" (Michener et al. 1990).

Stereotypes about ethnic (and other) groups may arise and persist in part because they help people to make judgments about others about whom they have little information. They provide a "mental map" (even though usually a flawed one), which helps to simplify an often complex world. Hogg and Abrams comment that stereotyping is a "product of a fundamental cognitive process which fulfills a basic human need for order and predictability" (1988: 84).

But people usually do not have stereotypes about ethnic or other groups simply because stereotyping helps them to simplify a complex world. Negative stereotypes about other groups also may serve to make people feel superior to others. As noted in chapter 2, people tend to compare the groups to which they belong with other groups. Because their self-esteem is tied to their identities as members of particular social groups, they are motivated to see their own groups as superior to other groups on characteristics (such as intelligence or courage) that they value. By overgeneralizing any differences that favor their own group, they can increase their own self-esteem (Tajfel and Turner 1986).

Stereotypes about ethnic groups are not perceptions that people hold as isolated individuals. Rather, they are widely shared within social groups (LeVine and Campbell 1972; Hogg and Abrams 1988). Generalizations about particular groups—that blacks are criminals, that Mexicans are lazy, that Jews are dishonest, that Catholics are dogmatic, and so on—are communicated to individuals by family, friends, and sometimes by the media. Thus, the overgeneralized images of other groups that individuals often hold are reinforced through the agreement of others. Stereotypes, then, often have a "social reality" despite their lack of matching objective reality.

Feelings

Studies examining people's reactions to other racial and ethnic groups have examined not only their images or stereotypes, but also their feelings toward members of such groups. For example, both whites and blacks in Indianapolis high schools were found to like members of the other racial groups less than members of their own racial group. Conversely, negative feelings such as dislike and anger toward the other group were frequent. Thus, many black students in the Indianapolis high schools were angry at what they perceived as attitudes of superiority and rejection by white schoolmates, while many whites were angered by what they perceived as offensive or inconsiderate behavior by blacks (Patchen 1982).

45

The negative feelings that members of one group have for another are not necessarily ones of hostility or hate. They may, instead, feel discomfort, uneasiness, disgust, and sometimes fear (Gaertner and Davidio 1986: 63). For example, most white children in one school were found to be fearful of physical attack by black classmates, while many of the black children were fearful of ridicule or rejection by the whites (Schofield 1982).

At times, people may experience a combination of positive and negative feelings toward another group. The work of Irwin Katz and his colleagues (Katz and Hass 1988) suggests that many white people experience friendly feelings toward black people (based in part on sympathy with blacks' experiences of discrimination) at the same time that they have negative feelings toward blacks (based on perceptions that many blacks do not live up to such values as hard work and thrift).

The feelings we have towards people tend to be consistent with the images we have of them (Rosenberg 1979). In accord with this general tendency, the absence of derogatory beliefs about black people is correlated with feelings of ease in interracial contacts and sympathetic identification with blacks (Woodmansee and Cook 1967). However, images (or stereotypes) of a group and feelings toward members of the group are not perfectly associated and sometimes may diverge considerably. Thus, some white people characterized as "aversive racists" (Kovel 1970) do not hold negative stereotypes about black people (at least consciously) but may nevertheless harbor underlying negative feelings toward blacks.

Relation of Attitudes and Behavior

How are attitudes (i.e., evaluations based on beliefs and feelings) toward members of another group related to behavior towards that group? Most of the research on this issue has concerned the relationship between whites' attitudes toward black people and their behavior towards blacks. In general, attitudes and behavior tend to be associated (Wicker 1969; Schuman and Johnson 1976). For example, students with more positive attitudes toward blacks were more willing than those with negative attitudes to sign a petition urging Congress to pass legislation that would end discrimination against blacks in employment (Kamenetsky et al. 1956). College students with less negative attitudes were more willing than those with more negative attitudes to interact socially with blacks (Warner and DeFleur 1969). Among both white and black high school students in Indianapolis, more positive attitudes toward schoolmates of the other race were associated with less avoidance of, less fighting with, and more friendly contact with members of that group (Patchen 1983).

However, the association between people's attitudes and their behavior, including behavior towards members of other racial groups, often is not very strong (Wicker 1969; Eagly and Chaiken 1993). Sometimes, there is no corre-spondence at all between attitudes and behavior. For example, in a time of wide-spread anti-Chinese prejudice in the United States, a social scientist traveled with a Chinese couple and kept a list of hotels and restaurants where they were served (only once were they refused service). He then wrote to the two hundred fifty establishments on his list, asking if they would accept Chinese guests. Over 90 percent of 128 proprietors who responded said they would not serve Chinese, although they had all actually done so (LaPiere 1934).

A lack of correspondence between racial attitudes and behavior has also been found in other studies. Rokeach and Mezei (1966) found that persons who are high in racial prejudice, as well as those low in prejudice, tended to choose co-workers on the basis of similarity in job-relevant beliefs, rather than by race. A study of change in neighborhood racial composition in the Chicago area (Taub et al. 1984) found that racial tolerance scores had no significant effect on the intentions of white residents to move from a racially-changing neighborhood. (Concern about crime and safety was the strongest predictor of intentions to move.)

Among a representative sample of white Americans, support of govern-ment programs aimed at helping black people—in particular to provide more jobs and better schooling for blacks—was found to be largely unrelated to atti-tudes toward black people (Kluegel and Bobo 1993). (Other factors, including beliefs about reasons for poverty, were more related to support for such gov-ernment programs.) The authors of this study comment: "Our results also raise questions about the central importance claimed for racial affect in opposition to race-targeted and general social welfare policies. Prejudice has no indepen-dent effect on attitudes toward race-targeted opportunity-enhancing policies" (1993: 460).

Why is there often only a weak association, or no association at all, between attitudes and behavior directed toward a particular racial or ethnic group? The question of the relation between attitudes and behavior has received considerable attention by social psychologists.

One explanation is that racial attitudes are often not measured adequately. A variant of this type of explanation has been advanced to explain the fact that many white people who do not express negative racial attitudes and who support the principle of equal opportunity nevertheless do not support some specific poli-cies (such as school busing to achieve racial balance and affirmative action pro-grams) aimed at promoting equality for minorities. Such a pattern of attitudes and policy preferences among whites has been interpreted by some writers as "symbolic racism." "Symbolic racists" are seen as harboring hostility towards blacks, which they are reluctant to express directly but which they express instead towards indirect symbols such as school busing and affirmative action (Sears 1988). However, the concept of symbolic racism has been criticized for its lack of clarity. Moreover, policy preferences on race-related issues sometimes can be explained without invoking racism at all (Pettigrew 1985). For example, the extent to which whites support preferences for blacks as part of affirmative action programs appears to depend on their beliefs about fairness rather than on their attitudes toward blacks (Sniderman and Piazza 1993).

There is broader agreement that one reason for a discrepancy between ethnic attitudes and behavior may be differences in the specificity of the attitudes and behavior that are being compared. A person's attitudes toward a broad group of people (say, Japanese) may be compared to her behavior towards a specific Japanese in a specific situation. But a person's attitudes toward Nurio Suzuki in a situation where she is cooperating with Nurio on some project may be quite different from her attitudes toward a "prototype" Japanese in a "prototype situation" of business competition (Stroebe and Insko 1989). Evidence from a study of a racially-mixed middle school (grades six through eight) illustrates this point (Schofield 1982); after the school was racially desegregated, there was little change in the attitudes of black students or of white students toward the other racial group as a whole. However, there was considerable increase in friendly contact with specific classmates of the other race. Janet Schofield, the author of the study report, states:

> One final factor that made possible the simultaneous reduction in fear and maintenance or buttressing of stereotypes about black aggression was the growing ability of white children at Wexler to differentiate between individual blacks. Thus, even students who continued to deal with their fears by avoidance had the opportunity to learn that general avoidance of all blacks was unnecessary. Black students, too, seemed more inclined to think of white students as individuals as they got to know many of their white classmates. (1982: 166–168)

Note that students in this school did not abandon generalizations about the other racial group; they merely recognized that not all of the others fit the generalization and became more able to distinguish between those who did fit and those who did not.

Generalizations about other groups tend to be probabilistic. "The perceiver expects that there will be 'exceptions to the rule'" (Miller and Brewer 1986). Thus, general attitudes toward a group and behaviors toward specific members of that group will not necessarily coincide (Horwitz and Rabbie 1989).

There are also a number of other reasons why attitudes and behaviors may not coincide (Eagly and Chaiken 1993). A person may behave in a way that appears inconsistent with her attitudes because of competing attitudes and values (e.g., about equality or private property), social norms (fear of disapproval), competing incentives (e.g., acting in accord with one's attitudes might risk legal penalties), or lack of the ability or opportunity to take actions consistent with an attitude (e.g., a person who dislikes having her children attend school with blacks may lack any affordable alternative).

Overall, we may conclude that people's attitudes toward a racial or ethnic group have only a weak relationship to their behavior towards members of this group. In part, this occurs because general attitudes may not be applied to specific individuals; in part it is because certain conditions—such as ability to act in accordance with one's attitudes—need to be present; and, in part, because other factors—such as social norms—also affect behavior.

But the limitations of explaining discriminatory actions in terms of negative or prejudiced attitudes may be more fundamental. People do not generally display negative feelings (such as anger) and behavior (such as aggression) toward members of another group simply because they have a negative evaluation of that group in some general way. Rather, they are most likely to display negative feelings and behavior toward another group when they see that group as affecting themselves or their own group adversely. Such perceptions, and their impact on emotions and behavior, are the subject of appraisal theories of emotion.

Appraisals, Emotions, and Behavior

Appraisal theories of emotion have attempted to establish the linkages between specific types of perceptions, termed "appraisals," the emotions that are elicited by those appraisals, and the types of behavior that result from certain emotions (Roseman 1984; Scherer 1988; Frijda et al. 1989).

Appraisals include perceptions both of a target person and of the situation or context. Appraisals that evoke emotions typically have important implications for the self. Frijda and his colleagues state that an event will lead to an emotion if it "appears to favor or harm the individual's concerns: his or her major goals, motives, or sensitivities" (1989: 213).

The type of appraisals that we make of a person affects our feelings or emotions toward that person, which, in turn, tends to affect our behavior. This is true also with respect to people's reactions toward members of other ethnic groups (Dijker 1987; E. Smith 1993; Patchen 1995).

Table 3.1 indicates some of the major types of negative appraisals of other ethnic groups and the perceptions that contribute to each type of appraisal. This table also indicates the primary emotions that result from each type of appraisal and the action tendencies that result from those emotions.

Three types of negative appraisals that are often important in interethnic relations are: 1) the other group has harmed, or is a threat to harm, oneself or one's own group; 2) the behavior of members of the other group is reprehensible (should be condemned); and 3) the other group has undeserved advantages relative to one's own group.

49

TABLE 3.1

Negative Appraisals of and Emotions and Action Tendencies Toward Members of
Another Ethnic Group

Appraisals	Primary Emotions	Primary Action Tendencies
A. Other group harms self (own group)		

1. Has Inflicted Harm in Past
   a. Actions reduce own (group's) material
     or physical well-being, status, power
   b. Actions were illegitimate
   c. Harm caused was intentional
	Anger
	(Dislike)
		Harm, punish others
		(Avoidance)

2. Poses Threat of Future Harm
   a. Is disposed to take actions harmful to
     self (own group)
   b. Is powerful
   c. Is active
	Fear
	(Anger)
	(Dislike)
		Avoidance;
		Protect self, own
		group; attempt to
		thwart or suppress
		other group

B. Behavior of members of other group is reprehensible		
1. Actions of other group are unpleasant and/or inconsistent with perceiver's own values	Dislike Disgust	Avoidance Refuse help
2. Actions of other group are freely chosen; oneself or one's own group are not responsible		
C. Other group has undeserved advantages		
1. Other group has high material benefits, status, or power relative to self or own group	Anger (Dislike) (Envy)	Protest Harm other group
2. Relatively advantaged position of other group is illegitimate		
3. Other group is (at least partly) responsible for the "unjust" situation		
4. Other group intends the "unjust" situation		

Other Group Seen As Doing Harm

Members of one ethnic group often see those in another group as
being harmful to their own interests or well-being. They may perceive that the others have already harmed them in some way or they may believe that there is a threat

of future harm from this group. They may also be concerned about harm done to themselves personally or harm done to the ethnic group to which they belong.

Personal Harm

We will first consider people's appraisals that they have been harmed personally by or are threatened with personal harm from members of another ethnic group. The specific personal interests that they may see as harmed or threatened include their economic prosperity, their safety, and their social status.

a) Economic Well-being

Sometimes members of one group believe that their jobs or business successes are being harmed or threatened by members of another group. When white workers were striking in East St. Louis in 1917 and heard that blacks were being brought in as strikebreakers, they rampaged through black areas, beating blacks and burning their residences (Rudwick 1964). Hostility to Mexican immigrants among many blue-collar Americans sometimes has been fueled by their perception that these immigrants are taking jobs that are rightfully theirs or depressing their wage rates (Lamm and Imhoff 1985). Hostility and violence against Chinese immigrants in the mid-nineteenth century came both from workers who believed that Chinese labor was lowering wages and from merchants who believed that "unfair" competition from Chinese was driving them out of business (Lyman 1974).

Another perceived economic danger is the threat to the value of one's home. Opposition by whites to having blacks in their neighborhood is often based in part on the perception that the presence of blacks will cause the value of their property to decline. Thus, a study of housing preferences in the Detroit area found that among whites who said they would move away from a neighborhood if more than a certain proportion of blacks were present, 40 percent explained this intention in terms of their concern about property values. Their belief that the entry of blacks would decrease property values often stemmed from images of blacks as not caring about or taking care of their homes (Farley et al. 1978).

b) Safety

Another important factor often affecting behavior is the perceived threat that many people feel to their personal safety. One of the main reasons why white people often avoid blacks is that whites are physically afraid of them. A study in the Chicago area found that fear of crime was the most important determinant of white residents' intentions to move out of racially mixed areas (Taub et al. 1984). Writing of the Hyde Park-Kenwood area of Chicago, Taub and his co-authors say: "It was crime and fear of crime that initially brought the University [of Chicago] into direct rather than passive action in the community. The robbery and attempted rape of a faculty wife started the committees that led to the formation of the South East Chicago Commission. Prior to this time, it had become more and more difficult to attract both students and faculty because the area was considered so unsavory" (1984: 101).

As is true more generally (Pettigrew 1973), fear of crime in that neighborhood was tied not only to race but to social class. Taub and his colleagues comment: "To the extent that blacks bear the symbols of being middle class, whites feel less wary; middle-class blacks feel pressure to bear those symbols so that they are not confused with the dangerous poor" (1984: 116).

Concern about crime and personal safety was also a frequent reason given by whites in the Detroit area in explaining their intentions to move away if blacks entered their neighborhood (Farley et al. 1978). Such fears are linked to perceptions of blacks as having "lower-class" traits, such as being rowdy and dangerous (Deutsch and Collins 1951). Whites are somewhat more willing to accept blacks as neighbors if they are described as having "about the same education and income" as the white respondents (Schuman and Bobo 1988).

Several studies of interracial contacts in schools also highlight the importance of perceived physical threats in affecting behavior. A study of high school students in Indianapolis found that many white students were afraid of their black schoolmates and that those who felt most fearful of attack (small white boys) were the most likely to avoid blacks (Patchen 1982). The major impact that fear of physical attack by black schoolmates had on the behavior of white schoolchildren—especially boys—is also emphasized in the study of a middle school (grades six through eight) in Pittsburgh (Schofield 1982). Describing the strategies employed by fearful white children, Schofield writes: "Another widely used strategy was withdrawal from the situation. The child did not object to perceived physical intimidation. He or she did not overtly refuse to do what was requested but just disappeared. . . . White children sometimes took flight at the approach of black children, even if the latter approached in a very ordinary and nonthreatening manner" (Schofield 1982: 119). The white children's fears were not wholly imaginary, however; Schofield reports evidence that aggressiveness was in fact, higher among black students than among whites (see also Patchen 1982).

Fear of physical harm may, of course, also be experienced by members of minority groups. Thus, while most black people in the United States prefer to live in racially mixed neighborhoods, the majority do not wish to live in otherwise all-white neighborhoods (Pettigrew 1973; Farley et al. 1978; Farley et al. 1994). Among the reasons given for this reluctance is a fear of physical attack by hostile whites. "They would probably blow my house up" is the kind of response given by many blacks (Farley et al. 1978: 331).

c) Social

There are other types of threats that may be perceived by members of one ethnic group when viewing another group. These include threats of social rejection,

52

threats to one's social status, and threats to the amenities of one's life (such as attractive streets and parks).

Minority group members in particular often are fearful of rejection by members of the higher-status or majority group. Thus, blacks are often reluctant to move into primarily white neighborhoods because they feel they would be unwelcome by white neighbors (Farley et al. 1994). Black students at primarily white schools are often concerned about being ridiculed or rejected by white schoolmates. Often, their reaction is to avoid or to be aggressive towards the whites (Patchen 1982; Schofield 1982).

Avoidance of black people by whites is sometimes increased by a perception among whites that living with and associating with blacks will endanger their social status among other whites (Pettigrew 1973; Jaynes and Williams 1989). Also, whites' resistance to having more blacks in their neighborhood is increased by the frequent perception that this will lead to greater physical deterioration of the neighborhood (Taub et al. 1984).

A threat to one's personal welfare usually evokes not only fear but anger. The relative salience of these emotions may depend on the power one feels relative to members of the other group. If the person feels weak and vulnerable, fear may predominate. When the person feels strong relative to the threatening group, then anger may predominate.

Emotions of fear and anger are linked to behavior (Dijker 1987; Frijda et al. 1989). When fear is strong, a person's natural tendency is to move away from or avoid members of the other group—as where whites move out of racially mixed neighborhoods because of their fear of crime. An alternative reaction (especially where other considerations prevent leaving or avoidance) is to try to protect oneself from the threat—for example, by carrying a defensive weapon. Still another reaction to fear—especially when a person feels weak—is to submit to the demands of the person(s) who threaten that person. For example, blacks in the American south during the era of segregation often complied out of fear to the demands of the more powerful whites.

Anger at a group that is perceived to be threatening (perhaps combined with fear) will dispose a person to take actions intended to thwart or perhaps even harm members of the other group (Dijka 1987; Frijda et al. 1989). For example, whites who feel threatened by a possible black "invasion" of their neighborhood may try to thwart the influx by various means (e.g., zoning changes, "steering" by realtors) or even may resort to violence against new black homeowners.

Harm To Own Group

A person may perceive that another ethnic group has done harm, or threatens harm, not only to himself personally but to his own ethnic group as a whole. Individuals anchor their personal identities to a considerable extent in the social categories which are applied to them by others—for example, as a black, a Japanese, a Catholic, a Jew, a Serb (Hogg and Abrams 1988). Because ethnic group membership is part of an individual's self-identity, she will experience the gains and losses, the triumphs and disasters of the ethnic group as her own.

For example, several studies found that most white opposition to school busing was based not on personal interest but on perception that black leaders were pushing "too fast" and that this was resulting in disadvantage for members of their own racial group (Bobo 1983). Bobo asserts that this research demonstrates that white opposition to busing reflects group conflict motives (1983: 1196). He states further: "People can form an opinion about ongoing and controversial issues like busing simply by thinking in terms of interests of 'myself and people like me.' People need not be touched by busing directly . . ." (1983: 1208). Consistent with this perspective, opposition to busing by working-class whites in Boston in the 1970s was related to their perceptions that other groups (such as blacks) were outdoing their own group, rather than to problems in their personal lives.

A study of negative attitudes toward immigrants and minority races in twelve European countries (Quillian 1995) assessed the extent to which such attitudes were related to the position of the responding individuals versus the position of their groups. An individual's own position was indicated by such variables as his education, income, change in economic status, and life satisfaction. Threat to the person's own group was indicated by the relative size of the out-group and economic conditions in the country. Results showed that individuals' characteristics had little effect on their prejudice and explained none of the difference in levels of prejudice among countries. However, the measure of perceived threat to the respondent's own group was a good predictor both of the level of prejudice within countries and of different average levels of prejudice among countries. Quillian asserts: ". . . it is the *collective* feeling that the dominant group is threatened that leads to prejudice against the subordinate group. This group-threat theory highlights the crucial importance of a notion of 'our' race or nationality and the 'other' race or nationality in the formation of prejudice" (1995: 592).

Appraisals that one's own ethnic group has been harmed, or is threatened with harm, from another group often stem from the competition that occurs among groups. Sometimes competition occurs over pragmatic concerns, such as land, jobs, housing, or recreational space. In addition, there is often competition among ethnic groups for prestige and power.

After reviewing conflicts between ethnic groups all over the world, Donald Horowitz writes: "Group attributes are evoked in behavior and subject to evaluation. The groups are in implicit competition for a favorable evaluation of their moral worth" (1985: 142). In African, Asian, and Caribbean countries, a frequent basis for comparison is the extent to which the group is "backward" versus advanced (in the sense of adjustment to modern technology and customs). The more advanced ethnic groups feel superior to the more backward groups, while those that are more technically backward may claim a moral superiority on the grounds of living a more "pure" and uncorrupted life. However, members of the more backward groups are likely to feel weak and powerless with respect to the more advanced groups. Attempts to not yield prestige and power to another group sometimes lead members of a more backward group to become aggressive. Discussing the reaction of so-called backward people to having been surpassed in modern skills by another group, Horowitz writes:

> It seems to have made such groups determined not to yield prestige or power whenever it was in their grasp, and it has shaped their behavior in myriad ways. There is much evidence that so-called backward groups are more frequent initiators of ethnic violence and advanced groups more frequent victims. To see this, one needs only to examine repeated sequences of rioting involving Hausa and Ibo, Malays and Chinese, Assamese and Bengalis, Sinhalese and Tamils, Lulua and Baluba. More than this, however, the sense of backwardness is a profoundly unsettling group feeling. It means that strangers are "wresting from one's people mastery over their own fates." (1985: 166)

The most important symbol of group prestige and control is that of language. Bengali and Assamese speakers in Assam, Sinhalese and Tamil speakers in Sri Lanka, French and English speakers in Quebec, Chinese and Malay speakers in Malaysia, as well as other ethnic groups all over the world, have disputed about when and where different languages may be used (e.g., whether or not one language would be "official"). The issue is important not only for practical reasons (such as job advantages to those speaking an "official" language) but because recognition of one's own group's language on an equal (or superior) plane with other groups' languages connotes high prestige and a sense that one's group "owns" (or at least shares ownership of) the country.

In the United States, as in other countries, people often are concerned not only with threats to the pragmatic interests of their ethnic group but to the prestige of their group and its ability to have control of its own affairs. Members of various ethnic groups—especially minorities—often react with anger to perceived slurs on their ethnic group. Many minority group members have been resentful of "outside" control of their communities by police, school teachers, landlords, merchants, and politicians from other ethnic groups (Blauner 1969).

In schools, members of minority groups are often resentful about what they see as lack of sufficient representation of their group in positions of prestige—for example, cheerleaders, Prom "queens," and student council members (Patchen 1982). Such concerns usually have little to do with any pragmatic benefits associated with holding such positions; rather, holding such positions is symbolic of group prestige and belonging.

Members of high-prestige majority groups sometimes have similar concerns about the position of their own group. Many non-Hispanic whites in parts of California, Texas, and Florida have been resentful of the numerical and cultural "dominance" of their areas by Spanish speakers (e.g., Cubans in the Miami area). Many feel as though they have become "outsiders" who no longer "belong." Hostility to immigration and to immigrants among some Americans, as well as among many Europeans (e.g., Germans, British and French) is based in part on such feelings of being overwhelmed by outsiders (Morgenthau 1993). Some Americans want to make English the only official language, in part out of a desire to keep the country "theirs." But many Spanish speakers in the United States feel that the right to use Spanish in such official locations as schools and government offices provides dignity and recognition to their own ethnic group.

The overall appraisal that another group represents a threat is likely to produce a combination of anger and fear. Because the individual identifies with her own group, she is likely to fear damage to its interests by another group. But because the possible damage is not necessarily to her own person or family, fear is likely to be weak—not the kind of visceral fear that comes, for example, from a salient threat of personal physical attack (Scherer 1988). The dominant emotion experienced is likely to be anger—anger at what is seen as an illegitimate threat to the welfare of the group to which one belongs.

Such anger is apt to lead, in turn, to tendencies to try to thwart and perhaps harm the threatening group and to defend her own group. Thus, for example, a person may make contributions to, and vote for, candidates who oppose the claims of the other group. More actively, she may join in demonstrations and even riots against the other group and its claims and in defense of her own group's claims.

Legitimacy and Power

If a person thinks that members of another ethnic group have harmed him in some way, he will tend to be angry at them. His anger will be enhanced when several other perceptions also are present (see table 3.1, Part A). One perception is that the harm that the others inflicted on him was intentional. Especially important are his perceptions of the legitimacy of the others' actions. Many times members of another ethnic group are seen as violating rules of custom, decency, or fair play. For example, a businessman who sees Jewish or Korean competitors as having reduced his profits may believe they use unfair business practices. A physical attack by a member of another ethnic group usually is seen as unprovoked and morally outrageous. Therefore, anger at, as well of dislike of, members of the other group is likely to be great.

Similarly, when people think that members of another ethnic group are acting in ways that harm their own group, their reactions depend in part on whether they see the other group's actions as legitimate or illegitimate. For example, demands by a non-English speaking group that its language be used in schools or public places are most likely to cause anger among those native English speakers who believe that it is right and proper that only their language should be spoken in "their" country.

If a person thinks that members of another ethnic group intend him harm in the future, his reactions will depend in part on how powerful he believes that group to be. As table 3.1, Part A-2 indicates, seeing some group as a personal threat requires that they are not only "bad" but also powerful enough and active enough to do harm (Osgood et al. 1957). For example, the personal fear that many white children have of black schoolmates is based in part on their perception that blacks are good fighters and often initiate fights (Patchen 1982; Schofield 1982). If a person perceives both that another group has harmful dispositions and relatively high power, his overall appraisal will be that a threat is present. Therefore, the person will be fearful.

Similarly, an appraisal of threat to one's own group arises not only from the perceived intent of the other group but also from its perceived strength and activity. If another ethnic group is seen not only as intending harmful actions, but also as having the power to carry out such actions, and as being active in pursuing its designs, then it will be seen as a threat. For example, Jews have been considered as a threat by those who see Jews as not only having undesirable aims but also as wielding great influence in business and government (Martire and Clark 1982). Similarly, Catholics have been seen as threats by Protestants in Northern Ireland and sometimes elsewhere not only because of their "undesirable" religious and political aims but because their large and growing proportion of the population in these areas gives them great potential power (Stark 1964; Darby 1976).

Behavior of Other Group As Reprehensible

A second major type of negative appraisal of another ethnic group (as shown in table 3.1, Part B) is that the behaviors of members of this group are reprehensible; that is, that their behavior should be condemned as improper or immoral. This type of appraisal is based first on cognitions that their actions are inconsistent with the values or norms that one accepts as proper.

With respect to perceptions of blacks, Katz and his colleagues point out: "It is well known that black rates of unemployment, welfare dependency, school failure, illegitimate birth, crime, and drug addiction are much higher than white rates" (1986: 44). For many white Americans such behaviors are discrepant with their individualistic values of hard work and personal responsibility (Katz et al. 1986; Jaynes and Williams 1989).

On the other hand, many blacks have been upset by perceptions of white police brutality, exploitation by merchants, discrimination by landlords and employers, and bias in the media (Turner and Wilson 1976; Farley et al. 1979). Such behaviors are seen as violating standards of equality and justice.

In addition to such perceptions of undesirable behaviors in the larger society, behavior by out-group members in specific settings may be seen as unpleasant or contrary to one's own values and norms. For example, a study of interracial contact among high school students found that white students tended to see black school-mates, compared to white schoolmates, as loud and noisy, not obeying school rules, starting fights, and not trying hard in school. Black students tended to see white schoolmates, compared to black schoolmates, as less willing to help blacks, acting superior or "stuck up," and expecting special privileges (Patchen 1982).

A second cognitive element contributing to an overall appraisal of reprehensible behavior is that members of the other ethnic group are fully (or at least primarily) responsible for their own behavior. They are seen as freely choosing to behave in undesirable ways, rather than having been forced to behave in these ways by another ethnic group, or by other outside agencies or circumstances. For example, high rates of illegitimacy and welfare status among American blacks are seen by many whites as due to the personal choices of people whose standards of morality and responsibility are lax (Apostle et al. 1983).

If the individual's overall appraisal of a minority group is that the behavior of (most of) its members is reprehensible, then she will experience emotions of dislike and perhaps even disgust. Such feelings will lead people to avoid members of the disliked group (Frijda et al. 1989). In addition, they may have the tendency to refuse help to members of the group (Piliavin et al. 1969). Similar feelings of dislike and disgust toward the dominant group lead some minority members to withdraw from contact with the dominant group as much as possible.

Other Group Has Undeserved Advantages

A third important type of negative appraisal is that members of the other ethnic group have a position relative to oneself (or one's own group) that is undeserved (see table 3.1, Part C). A first cognitive component of this appraisal is that other group members possess material benefits, power, or status that are greater than one's own or higher than some accepted standard (such as their traditional position). For example, Asians in East Africa may be seen as wealthier than native Africans, Protestants in Northern Ireland seen as having greater wealth than Catholics, and those of Indian descent seen by native Fijians as enjoying higher economic status than themselves.

A second key component of this negative appraisal is that the position of the other group is illegitimate. The bases of this judgment will vary. The other group may be seen as foreigners in "our" land (e.g., Fijian perceptions of ethnic Indians). A dominant group may be seen by a weaker minority as being favored by law, custom, or government preference (e.g., African Americans seeing European

Americans as favored by employers, banks, and the criminal justice system). A majority group may perceive a minority group as receiving unfair advantage in school admissions, jobs, and so on, because of government programs that favor minorities (e.g., affirmative action programs in the United States, India, Malaysia, and other nations).

An appraisal that members of another ethnic group have an undeservedly high position relative to oneself or one's own group will tend to result in anger at this "unjust" situation, as well as dislike and, perhaps, even envy of the other group. Anger will tend to be directed at the advantaged group, as well as at agencies (usually government) perceived to be responsible for the "unjust" situation. However, anger at the other ethnic group will be greater if it is seen as having some responsibility for the unjust situation (e.g., if it had lobbied successfully for preferential treatment).

Anger will tend to result in protest against the perceived injustice. For example, Brahmins in India have protested vigorously against preferential school admissions for the lower castes. Anger, and accompanying envy, may lead also to aggression against the group perceived to be unjustly advantaged. Thus, in the Philipines, ethnic Filipinos have at times physically attacked the commercially successful ethnic Chinese.

Positive Appraisals, Emotions, and Behavior

While the focus of this chapter and most work in ethnic relations is on negative reactions toward those in other groups, people's reactions may, of course, be positive instead. Table 3.2 lists three major types of positive appraisals of another ethnic group, the cognitive components of such appraisals, and the emotions and behaviors that are likely to result from the appraisals.

The three major types of appraisals listed are: 1) the other group is beneficial to oneself or to one's own group; 2) the behaviors of members of the other group are praiseworthy; and 3) the relative position of the other group is deserved or less than deserved. Note that these appraisals parallel the three major types of negative appraisals discussed previously, except that each is positive rather than negative—for example, that the behaviors of members of the other group are praiseworthy rather than reprehensible.

Other Group Seen as Beneficial

To illustrate appraisals of another group as beneficial to oneself, Americans tend to see Jews as intelligent and hardworking. If they see Jews as directing their talents and energies (e.g., as writers or political

TABLE 3.2

Positive Appraisals of and Emotions and Action Tendencies Toward Members of
Another Ethnic Group

Appraisals	Primary Emotions	Primary Action Tendencies
A. Other group benefits self (own group)		
1. Has helped in the past Actions increase own (group's) material or physical well-being, status, or power	Liking Gratitude	Approach

2. May help in the future
   a. Is disposed to take actions helpful to self (own group)
   b. Is powerful
   c. Is active
	Liking	Approach

B. Behaviors of members of other group are praiseworthy 1. Actions of others pleasant and/or are consistent with perceiver's own values	Liking Admiration	Approach Help
2. Actions of others are freely chosen; they are responsible		
C. Other group's economic or social position is deserved, or less than deserved	Satisfaction (Sympathy, pity)	No action tendency, if position deserved (Help other group, if their position less than deserved)

activists) toward desirable goals, they may see Jews as benefiting themselves personally or benefiting the community. Such an appraisal would lead to them to like Jews and perhaps feel gratitude towards Jews. Such positive feelings would dispose them to have contact with Jews and to support Jews if and when Jews needed help.

Some white students may see blacks as a group that benefits them personally. Many students see black schoolmates as "fun to be with" (Patchen 1982). If white students also believe that the "fun" activities of the blacks are legitimate (not disruptive), then they are more apt to like their black schoolmates and, perhaps, to be grateful to them for bringing enjoyment to a setting generally considered boring. If so, they are likely to seek out the blacks' company and to help them if, say, blacks need support of a request to the principal for certain school entertainments.

Black schoolmates may also be seen by white students as a group that is beneficial to the total student body to which the whites, as well as the blacks, belong. Blacks are seen as good athletes by most whites. If black students are active and make contributions in school sports and their actions in those activities are seen as legitimate (e.g., not unfairly competing with whites), then blacks are likely to be seen by whites as benefiting the whole school (including white students) by helping to bring prestige to the school. Because of this understanding, whites are more apt to like, feel gratitude, and be friendly towards black schoolmates.

Americans' appraisals of Afghans when Afghanistan was struggling against Soviet occupation is another example of an appraisal of another group as beneficial to one's own group. The resistance by the Afghans against Soviet control was seen by many Americans as helping our own country's battle against communism and against Soviet expansion at that time. The actions of the Afghans were seen by Americans as completely justified and the guerrilla fighters were also seen as being strong and active enough to seriously impede the Soviets. The overall appraisal that Afghan actions were benefiting the United States led most Americans to like and to feel gratitude towards the Afghans, as well as to give considerable material support to their efforts.

Sometimes, a particular ethnic group may be seen as being beneficial to oneself (one's own group) in the future. For example, some Americans view the growing Asian American population in the United States as likely to provide a source of needed scientific and technical talent, as well as cultural ties to important trading partners in Asia in the future. Those who see a particular group as likely to be helpful in the future will tend to like and to welcome that group.

Other Group as Praiseworthy

An appraisal of the behaviors of other group members as praiseworthy is exemplified by the appraisals of Asian Americans (Chinese, Koreans, Japanese, Vietnamese, and so on) made by many other Americans. Images of Asian Americans as hardworking, law-abiding, and having a close family life are widespread (Takaki 1994). Such characteristics are consistent with values that Americans traditionally have held. Moreover, such value-consistent behaviors generally are seen as due to the importation of Asian culture by the Asian Americans; thus, the desirable behaviors (such as hard work) generally are seen as behaviors that Asian Americans have freely chosen rather than forced upon them by harsh circumstances or by outsiders. These perceptions, when present, tend to produce feelings of liking and admiration for Asian Americans. Such feelings, in turn, dispose many people to be willing to have social contact with Asian Americans and to help them if necessary (e.g., against acts of bigotry by some persons). Of course, negative appraisals of Asian Americans occur frequently as well, as they do for many other ethnic groups.

Position of Other Group (Less Than) Deserved

Sometimes members of another group are seen as getting no more than (or even less than) they deserve. An ethnic group that tends to be more economically successful than other groups (e.g., Jews, Mormons, and Japanese in the United States, ethnic Chinese and ethnic Indians in other countries) may be seen as having earned its advantage by ability and hard work. This group even may be seen as having a position less than what it deserves, because of discrimination against its members.

Even when members of an ethnic group have been given some special advantages by the government, their position still may be viewed by some others as being deserved. For example, many white Americans believe that African Americans generally have been deprived of a fair share of the wealth and privileges of American society as a result of centuries of slavery and then discrimination. Such white Americans may see programs that give preferences to African Americans in jobs or university admissions as an effort to make up (only partially) for this history of oppression. In this historical context, they would appraise the present economic position of some blacks who have benefited from preferences as no more than, and perhaps even less than, they deserve.

Many white Americans also view the special privileges enjoyed by Native Americans in some areas (e.g., to fish and hunt in otherwise restricted areas, or to carry on otherwise prohibited gambling operations) in a similar way. They recall the way in which whites seized land occupied by Native Americans and killed or expelled the occupants. Against this historical background, some special group advantages for Native Americans may be seen as deserved.

When people see the economic or social position of another ethnic group as deserved, they are apt to be satisfied with the situation. When they see those in another group as getting less than deserved, they will tend to feel sympathy for this group and to support efforts to correct the injustices done to the other group.

Behavior as Influence on Attitudes

While we have focused on the effects that attitudes may have on behavior, sometimes the direction of causation may be reversed; that is, a person's behavior may affect her attitudes. This has been found to be true with respect to a number of social attitudes and behaviors; for example, although attitudes toward divorce do not predict such action, those who get divorced have more positive attitudes toward it (Wiggins et al. 1994).

Behavior towards those from another ethnic group sometimes affects attitudes toward that group. For example, white people who moved into racially mixed public housing, primarily because this was the best housing available to them, generally developed more positive attitudes toward blacks than they had previously (Deutsch and Collins 1951). After whites in the American south desegregated their schools and other public facilities, as a result of court decisions, there was a large shift in their attitudes toward acceptance of racially mixed facilities (Bem 1970; Rossell 1983).

Commenting on these changes in attitudes toward racial integration, Daryl Bem states: "We can now see one of the reasons why legislation and court decisions *can* change the 'hearts and minds of men,' why 'stateways can change folkways.' They do so, in part, by effecting a change in behavior; then when behavior has been changed, attitudes often follow" (1970: 69).

Changes in a person's behavior may lead to changes in her attitudes for several reasons. New behavior may result in new experiences that produce different perceptions and feelings. For example, whites who moved next to blacks often found their black neighbors to be respectable and helpful people; therefore their perceptions of and feelings toward blacks would become more positive. In addition, people often feel uncomfortable if their behavior and attitudes are inconsistent; that is, they experience what has been termed "cognitive dissonance." Eliot Aronson observes: "If I know that you and I will inevitably be in close contact, and I don't like you, I will experience dissonance. In order to reduce the dissonance, I will try to convince myself that you are not as bad as I previously thought" (1972: 194).

However, inconsistency between a person's attitude and behavior will produce a change in attitude only under certain conditions (Wiggins et al. 1994). One necessary condition is that the person is committed to a certain behavior or that it would be difficult for him to act differently. Otherwise, he could change the behavior, rather than the attitude. For example, whites in the racially mixed housing project studied by Deutch and Collins would have found it difficult to move to equally attractive and affordable housing.

A second condition that facilitates attitudes changing to match behavior is that the person believes he is acting voluntarily. If a person believes that he has been forced to act contrary to his attitudes, he may accept this inconsistency without great discomfort because he sees himself as having had no choice. Those whites who moved into a racially mixed housing project clearly acted voluntarily. The situation is not quite as clear for southern whites who attended (or sent their children to) racially mixed public facilities, but those individuals often had some alternatives (e.g., not patronizing integrated restaurants, sending their children to private schools).

In sum, a person's behavior sometimes affects his attitude—especially when he voluntarily commits to a behavior inconsistent with his initial attitude. To change attitudes, trying to *induce* changes in behavior (perhaps by offering incentives) is more effective than to *coerce* people to change their behavior.

Other Subjective Determinants of Intergroup Behavior

In addition to attitudes and the closely-related phenomena of appraisals and emotions, a number of other subjective factors have been identified as having an effect on intergroup behavior. These factors include general beliefs and values; explanation of differences in group outcomes; perception of social norms; perception of external incentives; identification with one's own ethnic group; and personality characteristics. We will briefly consider some of these additional influences on behavior towards other ethnic groups.

General Beliefs and Values

Attitudes and behavior toward members of another ethnic group may be affected by a person's more general beliefs and values. Beliefs about how people and government should behave, about what kind of society is desirable, and about why inequality between groups exists are all relevant.

One relevant set of beliefs concerns individuals' responsibility for their own welfare. Americans traditionally have believed in the importance of hard work, thrift, and self-reliance. There is widespread belief among white Americans that many members of minority groups—especially blacks—do not share or live up to these values of individualism. Such beliefs tend to lead to hostility towards minority groups and towards the lack of support for government programs designed to improve the welfare of minorities (Jaynes and Williams 1989).

A different set of beliefs widely held by Americans may be labeled "communitarian" values. These beliefs concern the desirability of helping others, especially those in need and those who have been mistreated. Holding such values may lead to more positive feelings toward minority group members, such as disadvantaged black people, and to support efforts to help such groups. Irwin Katz and his colleagues have found evidence that many Americans hold both individualistic and communitarian values. Katz and his colleagues believe that this leads many white Americans to have ambivalent feelings toward blacks (1986).

Another value that is relevant to intergroup relations is the belief or non-belief in equality as a social goal. Attitudes toward "affirmative action" programs aimed at producing more equal outcomes among racial groups tend to be correlated with whether or not Americans hold egalitarian values (Jaynes and Williams 1989).

Also important is whether or not people are willing to accept government mandates to achieve outcomes such as non-discrimination. After the influence of prejudice towards black people is taken into account, those who generally oppose government restrictions on behavior (as reflected, for example, in opposition to mandatory seat belt laws) are more opposed than others to open-housing laws (Schuman and Bobo 1988).

Explanation of Group Differences

Attitudes toward disadvantaged minorities and toward government programs aimed at helping such minorities are also affected substantially by the way in which people explain differences between groups, such as differences in income. A survey of the general population of the Bay area in northern California (Apostle et al. 1983) asked whites a series of questions concerning "why on the average white people get more of the good things in life in America than black people;" "why whites tend to do better on intelligence tests than blacks;" whether they thought "the fact that the average black person is less well off than the average white person" is mostly, partly, or not at all "the fault of white people living today;" and how much they thought that "it is the fault of black people that they don't do as well as whites." On the basis of their answers to these questions, the researchers categorized explanations of racial differences into a number of categories, several of which focused on perceived characteristics of black people (such as lack of effort or genetic inferiority) and some of which emphasized environmental reasons (such as discrimination and racism in American society).

People who explained differences in outcomes between the races as due to characteristics or behavior of blacks themselves (or to God's will) were much less likely than those who ascribed these differences to societal barriers to support government actions aimed at increasing the proportion of blacks in all types of jobs and at eliminating barriers to open housing. They were also less likely to approve of marriage between blacks and whites. The authors of the study conclude: "The results, in sum, lend strong support to a conclusion that how racial differences are explained is crucial to how whites respond to the black condition in America and what they are willing to do to improve it" (1983: 215).

The importance of beliefs regarding the causes of income differences is also shown in a study based on a national sample of Americans (Kluegel and Bobo 1993). Respondents were asked to rate the importance of four possible reasons "why there are poor people in this country": the lack of good schools; the failure of industry to provide enough jobs; loose morals and drunkenness; and lack of effort by the poor. The more that Americans attributed the existence of poverty to problems of individuals (morals and effort), rather than to the society (lack of good schools and jobs), the more likely they were to oppose government programs aimed at providing more jobs, spending more on schools and scholarships, and generally trying to raise living standards—regardless of whether such programs are targeted at the poor generally or at blacks in particular. When reacting to race-targeted programs, those giving individualistic explanations of poverty preferred programs that focus on equalizing opportunities, not on equalizing outcomes. Another study (Kluegel and Smith 1986) also found that white people's attitudes toward government programs aimed at helping minorities are influenced substantially by their beliefs about why some people do better economically than do others.

Measures of general conservatism have also been found to be related to attitudes on racial issues (Weigel and Howes 1985). For example, people who rated themselves as conservative on a liberal-conservative continuum were more likely than those rating themselves as liberal to oppose busing for racial balance (Bobo 1983). Since conservatives are more likely to hold such beliefs as the necessity of social inequality and the responsibility of individuals for their own outcomes, and to oppose the use of government coercion, the association of general conservatism with opposition to public efforts to redress racial inequalities is consistent with the other evidence presented earlier.

Perception of Social Norms

A considerable amount of research documents the effect of perceived social norms on behavior in general and on intergroup behavior in particular (Hogg and Abrams 1988). Studies in workplaces, schools, and other settings have found that the amount of social contact that blacks and whites have with members of the other racial group is affected strongly by their expectations of approval or disapproval for such behavior (Minard 1952; Warner and Defleur 1969; Patchen 1983; Gaertner and Dovidio 1986). For example, some white students avoid associating with black schoolmates because they expect that such associations will cause them to be rejected by peers, while some black students avoid association with white schoolmates because they fear such behavior will cause them to be labeled by peers, as "Uncle Toms" or "Oreos" (Patchen 1982). Both the actual and perceived norms of peers affect behavior towards people of another race, even when the effect of personal racial attitudes are controlled (Patchen 1983).

Perception of External Incentives

A person's actions toward members of another ethnic group may be influenced by his perception of external incentives—that is, of practical benefits or costs that may follow his actions. Sometimes this leads to actions that avoid contact with another group. For example, some white people prefer not to send their children to racially integrated schools because they believe this interaction will result in a poorer education. They may oppose school busing for racial balance because they are genuinely concerned about long bus rides for their children.

In other cases, perceptions of practical advantage may contribute to actions that increase intergroup contact. For example, whites may choose to live in a racially mixed housing complex because this provides them with a desirable apartment at a lower cost than is available elsewhere.

Similarly, perceptions of possible benefits and costs may lead either to discrimination or to non-discrimination against members of a particular ethnic group. Employers may choose to pay minority workers less than they pay others because they recognize that such actions will lead to larger profits (Grebler et al. 1970: ch. 10). Conversely, an employer may hire or promote people from minority groups because they wish to avoid legal penalties resulting from charges of discrimination.

Identification with One's Own Ethnic Group

There is a long history to the idea that persons who glorify the groups to which they belong tend to be hostile towards out-groups (Sumner 1906; Adorno et al. 1950). Social identity theory predicts that the more strongly a person identifies with an in-group, the more she will tend to show bias against out-groups (Hogg and Abrams 1988).

Attachment to an in-group does not always result in bias or enmity towards out-groups; some studies have found little relationship (Abrams 1992). However, an association between identification with an in-group and bias towards out-groups has been found both for some laboratory groups (Grant 1991) and for some real groups (Hollister and Boivin 1987; Kelly 1988). Stronger attachment to an in-group appears most likely to lead to bias and hostility towards out-groups when invidious comparisons between the groups are salient and especially when there is conflict of interest between the groups (see chapter 2 for a more extensive discussion of this subject).

Personality Characteristics

There is considerable evidence that certain personality characteristics (e.g., authoritarianism and dogmatism) dispose some people to be hostile to out-groups (Simpson and Yinger 1985). Moreover, those individuals who are hostile to one particular out-group (e.g., blacks) also tend to be hostile to other out-groups, such as Jews, homosexuals, and even the elderly (Weigel and

Howes 1985).

A person's personality traits may dispose her to make certain kinds of appraisals (e.g., a suspicious person may be more likely to perceive threat from another group); to magnify certain kinds of emotions (e.g., the person with much general anger may displace it onto other ethnic groups); and to value certain outcomes highly (e.g., the authoritarian may value conformity highly and thus be greatly influenced by perceived social norms).

However, while personality traits sometimes are an important influence on the ethnic attitudes and behavior of particular individuals, personality differences do not explain why specific ethnic groups generally are the target of hostility by other ethnic groups at certain times, nor why levels of hostility towards a specific group vary widely over time and location (Pettigrew 1958). (Because the focus of this book is on differences in relations among ethnic groups in society as a whole, we will not devote great attention to the effects of individual personalities.)

A Model of Subjective Determinants of Behavior

As the previous discussion indicates, many subjective factors—including appraisals, emotions, beliefs, values, identifications, and perceptions of norms and incentives—appear to have some impact on behavior towards members of another ethnic group. How do all of these subjective factors combine to affect behavior?

Figure 3.1 shows a model of the subjective factors that affect behavior towards members of a given ethnic group. (For other related models, see Ajzen and Fishbein 1980, and Eagly and Chaiken 1993: 209.) Two major direct influences on behavior are indicated in figure 3.1. One is the effect of *emotions*, which has already been considered at some length in this chapter. Emotions result directly in behavioral tendencies that are consistent with the specific emotions (Frijda et al. 1989). For example, fear leads to tendencies to move away from and anger results in tendencies to move against the target of these emotions.

A second major influence on behavior shown in the model is a person's *expectancies* about how desirable the outcomes of her actions will be. Here, one may draw on the general theoretical framework known as decision theory, a theory that has been used with some success to explain and to predict people's behavior (Abelson and Levi 1985). In its most general form, decision theory postulates that a person's actions are determined by 1) her expectations about the outcomes of her possible actions; and 2) the values (positive and negative) that she places on each of these various outcomes. For example, the likelihood that a

FIGURE 3.1

A Model of the Effects of Subjective Reactions on Behavior Towards
Ethnic Out-Groups

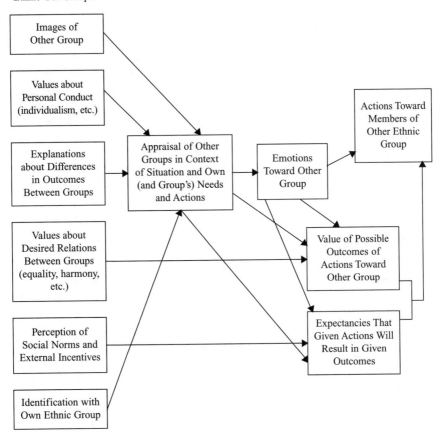

white person will buy a house in an interracial neighborhood might depend on 1)
her expectancies that this action will result in frequent contact with black neighbors, that the re-sale price of the house will stay high, that she will be a victim of
crime, and that family and friends will approve; and 2) how important each of
these possible outcomes is to her.

As shown in figure 3.1, a person's expectancies about the outcomes of her
actions and the value she places on those outcomes may be affected by her feelings or emotions. For example, if a person likes (most) black people she knows,
the value of social contact with black neighbors will be fairly high; if she is fearful of blacks, her expectancy of being attacked in an interracial neighborhood
would be raised. Thus, emotions toward a particular ethnic group may affect

behavior towards members of that group in two ways: directly, by the natural tendency to act consistently with one's feelings, and indirectly, through the effect of a person's emotions on her expectancies about the consequences of her behavior.

Figure 3.1 also indicates the important role of people's *appraisals* of other ethnic groups. A person's appraisals of an out-group will influence her emotions toward members of that group (as discussed earlier in this chapter). For example: an appraisal of threat to one's self or one's group will lead to fear and usually will lead to anger as well; appraisals of the other group as behaving in a reprehensible way will result in feelings of dislike and, perhaps, even disgust.

A person's appraisals of the other group also may affect her expectancies of, and values for, possible outcomes of action. For example, a person who sees blacks as a threat to her personal safety will have a high expectancy that neighborhood contact with blacks will result in personal injury. If she sees (most) blacks as behaving in a reprehensible way, then she may put a low value on contact with black neighbors.

Other subjective factors shown in figure 3.1 are viewed as affecting interethnic behavior primarily in either or both of two ways: 1) through their effects on people's appraisals of members of another group; and 2) through their effects on people's expectancies and values regarding the outcomes of their behavior.

A person's *images* of another group, his *values about personal conduct,* and his *explanations for differences in outcomes between groups,* all may affect his appraisals of the other group. For example, if a person 1) sees many Mexican Americans as having little education; 2) has individualistic beliefs about each person's personal responsibility for his own welfare; and 3) explains differences in educational and occupational outcomes between groups in terms of personal traits, such as motivation, he is likely to appraise Mexican Americans as having a deservedly low socioeconomic position.

Identification with one's own ethnic group may also affect appraisals of an out-group. As noted earlier, a person who identifies strongly with an in-group may tend to have more negative appraisals of an out-group, especially if the groups are competing for status or have a conflict of interest.

Perceptions of the social norms of those within one's own racial group may affect a person's expectancies of approval or disapproval for particular kinds of behavior toward members of an out-group. Similarly, the presence and nature of *external incentives,* such as the legal penalties that exist for discrimination, and how strictly these are enforced, would affect expectancies of external rewards and costs, as well as the value of such outcomes for particular kinds of actions.

Beliefs and values concerning desirable relations between groups also may affect a person's values regarding outcomes which may follow his behavior towards members of an out-group. For example, a belief that racial integration and harmony is a desirable goal (see Saltman 1990) would raise the value of contact with black neighbors. Beliefs in helping others (communalism) and in

70

equality would raise the value of improved opportunities for minorities, a possible outcome of supporting programs to help minorities. The individual might expect to feel personal satisfaction as a result of doing things that contribute to such valued goals.

The model presented in figure 3.1 includes only subjective factors that may affect attitudes and behavior toward those in other ethnic groups. It does not take account of the possible effects of contact between members of different ethnic groups, nor of the more indirect effects of conditions, such as economic inequalities, in the larger society. The impact of interpersonal contact between members of different ethnic groups is discussed in the next chapter. Later chapters consider the effects of conditions in the larger society.

Summary

A person's behavior towards members of another ethnic group may be affected by her attitudes, which are composed of her perceptions of and feelings toward that group. However, attitudes and behavior are not necessarily consistent.

Specifying the ways in which people see another ethnic group as significant for themselves (their appraisals of this group) and how specific types of appraisals affect their feelings and behavior toward the other group is important. One important type of negative appraisal of another ethnic group is that this group has harmed, or is a threat to harm, oneself or one's own group. People may be concerned about harm to themselves personally (e.g., that they may be physically assaulted). They may be concerned also about harm to the prestige, power, or welfare of their own group (e.g., that the language of their group will no longer be dominant).

When people see themselves or their group as having been harmed by another group, they are likely to be angry and will tend to retaliate by harming the others. When they see a threat of future harm from the other group, they are likely to be fearful, as well as feeling some anger and dislike of the others. They will tend then to avoid the others, to protect themselves, or to attempt to suppress the group seen as a threat. Which feelings and which actions are dominant will depend, in part, on the relative power of the groups involved.

A second type of negative appraisal of another ethnic group is that the behaviors of members of this group are reprehensible (should be condemned) because they do not meet one's own standards of proper behavior. For example, members of the other group may be seen as lazy, dishonest, and sexually promiscuous. If people appraise the behavior of those in another group as reprehensible, they are likely to feel dislike for, and to try to avoid, members of this group.

A third type of common negative appraisal of another ethnic group is that it enjoys undeserved advantages relative to one's own group. For example, members of the other group may be seen as generally having better jobs and higher incomes than those in one's own group because they have been unfairly favored by employers. When a person sees another group as enjoying undeserved advantages, he is apt to be angry and will tend to protest and/or to harm members of the other group.

Positive appraisals of another ethnic group are the mirror opposite of the negative appraisals: that the other group benefits oneself or one's own group; that the behaviors of those in the other group are praiseworthy; and that the economic and social position of the other group is deserved, or less than deserved. Such positive appraisals are likely to lead to positive feelings, such as liking, and to positive behaviors, such as approaching members of the other group.

In addition to appraisals and emotions, a number of other subjective factors may affect people's actions toward members of another ethnic group. Willingness to help disadvantaged minority groups is affected by people's general beliefs and values, such as those concerning individual responsibility, community responsibility, equality of outcomes, and government activism. Attitudes toward programs aimed at helping disadvantaged minorities are affected also by the way in which people explain differences between groups, such as differences in income. People who see group differences as due to personal characteristics, such as amount of effort, are less likely to support remedial programs than are those who see group differences as due to the situation (e.g., to lack of opportunity). Behavior towards members of another ethnic group may be influenced also by social approval and by practical incentives, such as legal penalties.

A model of the subjective factors that affect behavior towards members of an ethnic out-group was presented. Two major influences on a person's behavior may be distinguished. One is her emotions toward the out-group; the second is her expectations about the outcomes of her actions. Both emotions and expectancies, in turn, are affected by the person's appraisals of the out-group and by her other values, beliefs, and perceptions.

In the next chapter we look at how attitudes and behavior toward another ethnic group are affected by contact with members of that group.

4)   Contact Between Groups

How are the attitudes that people have towards another racial or ethnic group affected by contact with members of that group? Are hostile attitudes the result of unpleasant experiences with members of a certain group? Or are they more often the result of ignorance about another group that stems, in part, from a lack of contact with members of that group?

Many people have urged that members of different racial and ethnic groups should be brought together into closer contact. Such proposals have been based on beliefs that when people get to know those in another group better, their negative stereotypes will be weakened or eliminated and they will learn more positive attitudes and behavior.

Drake and Cayton observe, "Almost mystical faith in 'getting to know one another' as a solvent of racial tensions is widespread" (1962: 281). More recently, Sigelman and Welsh state: ". . . students of race relations have long assumed that close contact between members of different races promotes positive racial attitudes and that the lack of such contact fosters prejudice and ill will. . . . Adherents of this contact hypothesis view racial segregation as a source of ignorance and ignorance as a breeding ground for derogatory stereotypes and racial hostility" (1993: 781).

How much evidence is there for this view? In this chapter, evidence concerning the effects of intergroup contact on people's attitudes and behaviors is examined. After considering first some of the overall effects of contact, we will examine the conditions under which intergroup contact is most likely to lead to more positive attitudes and behavior. We will also discuss the psychological and social processes that affect the outcomes of intergroup contact.

Evidence of Positive Effects

There is evidence that, in many instances, increased contact between people from different racial or ethnic groups has resulted in their having more positive images of and feelings toward each other and their behaving in more friendly ways. Much of the research on this subject has concerned contact between blacks and whites. Many of these studies have been done in schools, especially in the period of extensive desegregation of American schools (St. John 1975; Prager et al. 1986).

These studies indicate that many students of both races change their racial attitudes and/or behavior in a positive direction as a result of interaction with schoolmates of a different race. An extensive study of eleven racially mixed high schools in Indianapolis provides some illustrative evidence. Thirty-six percent of all white students in these schools stated that their "opinion of most black people" had gotten better since coming to their school and 44 percent of black students said that their opinions of whites had changed for the better (Patchen 1982). A white student interviewed in one school about her experiences with black schoolmates said: "I used to feel like I'm not going to associate with them because from what you've learned, they're mean and everything. The ones I know are really great people. They just seem to fit into the school like anybody else and I'd say they are not any different than we are." A black student at another Indianapolis high school said: "I have a friend that's a white student here, and me and her, both of us, we're just almost the same. There's no difference" (Patchen 1982: 42, 53). Other studies of contact between black students and white students in racially mixed schools also have reported positive changes in attitudes and/or behavior among many of the students (Stephan 1991).

Contact between adult whites and blacks has been studied in residential settings. A number of such studies were done when public housing projects around the country began to be desegregated (Deutch and Collins 1951; Wilner et al. 1955; Works 1961). These early studies found that such contacts often resulted in positive attitude changes among whites and also among blacks. For example, a white housewife in a housing project located in a New England city said of her black neighbors: "They are just like the white people as far as I am concerned. . . . They are always dressed neat; as clean as the white; polite, as far as I can see. . . . They are friendly. . . . Why should there be all this talk about differences, or what you like or not? They seem to me the same in every way as the whites" (Wilner et al. 1955: 52). This study found that the closer the proximity of whites to black people in their housing project, the more positive were the whites' attitudes toward black people.

More recent studies in residential neighborhoods also have found that contact between blacks and whites may lead to friendly relations and more positive interracial attitudes (Taylor 1979; Helper 1986; Rosenbaum et al. 1991; Sigelman and Welch 1993). For example, low-income inner-city blacks who moved into better housing in mostly white suburbs were more likely to form interracial friendships than were a comparable group of blacks who moved to another predominately black part of the inner city (Rosenbaum et al. 1991).

Studies of contact between blacks and whites in other, often newly desegregated, settings, including work situations and the military, also provide evidence that such contacts can result in people having more positive racial attitudes. In the years following World War II, when black workers were admitted to previously all-white job settings, a number of studies found that the racial attitudes of many whites in such work settings became more positive (e.g., see MacKenzie 1948; Harding and Hogrefe 1952). For example, 55 percent of white college students who had worked with blacks occupying equal or higher-level jobs expressed favorable attitudes toward blacks as compared to only 13 percent expressing favorable attitudes among those who had not worked with blacks (MacKenzie 1948). Similar findings come from more recent studies (Jackson 1992). For example, a study of racial attitudes in South Africa concluded that positive contact experiences at work can contribute to improved intergroup relations (Bornman and Mynhardt 1991).

The military is another setting in which race relations are important. During World War II, white American soldiers had much more positive attitudes toward black soldiers after the two groups had fought together in combat situations. Only 18 percent of whites in segregated units thought it was a good idea to have both blacks and whites in the same combat companies, while 64 percent of whites who had served in racially mixed companies approved (Mannheimer and Williams 1949). In 1948, U.S. military services were desegregated by order of President Truman. Since that time, there has been little systematic research on relations between white and black personnel. But since the 1970s, the military services have devoted considerable effort to race relations training, which has had some success in encouraging positive relations across racial lines (Landis et al. 1984).

In church settings, too, there is evidence that contact between blacks and whites sometimes leads to more positive relationships. Among clergymen from five denominations, both white and black ministers with more experience in racially integrated educational institutions and churches had more favorable attitudes toward the other race than those with less experience of this type. Also, the entrance of black people into previously all-white denominations frequently has resulted in greater understanding between the racial groups (Reed 1991).

While much of the study of intergroup contact has focused on black-white relations, there has been some study of the effects of contact between other racial and ethnic groups. Positive changes in attitudes and/or behavior have been found to occur following contact between other groups—Hispanics and Anglos, Asian Americans and European Americans, Jews and Arabs, Catholics and Protestants, and many others.

Several studies have found that after Mexican American and Anglo students work together on cooperative projects in school, they usually get along better than they did before (Aronson and Gonzalez 1988). Aronson and Gonzalez describe the changed relations between Anglo students and a Mexican American boy named Carlos, as a result of their participation in such an ethnically mixed group: "After a couple of weeks, the other students concluded that Carlos was a lot smarter than they had thought he was. They began to see things in him they had never seen before. They began to like him. Carlos began to enjoy school more and began to see the Anglo students in his group not as tormentors but as helpful and responsible people" (1988: 307).

In Northern Ireland, great hostility has long existed between the Protestant and Catholic communities, accompanied by frequent interreligious violence. Despite these tensions, Catholic and Protestant children who have attended non-sectarian schools (e.g., the physically handicapped) frequently have developed friendships "across the divide" (Trew 1986). Similarly, Catholic and Protestant farmers living in close proximity on farms in Northern Ireland were found to share equipment and work closely together in their farming, though they remained segregated in other activities (Harris 1972).

Another area of the world in which great interethnic tensions exist is Israel, where the Jewish and Arab communities have long been hostile to each other. When members of these groups have been exposed to greater contact with members of the other group, their attitudes sometimes have changed for the better (Bargal 1992). For example, many West Bank Arabs working in industrial plants in Israel formed positive views of Jewish co-workers, as well as of Jews outside their work setting (Amir et al. 1982).

Contact between a variety of other ethnic groups, including Japanese Americans and European Americans (Irish 1952); Australians and Vietnamese immigrants (McKay and Pittam 1993); and ethnic Germans and Turks living in Germany (Wagner and Machleit 1989) has also been found sometimes to result in more positive intergroup attitudes. For example, children from different nationalities who attended international summer camps became more accepting of those from different nations (Bjerstadt 1962). One American girl stated: "I have changed my ideas about other peoples. I thought I would dislike them. I find them to be different in customs, but all the same, the same kind of people" (1962: 26). Contact between people of different nationalities as a result of tourism or exchange programs also can result in more positive views of other groups (Amir and Ben-Ari 1985; Kemal and Maruyama 1990).

Contact May Not Have Positive Effects

While contact between people from different racial or ethnic groups may result in more positive attitudes and more friendly behavior, such positive outcomes do not necessarily occur. Often intergroup contact has little effect, and sometimes even has a negative effect, on the way people feel and act towards those in another ethnic group. Amir observes: ". . . there is evidence that intergroup contact does not necessarily reduce intergroup tension or prejudice. At times, it may even increase tension and cause violent outbreaks and racial riots" (1976: 246).

The many studies of contact between black and white students in racially mixed schools report many instances in which intergroup contact results in negative perceptions of the other group and less favorable attitudes than existed before the additional contact occurred. For example, the study of interracial contact in the Indianapolis high schools quoted a white boy who wrote at the end of his questionnaire: "I have never been against black people, but these last two years I have really changed my attitude, just because it seems the higher authority in this school seems to be afraid of them and they get by with more than whites" (Patchen 1982: 53). A black girl in the same school system said of white schoolmates: "I don't like their attitudes. Some of them are just plain mean. They treat you like you're nothing but something you can walk on. . . . Some white people act like they're better than everybody else" (1982: 43).

While over one-third of white students in this school system said their opinions of most black people had gotten better since coming to their recently mixed school, 38 percent said their opinions of blacks had stayed the same and 27 percent said their opinions of blacks had gotten worse. Among blacks, the ratio of positive to negative change was higher: 44 percent of black students said their opinions of whites had changed for the better, 14 percent said their opinion of whites had worsened, and 41 percent reported no change.

In general, studies of interracial contact in schools have found mixed results. After a careful review of a large number of studies following the legal end of school desegregation, St. John concludes: "This review of research on racial attitudes and behavior in schools indicates that desegregation sometimes reduces prejudice and promotes interracial friendship and sometimes promotes, instead, stereotyping and interracial cleavage and conflict" (1975: 85).

Stephan (1978) reported that desegregation reduced white prejudice towards blacks in only 13 percent of the school systems studied; moreover the prejudice of blacks towards whites increased in about as many cases it decreased. In a more recent review of research on school desegregation, Stephan states that ". . . desegregation as it usually occurs is not likely to lead to reductions in stereotyping or prejudice" (1991: 114).

Greater contact between blacks and whites in residential settings also does not necessarily lead to more positive racial attitudes. Several studies have found that, when black families moved into previously all-white neighborhoods, the attitudes of their white neighbors toward black people in general changed little (Northwood and Barth 1965; Meer and Freedman 1966; Zuel and Humphrey 1971). While some families reported positive interracial experiences, others described more negative reactions. One white family in a newly integrated neighborhood in Seattle said of blacks living close by: "Our kids are not on speaking terms. [They] have five children who are noisy and run through our house. They also had a cat who came in our house. Finally, I gave the cat away. Now we are not on speaking terms" (Northwood and Barth 1965: 42).

The negative experiences that black people sometimes have is illustrated by the experiences of a black family in a newly integrated neighborhood included in the Seattle study. This family was invited to participate in a local church. However, when the wife attended the morning Mass, she received a frosty reception. She never returned to the church (1965: 44).

While contact between blacks and whites in housing projects often resulted in more positive racial attitudes, this was not always the case. For example, one study (Ford 1972) found that, after they had entered a racially mixed public housing project, only 22 percent of white women reported that their attitudes toward blacks had become more favorable; 62 percent said their attitudes had not changed. Self-reported attitude change among black women in the same housing project was more positive: 34 percent said they became more favorable toward whites, 8 percent said they had become less favorable, and 58 percent said their attitudes had not changed.

Living in racially mixed neighborhoods does not necessarily make people more favorable to interracial mixing. An early national study compared attitudes toward racial integration of Americans who lived in neighborhoods of varying racial compositions. It found that white residents of substantially integrated neighborhoods were less favorable towards racial integration with blacks than whites who lived in racially segregated neighborhoods (Bradburn et al. 1971: 121). A more recent national survey found that whites who had more interracial contacts in their neighborhoods were slightly more likely than whites who had less contact to support racial integration (in neighborhoods, school, churches, socializing). But this was not true among blacks; blacks who had more contact with whites in their neighborhoods were not more supportive of racial integration than were other blacks (Sigelman and Welch 1993).

Similarly, contact between blacks and whites in work situations does not necessarily result in positive attitudes and behavior, either in the United States or in other countries (Jackson 1992). Thus, for example, when black workers of West Indian origin were employed in London public transportation, there was considerable hostility from white co-workers and de facto segregation in eating areas (Brooks 1975). Similarly, interaction between blacks and whites in U.S.

military units often has been accompanied by considerable hostility (Hope 1979).

As with contact between blacks and whites, greater contact between people of other ethnic groups does not necessarily lead to more positive attitudes. A study of desegregation in the grade schools of Riverside, California found that there were few friendships formed between Hispanic and Anglo children. The authors of that section of the study concluded that "little or no integration occurred during the relatively long-term contact represented by Riverside's desegregation program. If anything we found some evidence that ethnic cleavage became more pronounced over time" (Gerard and Miller 1975: 237).

Among a national sample of non-Jewish Americans, amount of contact with Jews at work or contact with Jewish doctors or dentists were unrelated to anti-Semitism (Martire and Clark 1982). Arabs and Jews who met in "sensitivity training" sessions in Israel had great difficulty understanding each other's viewpoints (Lakin et al. 1969). Some studies in Israel have found that Arabs who worked with Jews showed no significant decrease in hostility; the most negative attitudes were found among Arabs who had worked longest in Israeli industry (Ziegler 1993: 393). Also, people (Americans and others) who visit foreign countries often become more nationalistic than they were previously (de Sola Pool 1965; Kelman 1975; Brislin et al. 1986).

Conditions of Contact

Contact between people from different racial or ethnic groups sometimes leads to improved intergroup attitudes and relations, however, these positive outcomes do not necessarily occur. Whether contact results in more positive attitudes and behaviors, has no effect, or even has a negative impact depends on the conditions under which such contact between groups takes place. The relevant conditions include the nature of the contact, the contact situation, and characteristics of the individuals and groups involved (Allport 1954; Amir 1976; Cook 1978; Stephan 1987; Jackson 1993).

Nature of the Contact

Contact between individuals and groups may, of course, vary in sheer magnitude—how often they occur, for how long, how many people are involved, in how wide a variety of activities (Allport 1954). Sometimes, an assumption is made that there is contact between people from different racial or ethnic groups when, in fact, little or no interaction actually is occurring. Some studies of racially mixed American neighborhoods have found that most white people in these neighborhoods had little, if any, personal contact with their black neighbors (Molotch 1972; Hamilton et al. 1984). Similarly, a study of Israeli immigrants found that, despite living close to one another, those coming from different national backgrounds had little social contact with each other (Weingrod 1965).

In American schools that nominally are racially mixed, most students may have little contact with schoolmates of another race because the school divides students by previous academic performance ("ability groups") and/or by different programs (academic, general, vocational, and so on). Often, these groups tend to divide along racial lines (Epstein 1985). Even when students of different races have an opportunity to interact, they often segregate themselves, especially in lunch rooms and activities outside of class (Schofield 1982; Rogers et al. 1984). Interaction between people who are physically close to each other sometimes is further reduced by barriers of language, as among ethnic groups on Guam (Brislin 1968), or by religious barriers, such as those between those of different castes in India (Anant 1972). Clearly, unless physical proximity is accompanied by people talking often to each other and doing things together, it is not likely to have much effect on intergroup relations. (Williams [1964] found that work situations provide the best opportunities for intergroup contact.)

Even when there is frequent contact between members of different groups, such contacts may vary in what has been called "acquaintance potential." Cook used this term to refer to "the opportunity provided by the situation for the participants to get to know and understand one another" and maintained that it is an essential condition for contact to have positive effects (1962: 75). Using similar terms, other writers have stated that effective contact should be "intimate" rather than trivial and transient (Allport 1954), "intensive," and having "personal involvement," rather than being superficial (Amir 1976). Amir (1976) suggests that studies that found the most positive effects from intergroup contact were done in situations where participants were most fully involved in the contacts.

The Situation

The outcomes of contact between people from different racial and ethnic groups may depend also on the type of situation in which they interact. In the original statement of the "contact hypothesis," Gordon Allport wrote: "Prejudice (unless deeply rooted in the character structure of the individual) may be reduced by equal-status contact between majority and minority groups in the pursuit of common goals. The effect is greatly enhanced if this contact is sanctioned by institutional supports (such as, by law, custom, or local atmosphere), and if it is of the sort that leads to the perception of common interests and common humanity between members of the two groups" (1954: 267).

Note that Allport's statement specifies three aspects of the contact situation as especially important: the presence of common goals, the relative status of the groups, and institutional supports for intergroup contact. Related to the third of these situational conditions, many researchers have discussed the importance of social norms that may either encourage or discourage intergroup contact. Other aspects of the situation including the relative size of the groups, whether people enter into contact voluntarily or involuntarily, and whether there are procedures for resolving conflicts, may also affect the outcomes of contact. Let us consider the possible effects of these aspects of the contact situation.

Compatibility of Goals

Perhaps the single most important feature of any social situation is the extent to which the effort of one group to reach its goals is compatible with the goal-seeking efforts of another group. Two aspects of this compatibility may be distinguished: 1) how much the outcomes of each group correspond; and 2) how much the groups are required to work together in order for each of them to reach its own goals (Brewer and Miller 1984).

When the outcomes of individuals correspond, each does well when the other does well and each does poorly when the other does poorly; their fates are tied together (Kelley and Thibaut 1978; Johnson et al. 1984). When people share the same outcomes, rather than having different outcomes, they are more likely to be friendly and to help one another (Deutsch 1968; Johnson et al. 1984). Moreover, though some competition for status between groups may continue (Brewer 1979), people tend to see those who share a common fate as being in the same group as themselves (see chapter 2 for further discussion of this point).

Separate from the matter of how much individuals' outcomes correspond is the extent to which reaching one's goals requires joint effort with others. Sometimes one person, or a limited set of persons, is best able to reach her goals by working alone. In other situations, it may be necessary for individuals or subgroups to work together cooperatively in order for any of them to reach their goals. When people need to work cooperatively with others, they are likely to feel more positive towards each other. After all, the efforts of each are helping the others, such help is rewarding, and we tend to like and to feel gratitude towards those who enable us to reach our goals. We also are likely to feel empathy for those acting in ways that will help us. We are "pulling for them" because their successful efforts help us as well.

A classic experiment with boys at a summer camp demonstrated that situational pressures to work together can have a powerful impact on relations between hostile groups (Sherif et al. 1961). Hostility between groups of campers was created by having them compete in a series of athletic contests, such as a tug-of-war. Then they were placed in a series of situations that required joint effort

81

of the separate groups in order to reach common goals, such as pulling a truck out of a ditch so it could go to get lunch for them. The researchers called these "superordinate goals." After the boys worked together to reach a number of common goals (getting the "lunch truck" free, locating blockage of their water supply, getting an enjoyable movie, and so on), the changes in their attitudes and behaviors toward each other were dramatic. Negative stereotypes and feelings toward the "out-group" changed to positive attitudes; insults and fighting turned to friendship. The distinctions between the groups faded as boys mingled across group lines.

Ethnic Groups: Schools

Our discussion so far has indicated that people will get along better when they share common rewards and/or they need to work together to achieve their goals. (When people *both* share a common fate and have to work together to achieve their best outcomes, the positive impact on their relationship is likely to be the greatest.)

These general principles have been found to be relevant to relations between people of different ethnic backgrounds. Much of the research concerning different ethnic groups working together towards common goals has been done in schools, where there can be a great deal of competition among students. Students often compete with each other for the best grades and for the approval of teachers. Often, the competition acquires an ethnic aspect, as when black or Latino students are resentful of white schoolmates or white students are resentful of Asian American schoolmates who get better grades than they do.

A variety of specific methods have been devised to replace the competitive classroom situation with one in which students of different ethnic groups work together cooperatively (Sharan 1980; Slavin 1983; Aronson and Gonzales 1988). Some of these techniques induce cooperation by making the students dependent on each other to complete their tasks; that is, no child can fulfill his or her assignment without the help of others. Other techniques tie the grades of each child to the grades of others in his group; each child's grade depends partly on the success of other group members.

An interesting example of a technique that promotes joint effort across ethnic lines is the Jigsaw Method developed by Eliot Aronson and his colleagues (Aronson 1984). In this method, students are placed in six-person learning groups. The lesson of the day is divided into six segments so that each student has one part of the material. Thus, each student has a portion of the necessary information that, like the pieces of a jigsaw puzzle, must be put together before any of the students can learn the lesson as a whole. Each student must learn his part and then teach it to the other members of the group. In order to learn all the material (and, thus, to do well on a test), each student must listen to and make use of the contributions of the others. When the Jigsaw Method was used in a number of school districts throughout the United States, participating students showed significant increases in their liking for others in their group, whether the others belonged to their own ethnic group or to other ethnic groups (Blaney et al. 1977; Geffner 1978).

Many experiments in schools also have found that interethnic work groups are effective in improving relationships among those of different ethnic backgrounds. An example is a field experiment conducted in the seventh and tenth grades of the Denver public schools (Weigel et al. 1975). Each of ten English teachers taught one class in which the class was divided into mixed ethnic groups—usually three "Anglos" (European Americans), one African American, and one Mexican American. As a "control group," each teacher taught another class in which she lectured to the entire class. After several months, the classes that were divided into interethnic work groups, compared to the control classes, had much more cross-ethnic helping behavior for all ethnic groups and more respect, liking, and friendship choices by Anglos for Mexican Americans. Several studies have found that cooperative learning techniques are especially effective for Mexican American students, whose cultural background encourages cooperative activities (Fernandez and Guskin 1981; Aronson and Gonzales 1988).

Opportunities in schools for working together on common tasks are not limited to the classroom. Athletic teams, school plays, musical groups, school publications, and other school activities are settings in which students can and often do work together on common tasks. The rewards (success, praise, prestige) or costs (failure or ridicule) are often shared by all those who participate.

There is evidence that when students from different ethnic groups participate together in such cooperative activities, their attitudes and behavior toward schoolmates from other backgrounds become more positive. Participating together on interracial sports teams has strong and consistent positive effects on race relations (Miracle 1981; Chu and Griffey 1985). The study of race relations in the Indianapolis high schools (Patchen 1982) found that white students who participated in school activities—sports teams, clubs, musical groups, publications, and so on—had more friendly contact than others with black schoolmates; the more blacks were represented in their activities, the more interracial friendships the whites had. The friendly relationships that form in cooperative activities in the school sometimes extend outside the school, as illustrated by the comment by a black student about some white teammates: "I have a couple of good friends, they take me home from practice sometimes . . . during the summer, they will come pick us up, we'll pick him up. We call each other, we go over to his house, or they come over to our house. . . . We come to the football games together. He's almost like a brother" (Patchen 1982: 61).

The widespread importance of cooperative school activities is shown in a study of students from seventy-one high schools in both the northern and southern United States. For white students and black students alike, working together on class projects or playing together on athletic teams was associated with lower prejudice and more friendly behavior towards schoolmates of the other race (Slavin and Madden 1979). Cooperative activities contributed much more to good intergroup relations than did a variety of other school factors.

Researchers who have reviewed the evidence from the large number of relevant studies have been unanimous in concluding that having interethnic groups work on common school tasks is an effective way to build positive relations between the groups (Schofield and Sagar 1979; Johnson et al. 1984; Slavin 1985; Brewer and Miller 1988). John McConahay comments:

> The concept of interracial work groups as a means of reducing prejudice and improving race relations has a great deal of empirical support. It is rare in social science to find such robust results across such a wide range of techniques, empirical evaluation methodologies, grade levels, regions of the country, and curricula . . . it is clear that this is the most effective practice for improving race relations in desegregated schools that we know to date. (1981: 49–50)

Other Settings

People from different ethnic groups may, of course, work cooperatively together on common tasks in settings other than schools—in work organizations, community organizations, churches, the armed services, and so on. There is much less systematic research about the effects of cooperative tasks in such other settings than there is in schools. However, the available evidence supports the idea that working together on common tasks usually promotes good relations.

In a field experiment, Stuart Cook and his colleagues studied cross-racial attitudes and interactions among women from low socioeconomic backgrounds whose children attended an interracial preschool in Nashville. Mothers who participated in interracial groups that discussed common child-rearing problems were more attracted to each other than were mothers in mixed-race training groups that did not participate in this joint activity (Cook 1984).

In Northern Ireland, despite high tensions between Catholics and Protestants generally, farmers from the two religious communities who shared equipment and worked together frequently developed friendships (Trew 1984). In interracial housing projects, mutual aid in such activities as child care and shopping appears to have contributed to positive attitudes and friendly interaction between black and white neighbors (Deutsch and Collins 1951; Wilner et al. 1955). In military units where black and white soldiers fought together against a common enemy, whites' racial attitudes were much more positive than where this was not the case (Mannheimer and Williams 1949).

Conditions for Effectiveness of Cooperative Tasks

Working together on common tasks is not equally effective in all circumstances for improving relations between members of different ethnic groups. First, the effects are likely to depend on how successful the group is in completing its task and reaching its goal. If a group experiences failure and frustration, members are less likely to like and respect each other than if the group has been successful (Amir 1976; Brewer and Miller 1984; Cook 1984; Hewstone and Brown 1986). For example, interracial military personnel had less liking and respect for their teammates in experimental groups when those groups were unsuccessful in task competition with other groups than when the groups were successful (Cook 1984).

Regardless of the overall success of the group, an important problem in cooperative task groups is the possibly differing competence and contributions of the members. Group members who are perceived by other members as low in task competence and contributing less than others to task success are apt to be less liked and less respected by others (Fromkin and Sherwood 1974; Cook 1984). Those who need help from others may be less respected than others. In addition, if rewards for group success are given equally to all members (e.g., when all members of a classroom group receive the same grade for a group project), those who have contributed most to group success may be resentful towards those who they believe have done less, either from lack of ability or effort. (The Jigsaw method for cooperative school activities minimizes the latter problem by arranging for essential contributions from each group member.)

Thus, activities that require people to work together towards common goals need to be planned with due attention to possible pitfalls. Nevertheless, cooperative activities can be a powerful tool for improving relationships among members of different ethnic groups.

Relative Status

When people from different groups come into contact, they may be equal or unequal in status. Members of one group, usually a majority or dominant group, may have authority over those of another group, as where Anglos are supervising Latino field workers. Whether or not there are differences in power or authority, there may be differences in prestige, as where white workers occupy more skilled jobs and African American workers have less skilled jobs.

Regardless of differences in ethnicity, there tends to be social distance and often some tension and hostility between those who have power or authority over others and those who are subject to such authority (Homans 1974). Similarly, even among those who do not differ in ethnicity, people are more likely to associate with those of like economic status and social prestige than with those of different status (Michener et al. 1990: ch. 11). High-status people often look down on those of lower status and refrain from associating with them for fear their own status will be lowered if they do so. Those of lower status may feel resentful of and uncomfortable with those who they think may feel superior to them.

When people differ in their ethnicity, it is difficult to reduce barriers between them when they meet under conditions of unequal status. For example, in the southern part of the United States during the first half of this century, blacks and whites had a great deal of daily contact. But this contact usually was between people very different in power and prestige—for example, white men as bosses of black laborers and white women in charge of their black maids (Dollard 1957). Under these conditions of unequal-status contact, negative attitudes or attitudes of superiority by whites, feelings of resentment by blacks, and little social contact were the rule.

However, when people from different racial or ethnic groups come into contact as equals, attitudes toward the other group often become more positive. In work situations of many types (such as department stores, merchant ships, and coal mines), whites who worked with blacks who had a job at an equal level often formed positive attitudes toward the blacks (Amir 1976). In neighborhoods, whites who perceived the social class of blacks they knew as being the same or higher than their own had more favorable attitudes than those who saw the status of blacks as lower than their own (Morris 1970; Helper 1986). Mothers and children who participated in a school busing program that involved black children of relatively high social status reported more cross-racial friendships than did other mothers and children (Jonsson 1966), and the more equal in socioeconomic status and achievement were black students and white students in a middle school, the more cross-racial friendships were formed (Schofield and Sagar 1979).

Relative Power

Most of the research concerning the relative status of groups has focused on their prestige and/or their socioeconomic positions. There has been little direct research on how the relative power of groups may affect the results of contact between them. However, the study of race relations in the Indianapolis schools provides some relevant evidence about the effects of perceptions of relative power. Students were asked whether they thought that white students or black students were more or equally "able to affect what goes on at this school" in seven specific ways; for example, get the school to sponsor entertainment events, get students they like elected, and get teachers to do things in a different way.

Among both black and white students, attitudes and behaviors toward schoolmates of the other race were more positive when students saw the two races as being more equal in overall power. Seeing either the other group or one's own group as having more power was associated with more negative attitudes and more negative behaviors (e.g., more avoidance and more fighting) and with less friendship (Patchen 1982). These findings are consistent with the proposition that those who see their own group as low in status tend to be resentful towards the higher-status group, while those who see their own group as enjoying high status tend to be arrogant or condescending toward the lower-power group.

Status In and Outside the Contact Situation

The status of people in a particular contact situation may differ from their more general social status based on such factors as education, occupation, and income (or that of their family). For example, blacks, whites, Hispanics, and others who attend the same school may be of equal status as students in that school, having the same privileges, responsibilities, and prestige in their position as students. However, the occupations and incomes of most of the white students' families may be higher than those of the families of most black or Hispanic students.

Is status in the broader society or status in the immediate contact situation more important? Some have suggested that status in the contact situation has the greater effect on attitudes and behaviors (Jackson 1993). While there is little systematic evidence on this point, the study of racially mixed high schools in Indianapolis (Patchen 1982) supports the idea that status in the immediate contact situation is more important than general social status. Greater differences between the educational level of black and white students' parents were *not* associated with less positive race relations. However, differences in the proportion of each race in the higher-status academic program of each school were related to opinion change among white students. The greater the difference in proportions of blacks and whites in the academic program, the smaller the proportion of white students who changed their opinions of black people in a favorable way.

If peoples' status in the immediate contact situation matters more than their general status in society, as seems to be the case, this makes the creation of equal status much more possible. Matching the status of people from different racial or ethnic groups within, say, a particular business firm or a particular school is much easier than matching them in family socioeconomic status.

However, a person's or a group's status within and outside the immediate situation are not completely separate; they may be related and overlapping (Riordan 1978). A person's or group's status as judged by the larger society, based on, perhaps, family wealth and prestige, and, perhaps, directly on ethnicity may affect the prestige that they are accorded by others in school, on the job, in the neighborhood, in a civic organization, and so on.

In addition, even when people are formally of the same status, members of a majority group may expect those from some minority groups to have less ability and to be able to contribute less to the accomplishment of group tasks. Thus, studies of student groups composed of whites and African Americans, of Anglos and Mexican Americans, and of whites and Indians have found that whites were more active and influential (i.e., had higher status) when students discussed school assignments. Elizabeth Cohen (1984) attributes this phenomenon to the expectations of the white students (and sometimes of the minority students as well) that the whites would be more competent. Where such expectations exist, activities that ensure or demonstrate the competence of minorities, may be needed in order to create genuinely equal status in the situation (Cohen 1984). In Israel, precontact workshops were used to enable Jewish and Arab youths to develop similar communication and other skills, and thus more equal status when members of the two groups interacted in the workshops. This procedure appeared to contribute to more positive relationships between the Jews and the Arabs (Bargal 1990).

Negative Effects of Equal Status

While equality of status generally seems to facilitate good relations between groups, there are circumstances in which changes toward more equal status may lead, at least temporarily, to more hostility between groups. When a group that has been lower in status and privilege than another group begins to increase in resources and prestige, it may begin to compete more effectively with the higher-status group. Thus, the rising group may be seen as a threat and the higher-status group may react with hostility against the "upstarts" (Amir 1976; Jackson 1993). For example, in the late nineteenth and early twentieth century, as African Americans moved into jobs previously reserved for whites, hostility and often violence against blacks frequently erupted (Olzak 1992).

Support From Authorities

Another aspect of the situation that may affect the outcomes of contact between different ethnic groups is the extent to which friendly equal-status contact is encouraged and approved by those in authority.

Some of the reasons why support by authority figures is important are indicated by Yarrow, Campbell, and Yarrow: "The development of new norms is facilitated if what is 'appropriate' and 'expected' in the situation is clarified immediately by the persons in authority roles. The definition of the situation is established effectively by leader model, by leader direction, by manipulation of physical environment" (1958: 60).

Again, relevant evidence comes primarily from studies of interracial contact in schools. Both intensive case studies of individual schools and larger correlation studies have found that support by the school principal is an important contributor to good relations between black and white students (Schofield and Sagar 1979). For example, one study found that in schools where the principal felt that good intergroup relations are important, black and white children were more likely to interact in the lunchroom and at recess than were children in schools where the principal did not give this goal high priority (Wellisch et al. 1976).

Support by school principals and other school administrators for positive intergroup relations may have effects in several ways. First, administrators can affect the school structure—such as tracking, ethnic grouping within classes, curriculum, seating assignment policies, and extracurricular programs—that provide opportunity for students to come into contact and the conditions of such contact (such as the frequency of cooperative activities). Secondly, administrators can influence the attitudes and behavior of teachers (Forehand et al. 1976). Aside from any effects a principal may have as a persuader and as a model, teachers are likely to act in ways that will please and win rewards from the principal. If the principal clearly makes known that she cares about good race relations, there are clear incentives for teachers to act in ways that support this goal.

Support by teachers for positive ethnic group relations, in turn, is likely to contribute to good race relations. For example, the more that white students in the Indianapolis high schools perceived their teachers as having favorable attitudes toward black people and toward friendly interracial contact, the more positive were their own racial attitudes and the less likely they were to avoid or to fight with black schoolmates (Patchen 1982). Teachers probably affect their students' attitudes and behaviors because they may serve as models, because they may help to define what is right, and because students usually want to please their teachers.

School administrators and teachers also can affect relations among ethnic groups by whether or not they show favoritism toward any group. Those groups (e.g., African Americans, Hispanic Americans, Native Americans, and Asian Americans) that often have been subjected to discrimination in our society are especially sensitive to evidence that they are being treated unfairly. When students believe that school administrators and teachers show favoritism to members of another group, they tend not only to be angry at the school authorities but also at the students whom they see as the beneficiaries of the favoritism. Thus, the more that black students in the Indianapolis high schools saw administrators and teachers favoring whites, the less positive their attitudes toward and the fewer friendships they reported with their white schoolmates (Patchen 1982).

The negative effects of perceived favoritism are not limited to members of a minority. Those belonging to a majority or dominant group may also see favoritism on behalf of a group other than their own. Thus, white students may believe that black students get special privileges and white workers may believe that African Americans and Hispanics are given special preference in hiring or promotions. Such beliefs often result not only in hostility towards those in authority believed responsible for the alleged discrimination but also in hostility towards the minorities that are seen as benefiting (Schuman 1991).

What has been said about the importance of support by school authorities for positive intergroup relations also applies to support by authorities in other settings. Executives in business firms and commanding officers in the military, for example, can help to create the conditions for friendly interethnic relations among their subordinates. Like school administrators, they can create organizational arrangements and policies, such as interethnic task groups composed of people of equal rank, that are likely to result in friendly contact between members of different groups. They can attempt to eliminate discrimination in the treatment, hiring, promotion, assignment, and so on of any group, thus reducing resentment between groups. They can help to change attitudes by persuasion and personal example and they can make it clear that subordinates' support for egalitarian, friendly intergroup relations will be rewarded while opposition to this goal will be penalized. Finally, they can help to build strong ethnically-inclusive groups, with shared goals and common symbols.

While there is little systematic research on the unique impact of authorities on intergroup contact in work settings, experience both in business and the military is consistent with the proposition that strong support from authorities for good intergroup relations makes an important contribution to positive outcomes. In their review of race relations training programs in the military services, Landis, Hope, and Day comment: "As the Race Relations/Equal Opportunity (RR/EO) programs are reviewed and evaluated over the last eleven years, it becomes clear that the commander's support is critical to the success of these programs. . . . Attendance, involvement, and ultimately improved understanding between minority and majority, black and white, male and female, was directly influenced by the degree of participation of commanders at all levels of administration . . ." (1984: 267–268). Similarly, the support of top management has been found important for the success of programs designed to promote harmony between ethnically diverse workers in business firms (Morrison and Herlihy 1992). Support by authority figures also appears to have contributed to favorable attitude change in other settings, including the church (Parker 1968).

Norms of Social Groups

When people interact with those of another ethnic group, they may be concerned not only with the reactions of people in authority but also with those of others who are important to them. The approval or disapproval of family members, friends, neighbors, and co-workers may greatly influence their own attitudes and behaviors toward other ethnic groups.

Studies of residents of interracial housing projects showed that as white residents anticipated more favorable reactions to interracial contacts by their friends, their own attitudes toward their African American neighbors were more positive (Deutsch and Collins 1951; Wilner et al. 1955). Wilner, Walkey and Cook comment: "It is possible that perception of a climate favorable to racial interaction cuts down impediments to one's own contacts with Negroes, and with developing contacts comes appreciation of Negroes as human beings with good and bad qualities" (1955: 105).

Studies in schools provide extensive evidence that students' attitudes and behaviors toward schoolmates from other ethnic groups are greatly affected by the attitudes of their families and peers (Chadwick et al. 1971; Bullock 1976; Patchen 1982). In the Indianapolis high schools, the racial attitudes of students' parents were one of the strongest predictors of both white and black students' attitudes and behaviors toward schoolmates of other race (Patchen 1982). The strong influence of parents is illustrated by the remarks of a white female student: "My Dad is very prejudiced. If he for once saw me talking to Negroes, like I do, he would be very upset. . . . When your parents are prejudiced, it's hard for yourself not to be" (Patchen 1982: 171).

The Indianapolis high school study also found that the racial attitudes of students' peers (both actual and perceived) had a substantial impact on racial attitudes and behaviors, especially among white students. Asked how his friends would act if he became friendly with a black, one white freshman boy replied: "They probably wouldn't play with me anymore. They're always talking about them. They say they hate them." Many blacks reported similar social pressures. Asked about the reactions of black friends to friendship with whites, a black sophomore boy said: "They don't like it. Some of them call you a 'Tom'" (1982: 170–171).

Overall, the evidence indicates clearly that contact between members of different ethnic groups is more likely to result in friendly relations if such relations are approved, rather than disapproved, by the important social groups to which the participants belong.

Size of Groups

There are additional aspects of the situation that may affect the outcomes of contact between members of different ethnic groups. One such factor is the relative size of the groups. Evidence concerning how the relative size of groups is related to members' intergroup attitudes and behaviors is not consistent. Some studies have reported that members of different ethnic groups get along best when the groups are about the same size (Coleman et al. 1966; Gonzales 1979; Garza and Santos 1991). Other studies have found negative attitudes and behaviors, such as avoidance or fighting, to be more frequent when ethnic groups are about equal in size than when one group is numerically predominant (Longshore 1982; Patchen 1982; Hallinan and Smith 1985).

When groups are about equal in numbers, there is optimal opportunity for members of each group to interact with those from the other group; greater opportunity for contact leads to more friendly interactions (Patchen 1982). Being equal in number may also contribute to equal status for those in both groups. In addition, when no group is a small minority, ethnic group characteristics are less distinctive and ethnicity would tend, other things being equal, to be less a focus of attention.

On the other hand, members of one or both groups may be concerned that they will be dominated by the other. With relatively equal numbers, a real or perceived struggle for power may occur. This struggle is especially likely when the proportion of one group has been rapidly increasing and when this group is seen as having different goals or standards than the previously dominant group (see section on "Size of Ethnic Groups" in chapter 5). This situation has sometimes occurred, for example, where rapid increases in the proportion of lower-income blacks in a school has led whites to perceive a threat that "their" school is being "taken over" by blacks. For intergroup contact to have positive effects on attitudes and behaviors, it appears desirable to try to avoid perceptions that a struggle for power, related to numbers, is taking place. Assurances by authorities to each group that its interests will be protected may help to reduce the concerns of members of both groups.

Voluntary versus Involuntary Contact

Some theorists have maintained that intergroup contacts will have more positive outcomes when they are voluntary rather than forced (Blalock and Wilken 1979: ch. 14). Reviewing evidence on school desegregation, Norman Miller asserts that: "Desegregation plans should rely on voluntary action and choice as extensively as possible. Choice and voluntary behavior produces commitment and favorable attitudes. . . . Restricting choice and freedom can produce resistance and negative attitudes" (1980: 319).

There is some additional evidence to support the idea that voluntary contact between ethnic groups is preferable to involuntary contact. Thus, Wagner and Machleit (1989) found that only voluntary contact between Germans and foreigners was associated with reduced prejudice. However, there is also evidence that attitudes toward another ethnic group may change after contact that is induced by outside pressure. Most notably, the attitudes of whites toward racial integration in American schools and other settings became much more positive after court-ordered desegregation. In the south, the greatest reduction in resistance to racial mixing occurred between 1970 and 1972, soon after the greatest reduction in school segregation (Rossell 1983).

For most whites during this period, greater contact with blacks was not entirely forced, since most whites had some alternatives, such as staying away from racially mixed public places (restaurants, movie theaters, and so on) or sending their children to private schools. However, there were strong, practical inducements for most people to change their behavior; for example, sending children to a racially mixed public school was less expensive then sending them to a private school and patronizing racially-mixed public places met important needs, such as those for easily-accessible recreation. Once the changed situation had induced people to have contact with blacks in situations where they had not previously done so, their attitudes about such contact changed. There may have been a variety of reasons for such attitude change, including positive interracial experiences, a changed position by authorities, and a need to reduce cognitive dissonance between one's behavior and one's attitudes (see the section on "dissonance reduction" later in this chapter).

Procedures to Resolve Conflicts

Since disagreements are likely to occur between any groups that interact with each other, procedures for resolving conflicts may help to promote good relations. A large amount of literature on conflict resolution suggests that these procedures are helpful for individuals and groups in general (e.g., see Fisher 1994). For example, committees with representatives from both disputing parties and mediation by mutually respected third parties are often useful in devising solutions to conflicts. With respect to ethnic group relations specifically, one study found that the more that students perceived active efforts by school authorities and by student groups to solve problems (racial and otherwise) in the school, the more positive were their attitudes toward schoolmates of the other race (Patchen 1982).

Personal Characteristics

While the nature of the contact situation is very important, the outcomes of contact between different ethnic groups may depend also on the characteristics of the people involved. Relevant characteristics include the initial attitudes of those in each group toward other ethnic groups, similarities and differences between the members of each ethnic group, and individuals' personalities and social characteristics.

Initial Attitudes

When individuals come into contact with members of another ethnic group, their subjective reactions and behavior may depend in part on the attitudes they bring to the contact situation. Those who initially dislike another

93

ethnic group may make minimal use of the opportunity for contact. Moreover, because people have the tendency to attend to information consistent with existing attitudes (Oakes et al. 1994), they may selectively notice especially any negative traits or actions and disregard positive traits and actions of the other group. Conversely, those who initially have favorable attitudes may make more use of opportunities for intimate contact and may more readily perceive positive characteristics and actions by the other group.

Research on contact in a variety of settings (including schools, housing projects, and neighborhoods) has found two major effects of people's initial ethnic attitudes. First, their attitudes toward members of another ethnic group tend to remain in their initial direction (favorable or unfavorable), while becoming stronger than they were initially. In other words, those who liked the other group to begin with like them even more, and those who disliked the other group initially dislike them even more after contact. Secondly, those with more favorable initial attitudes are most likely to develop the friendly relationships that contribute to reductions in prejudice (Amir 1976).

Initial attitudes toward a particular ethnic group may be linked to the information, and sometimes misinformation, that people get from a variety of sources, including their families, their peers, and the media. Initial attitudes also may be influenced by prior experiences with members of another ethnic group. When prior relationships with members of another ethnic group have been friendly, people tend to exhibit positive attitudes and friendly behavior toward members of that same group in a new setting. When they have had bad experiences with members of this group (such as being the target of snubs, insults, or attack) they will tend to avoid members of this ethnic group in new situations.

For example, white high school students were asked how many black people they had "gotten to know" in grade school and in their neighborhoods before high school and how friendly or unfriendly their experiences with blacks usually were. Black high school students were asked similar questions about their earlier contacts with white people. For students of both races, the more favorable their earlier contacts, the less avoidance of, the more friendly contact with, and the more positive attitudes they had currently towards schoolmates of the other race (Patchen 1982). Thus, the chances of children developing positive attitudes and behaviors toward people from other ethnic groups are maximized if such contact occurs, under favorable conditions, at an early age.

While those with initially negative ethnic attitudes usually are resistant to change, contact with another ethnic group can result in positive change among such persons under the right circumstances. In some experimental studies, whites with prejudiced attitudes toward black people have been brought into contact with blacks in situations where they must work together towards common goals on an equal status basis. Under such conditions, which we have seen are likely to promote good intergroup relations, even those people who were initially prejudiced against blacks changed their racial attitudes in a positive direction (Cook 1984, 1990).

Similarities and Differences

How people belonging to different ethnic groups react to contact with members of the other group may be affected also by how similar members of these group are. Liking for others generally increases as there is greater similarity between oneself and the others, especially with regard to attitudes and values (Newcomb 1961; Neimeyer and Mitchell 1988).

The idea that contact between members of different ethnic groups is a good way to improve intergroup attitudes sometimes has been based on the assumption that such contact will lead participants to see that they share values and beliefs with those from other ethnic groups (Pettigrew 1971; Cook 1978; Stephan and Stephan 1984). Often, an assumption has been made that intergroup contact will show participants that their negative beliefs about members of another group are incorrect (Stephan and Stephan 1984; Rothbart and John 1985). But there may be real differences in the attitudes, values, and behavior patterns of different groups, for example, concerning effort in school or sexual behavior.

Several analysts of intergroup contact have seen the absence of important differences among ethnic groups as a condition for intergroup contact to have positive outcomes. Thus, Cook has stated, as one of five basic conditions necessary for contact to improve intergroup attitudes, that "the characteristics of outgroup members with whom contact takes place disconfirms the prevailing outgroup stereotype" (1978: 97). Writing about interracial contact in schools, Norman Miller includes among "harmful situations" one that "emphasizes moral and ethnic standards on the part of one group that are objectionable to the other" (1980: 317). Other writers have specified "high acculturation among minority groups" (Eshel and Peres 1973) as one of the conditions for successful intergroup contact.

Evidence supports the proposition that contact between different ethnic groups will have more favorable social outcomes when the groups' attitudes, values, and related behaviors are more similar. One study found that after contact with potential co-workers on a job, people chose those with whom they preferred to work on the basis of similarity in job-related attitudes more than on the basis of shared race (Rokeach and Mezei 1966). Among high school students in Indianapolis, the greater the similarity between blacks and whites with respect to average effort and amount of fighting in school, the less the unfriendly interaction that black students reported with white schoolmates. In addition, the smaller the average differences between the races in their academic aspirations (wanting to get good grades, wanting to go to college), the more favorable the change in whites' opinions of blacks (Patchen 1982). In a middle school in another city,

95

similarity in achievement between black and white students was found related to interracial friendship (Schofield 1982).

Studies in countries outside the United States also have found that relations among different ethnic groups are affected by the degree of similarity in their members' attitudes and behaviors. Among Jewish and Arab high school students and college students in Israel, personal contact that stressed similarities between members of the two ethnic groups has led to improved intergroup attitudes; but contacts that involved discussion of political issues on which the two groups have different views did not lead to more positive attitudes (Ben-Ari and Amir 1986). In contacts between Dutch and Turks in Holland and between Hindus and Moslems in India, actions by members of each group that violate norms of the other group often have led to intergroup hostility (De Ridder and Tripathi 1992).

Some have concluded that it is important to emphasize similarities between members of different ethnic groups (Stephan and Stephan 1984). Yet, arguments have been made that ignoring important differences between groups ultimately may cause problems. If people expect those of another ethnic group to be just like themselves, they may be shocked and dismayed when those expectations are not met (Hewstone and Brown 1986). Therefore, explaining the real nature of the differences that exist between groups and the reasons for these differences seems important. Sometimes, differences such as those in food and music may even be a source of interest and attraction.

But while it may be legitimate to "celebrate" diversity, the important link between similarity and attraction should not be forgotten. While people should be helped to understand and respect differences between groups, they should also be encouraged to recognize important similarities that people of all ethnic groups share. They may come to see that varying customs and practices usually are different ways by which different groups attempt to satisfy the same basic human needs and aspirations. In addition, people from different ethnic groups can be brought together on the basis of important similarities of values and interests. For example, students from different ethnic backgrounds can be grouped on the basis of their interests in a particular subject (e.g., in a "magnet school" for music, art, or science) and adults may come together on the basis of shared values and interests in sports, hobbies, religion, or politics.

Other Personal Characteristics

A number of other personal characteristics have been found to be relevant to intergroup attitudes and behavior. There is, for example, considerable evidence that personality factors play a role in attitudes toward other ethnic groups. One relevant personality variable has been labeled "authoritarianism." People scoring high in authoritarianism emphasize obedience to conventional rules and authority but show great hostility to out-groups, especially minorities. This pattern of attitudes has been explained as the result of repressed hostility towards strict authority figures, such as parents, being displaced against more socially acceptable targets, such as minorities (Adorno et al. 1950; Hassan 1987). Other personality traits that have been found to be associated with negative attitudes toward ethnic minorities or out-groups include dogmatism (Rokeach 1960) and low self-esteem (Sniderman 1975). Those who are dogmatic or closed-minded are more likely to hold on to fixed stereotypes about members of other groups, while those with low self-esteem may try to raise their self-esteem by derogating out-groups to whom they then can feel superior. People who have such personality traits, traits that dispose them to be hostile to or to feel superior to those in other groups, are less likely to be influenced positively by contact with other groups than are those with more open, accepting personalities (Stephan 1987).

Other personal characteristics that are associated with ethnic attitudes include age, education, and socioeconomic status. People who are younger, better educated, and occupy higher socioeconomic status generally have the most positive attitudes toward ethnic minorities such as blacks and Jews (Case et al. 1989; D'Alessio and Stolzenberg 1991). Since intergroup contact has more positive effects on those with more positive initial attitudes, one would expect younger, better educated people to show more positive changes in attitudes after contact with other groups. There is some evidence that this may be true (Williams 1964). Yet, if highly educated, high-status individuals come into contact with members of a poorly-educated ethnic group whose behavioral style is very different than their own, the differences between the groups may result in little or no positive attitude change.

The Search for Desirable Conditions of Contact

We have discussed a variety of situational and personal factors that may affect the success of intergroup contact in changing people's attitudes and behaviors toward other ethnic groups. This information is useful.

But there are several problems that arise when one tries to specify the conditions that will make contact between groups more successful. First, considering each relevant condition separately is not enough because the effect of one condition of contact depends on the presence or absence of other relevant conditions. For example, the effect of status differences on intergroup relations may depend on how much the groups differ in their values and patterns of behavior; and the extent to which participation in cooperative tasks results in more positive intergroup attitudes may vary with whether rewards for task success are given to all participants together or separately to different individuals or groups. With few exceptions (see Patchen 1982), there has been little research on the ways in which various aspects of situations and persons may combine to produce positive or negative changes in attitudes and behavior.

Another problem is that the list of possibly relevant conditions of contact may become quite long. We have mentioned only the conditions that appear to be most important and other conditions could be added; researchers have constantly suggested additional aspects of the contact situation or of the participants that may be relevant (Stephan 1987). After reviewing the many specific conditions of contact that have been proposed by various theorists and researchers, Stephan remarked: "At this stage in its development, the contact hypothesis resembled a bag lady who is so encumbered with excess baggage she can hardly move" (1987: 17).

One of the problems that stems from creating a long and ever-lengthening list of desirable conditions for intergroup contact is that the "proper" conditions may become unrealistic. Finding circumstances in which, for example, contact is intimate rather than superficial, status is equal, people work on cooperative activities, social norms and institutional supports are positive, initial attitudes are positive, and values are similar, may restrict us to a very small proportion of actual situations. By making the list of desirable conditions too long, the "contact hypothesis" may become irrelevant to the real world.

Secondly, an emphasis on specifying the "proper" conditions does not tell us just why contact with other groups does or does not lead to changes in attitudes and behavior. Pettigrew has said of the contact hypothesis: "Its specification more closely resembles a laundry list than a subset of corollaries of a single more generic theory" (1986: 171). What is needed, instead, is to go beyond a mere list of appropriate conditions to try to specify the crucial cognitive, emotional, and social processes that may result from intergroup contact. We turn next to these issues.

Effects of Intergroup Contact

Why does contact between members of different ethnic groups sometimes result in more positive attitudes and behavior and sometimes have less positive results? We will consider several ways in which contact with members of another group may affect a person: 1) by changing the ways in which she perceives those in the other group; 2) by changing her feelings toward those in the other group; 3) by changing her expectations about the probable outcomes of her actions toward members of that group; and 4) by directly affecting her behavior that in turn, may affect her perceptions and feelings. Figure 4.1 presents a schematic overview of these possible effects of intergroup contact. In the sections to follow we will discuss each of these possible effects and the psychological processes by which they occur (also indicated in figure 4.1).

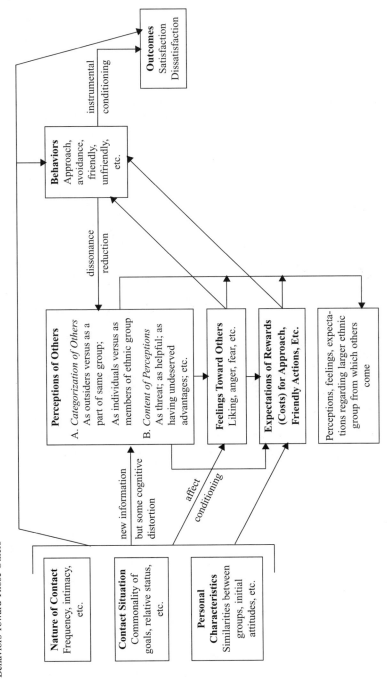

FIGURE 4.1

Effects of Contact with Persons of Another Ethnic Background on Individuals' Reactions and Behaviors Toward Those Others

Cognitive Effects

Those who have urged contact among groups as a way to improve intergroup relationships have often assumed that such contact will cause people to see that their perceptions, especially their negative stereotypes, of the other group are wrong. They also expect that people will form more accurate perceptions of the other group, thereby (by definition) having more positive attitudes, which in turn will result in more friendly behavior (Rothbart and John 1985; Ben-Ari and Amir 1986). This view assumes that more knowledge about the other group will lead to positive changes in attitudes and behavior.

Some support for this view comes from a study of Anglo and Chicano students in two junior high schools in New Mexico (Stephan and Stephan 1984). They found that greater contact with Chicano (Mexican American) students was related to Anglo students having more knowledge of Chicano culture, which, in turn, was associated with more favorable attitudes toward Chicanos. Stephan and Stephan argue that: "Ignorance of out-groups promotes prejudice because people tend to fear and reject what they do not understand. Ignorance of out-groups creates anxiety about how members of out-groups will behave and how in-group members should behave toward them" (1984: 238).

A large number of studies have examined the effects on prejudice of school programs that provide students with information about the history of different groups in our society, especially African Americans, and highlight the achievements of members of minority groups. A majority of the studies have found that such information can reduce prejudice; however, many of the studies found no effects on students' attitudes and one found an increase in prejudice (Stephan and Stephan 1984). Some writers have been skeptical that information resulting from intergroup contact is likely to have much effect on negative stereotypes of an out-group (Hewstone and Brown 1986). These writers have focused particularly on the cognitive mechanisms by which people often keep old stereotypes intact despite new information.

Effect of Expectations

People often tend to see what they expect to see (Rothbart 1993). For example, if a white person expects blacks to be aggressive, he will tend to notice and remember instances in which a black person is angry or belligerent and to ignore or forget instances in which a black person is pleasant or conciliatory. An ambiguous remark by a black person may be interpreted as an insult rather than a joke. This tendency of expectations to affect our perceptions makes the process of changing the images of an out-group more difficult.

In addition, when the behavior of others is recognized as contrary to expectations, the behavior may be explained as due to external forces rather than freely chosen by the others. For example, if a Mexican American works hard at a job, this may be interpreted as due to pressure from a boss rather than to high motivation. However, when there is a clear and undeniable evidence that the out-group is different than people's expectations, a positive change in attitudes is likely to occur (Rothbart and John 1985; Oakes et al. 1994).

Sub-types and Exceptions

One or more members of an out-group may behave in ways that clearly are not consistent with a person's stereotype of this group. Often, the person's reaction is not to change his group stereotype but, instead, to see individual members of the group as exceptions or as a special sub-category that is different from most of the out-group (Rothbart and John 1985; Rothbart 1993). Thus, if a person sees Mexican Americans working hard, he may see them as not typical of their ethnic group and thus not change his stereotype of Mexicans as lazy. In addition, an out-group member who displays positive behavior may be perceived mainly in terms of some identity other than her ethnic group. For example, a very intelligent black physician may be seen primarily as a physician rather than as a black so that a positive view of this person is generalized much more to physicians than to blacks.

Given the tendencies of people to maintain their stereotypes of other groups, contact that provides contrary information may change these stereotypes only under special conditions. Rothbart and John state that disconfirming information will modify stereotypes of out-groups only if the behavior is clearly inconsistent with expectancy, occurs repeatedly, occurs in many different settings, the out-group members are seen as typical of their group, and the disconfirming information is associated with the group label. Since this set of conditions is not typical of most situations, Rothbart and John suggest that "the conditions leading to change through contact are quite restricted" (1985: 100).

Changes in Perception of Group Boundaries

Contact between people from different ethnic groups may lead them to see less clear boundaries between themselves and the others. The line between "we" and "they" may become blurry. To some extent people from different ethnic backgrounds may be viewed primarily as part of the same group, for example, as fellow students, co-workers, residents of the same neighborhood, or members of the same church. Distinctions between groups that are seen as salient then may be based on differences other than ethnicity. If those of different ethnic backgrounds are seen as part of one's own group, feelings and behavior toward these others are likely to be more positive (Worchel 1979; Brewer and Miller 1984).

101

Contact between those from different ethnic backgrounds may lead to a reduction in the salience of group boundaries to the extent that it promotes relationships based on individual characteristics and not on ethnic group labels. Contact that leads to close interpersonal contacts may lead people to differentiate among members of another ethnic group, to see them primarily as individuals rather than as "all alike"—the essence of a stereotype. The salience of ethnic group categories will also be reduced if other important distinctions between people (e.g., in interests, in abilities, in occupation) do not coincide with the ethnic categories (Brewer and Miller 1984). If those other distinctions "cross-cut" rather than coincide with ethnic categories, ethnic categories will be less likely to be used, or to be useful, in dealing with others. Other similarities and differences between people will be more salient. When ethnic stereotypes are less useful in predicting how people will behave, they are likely to be altered or discarded (Rothbart and John 1985).

The extent to which intergroup contact will lead people to see the distinction among groups as less important will also depend on other circumstances of the contact, many of which we discussed earlier in this chapter. Contact that occurs under conditions that lead people to see themselves as sharing a common fate and a similar status will make intergroup boundaries seem less important. Having to work cooperatively on common tasks, being treated similarly by authorities, having similar roles and authority, and being assigned to roles or task groups on bases other than ethnic group identity are examples of such conditions. Circumstances that promote intimate, rather than superficial, interactions also are likely to contribute to the judgment of people as individuals rather than as members of a particular ethnic group. Such a weakening of boundaries between ethnic groups will be made more difficult if political tensions between ethnic groups in the larger society polarize people according to ethnic group.

Interpersonal versus Intergroup Contact

In discussions of the cognitive processes involved in contact, there are two somewhat contradictory emphases that appear: one is an emphasis on inter*personal* contact, where people interact primarily as individuals, and the other on inter*group* contact, where people interact with a focus on their group identity.

Some writers have emphasized the importance of contact leading people from different groups to see each other as individuals, rather than in terms of their group labels. They have emphasized that contact is likely to make people realize that they have important similarities with those in other ethnic groups, especially with respect to values and beliefs. Seeing others as individuals, most of whom are basically similar to themselves, is seen as leading people to place less importance on group categories or labels (Amir 1969; Cook 1978; Brewer and Miller 1984).

An alternative view is that when contact emphasizes interactions among people as individuals, with little or no attention to their ethnic group identity, any favorable changes in attitudes that occur are not likely to generalize to the ethnic out-group as a whole. Thus, for example, while a white person may have more favorable attitudes toward a few black individuals whom she has met at school or at work, she may continue to hold negative attitudes about blacks in general. And in order for favorable attitudes developed through interethnic contact to generalize, this argument continues, those from another group must be seen as typical of their group and attention must be focused on their group identity (Hewstone and Brown 1986; Pettigrew 1986).

There often appears to be a trade off, at least in the short run, between the benefits of emphasizing interpersonal contact in order to get people to see others as individuals, and the benefits of emphasizing group identity in order to change stereotypes of groups. In the long term, however, an emphasis on interaction between people from another ethnic group as individuals and a change in attitudes toward the other ethnic group as a whole may be entirely compatible. People may see particular members of another ethnic group as exceptions to the group stereotype or they may form ideas about various sub-types of the other ethnic group—for example, of middle-class Hispanics, friendly Hispanics, hard-working Hispanics, and so on, that permit them to maintain a stereotype of the rest of the ethnic group. However, as people come into contact with more members of the other group, they may have to create more and more exceptions and more and more subtypes of the broader ethnic category. The end result is a virtual destruction of the original stereotype.

In fact, the process of differentiation among individuals within a group is the very antithesis of stereotyping. Thus, seeing members of another group as individuals or special types may not change the nature of the image of another group as a whole. Rather, over time, it may destroy any unified image of the other group. As Rothbart and John state: "Continued subtyping should effectively lead to dissolution of the stereotype" (1985: 99).

Whether or not it is realistic to encourage interaction among people as individuals, or in terms of their ethnic identity, depends on the strength of ethnic identities and the relevance of these identities to the interaction situation. Where there are ongoing conflicts between different ethnic groups—for example, Serbs and Croatians in former Yugoslavia, Jews and Arabs in Israel—it would be unrealistic to expect members of either group to ignore their group identities. Similarly, where there is strong intergroup competition or clear differences in status or power between the groups in the contact situation, group identity necessarily is salient. However, where conflict between groups in the larger society is not at a high level and where group differences are not salient in the contact situation, the possibility that contact on a more interpersonal basis will lead to a deemphasis of group categories seems greater.

There may also be a possibility of promoting a two-step process by which contact between groups focuses first on relations between individuals and then

103

attempts to get people to generalize improved attitudes to the other ethnic group as a whole. Stuart Cook brought people from different ethnic groups into contact under conditions that emphasized relations as individuals and led to friendly interactions. Then a minority person who had become a friend initiated discussion of the discrimination experienced by other persons in her ethnic group. This discussion often led to a more general change in attitudes about treatment of all those in the minority group (Cook 1984).

Effects on Emotions

While great attention has been given to the ways in which intergroup contact may affect people's perceptions of other groups, much less attention has been given to the possible impact of contact on people's feelings toward other groups (Pettigrew 1986; De Ridder and Tripathi 1992). But contact with members of another ethnic group may have great effects on such emotions and the emotions, in turn, may change behavior (Smith 1993; Patchen 1995; Pettigrew 1996).

The effects of contact on feelings toward another group may be both indirect and direct. The indirect effects come primarily through changes in how people see the other group, discussed in the previous section of this chapter. More positive images of another group as a whole will lead to less negative and more positive feelings toward members of this group. For example, if whites see blacks in general as less aggressive, more honest, and more hardworking than they were seen before contact, then the whites will feel less anger, less fear, and more liking towards black people. If, as a result of their contact experiences, white people make more distinctions among various groups of black people (with respect to their friendliness, honesty, hard work, and so on) then they will have more positive feelings toward at least some black people; and to the extent that contact with blacks leads whites to see at least some blacks as part of their own group (e.g., as carpenters, as Baptists, as classmates), rather than only as an out-group, then the more positive feelings usually directed toward in-group members will be increased.

The effect of contact on feelings toward members of another group also may be more direct. Contact with members of another ethnic group may result in positive feelings because the contact has been associated with pleasant and rewarding experiences. Similarly, contact may lead to negative feelings because contact has been associated with unpleasant experiences (Amir 1976). For example, when students of different races experience success and reward in cooperative groups, liking between members of the different groups increases (Johnson et al. 1984). When the contact that Jews had with Arabs resulted in unpleasant experiences, hostility towards Arabs increased (Ben-Ari and Amir 1986).

A key psychological process involved when intergroup contact is pleasant or unpleasant is that of classical conditioning, sometimes called "affect conditioning" when applied to the conditioning of affect or emotions (Byrne 1971). In intergroup contact, members of another ethnic group may become associated with experiences (stimuli) such as success or failure in a task, reward or penalty from a teacher or a boss, fun or drudgery in an activity, laughter or stress, or praise or criticism, that typically produce positive or negative feelings. Members of the other ethnic group may or may not have been responsible for these pleasant or unpleasant events. By being associated with such stimuli they elicit the same feelings as those events elicit.

Whether contact with another ethnic group is associated with pleasant or with unpleasant events depends to a large extent on the nature of the contact situation. Many of the conditions of contact (discussed in previous sections of this chapter) are important in part because they make contact more or less pleasant and rewarding. For example, contact with members of another group usually is more pleasant when people are engaged in cooperative tasks, when they are of equal status, and when other important people are supportive of the interaction. The pleasantness of contact also will be affected by the characteristic of the people involved. For example, intergroup contact is apt to be more pleasant and rewarding when those from different groups share similar values and beliefs, especially those that are relevant to the contact situation (such as whether success in an activity is important).

When contact with members of another ethnic group changes a person's feelings toward that group, either directly through affect conditioning or indirectly through changes in his perceptions of this group, then his behavior is likely to change as well. Emotions affect behavior directly—for example, fear of a group may cause a person to avoid that group. Emotions may also affect behavior through their effects on expectations—for example, a person who likes Japanese may expect that he will enjoy spending an evening with a Japanese family and so invite them to his home. Thus, by leading to more positive feelings toward members of another ethnic group, contact may result in more positive behavior.

Expectations About Outcomes of Behavior

The way in which a person acts is usually affected by his expectancies about the outcomes of his behavior (Ajzen and Fishbein 1980; Abelson and Levi 1985). We have noted that, by changing a person's perceptions and feelings concerning members of another ethnic group, contact with this group may change his expectations about the outcomes of his actions toward members of the group.

Intergroup contact also may affect a person's expectancies more directly. For example, if such contact involves working together on common tasks, he may expect that, by acting in a cooperative way, he will help to bring rewards to himself (success in the task, a good grade, a good performance rating, and so on).

105

If the intergroup contact is with members of another ethnic group whom he finds are similar to himself in interests and attitudes, he might expect that engaging the other in conversation or joint activities would be pleasant for him. If he detects that members of the other group initially have friendly attitudes toward his group, he would expect that a friendly approach by himself towards the others would be reciprocated.

If an authority figure (such as a boss, a teacher, or a clergyman) has said that she wishes people to act in a friendly, egalitarian way towards members of other ethnic groups, the person may expect that, by acting in such ways, he will win approval from the authority. Similarly, if contact occurs under circumstances in which friends, family, co-workers, or neighbors endorse friendly interaction between those from different ethnic groups, the person can expect that his friendly behavior will be followed by social approval.

If intergroup contact occurs under more unfavorable conditions, the person may form different expectations concerning the likely outcomes of his actions toward those in another ethnic group. The ethnic groups may be competing with each other; the members of the two groups may hold different attitudes about issues relevant in the contact situation (e.g., about conformity to the rules of a school); members of both groups may enter the contact situation with negative attitudes toward the other group; authorities in the situation may not promote positive intergroup relations; and peers or family members may express disapproval of friendly relationships with members of the other group.

The more that these and other unfavorable conditions exist, the more a person is likely to expect that approaches or friendly actions by himself toward members of the other group will bring unpleasant results (rebuff by or unpleasant encounters with the other group, disapproval by members of his own group, lack of success by his own group in competition with the other group, and so on). He may expect, instead, that avoiding members of the other group or even acting in unfriendly or domineering ways toward members of that group will be rewarding. For example, he might expect avoidance to reduce his anxiety about interaction, and "putting down" members of the other group to win approval from his friends.

Direct Effects on Behavior

In addition to affecting behavior indirectly, through its effects on perceptions, feelings, and expectations, intergroup contact may affect behavior more directly. Such behavioral changes may lead to attitudinal changes as well.

Instrumental Conditioning

Situations that bring people from different groups into contact often require the people to behave in new ways. Employees may have to work with each other cooperatively on the job. Students may have to cooperate on a group project. Those on a baseball team have to coordinate their play. Behaviors in these situations will have consequences, either good or bad. For example, friendly actions toward members of another ethnic group may bring reciprocal friendly actions by the others, help from them, approval from third parties, success, fun, or other rewards. On the other hand, sometimes these behaviors may lead to negative outcomes (rejection of friendly overtures, disapproval by peers within one's own ethnic group, and so on).

When a certain type of action is repeatedly rewarded, it will tend to recur. This process has been called "instrumental conditioning" (Michener et al. 1990: 61–65). Behavior learned in this way is slowly strengthened. This type of conditioning is one of the reasons that the effects of contact are likely to be cumulative over time (Pettigrew 1986). For example, in the Robber's Cave experiment with boys at a summer camp, members of formerly conflictive groups were induced to work together to reach a series of common goals; as their more friendly behavior was rewarded repeatedly, it became stronger (Sherif et al. 1961).

In addition to learning behavior directly by experiencing rewarding or punishing consequences, people also learn by watching the behavior of others, who become models (Bandura 1977). In many intergroup contact situations, people get a chance to see models acting in friendly or unfriendly ways toward another ethnic group. They also can observe any positive or negative consequences of this behavior for the model. Such observations may encourage them to imitate the observed behaviors. For example, a black student may observe that a black schoolmate or a black teacher having lunch with whites is apparently enjoying the experience. Therefore, she may be encouraged to do this herself.

Whether a particular situation of intergroup contact will provide rewards for friendly behavior will depend on some of the conditions of contact discussed earlier. For example, where members of different groups must work together in activities, such as the "jigsaw" program (Aronson and Gonzalez 1988), cooperative actions usually result in a variety of rewards, including help, success, acceptance by others, good grades, and a belief that one is liked by others. Other examples of contact conditions that are likely to make friendly behavior rewarding are supportive norms by important others (family, peer groups, and so on) and greater similarities between the members of the different groups. Such favorable conditions of contact will lead people to expect either more extrinsic rewards (such as good grades or money) or more intrinsic rewards (such as pleasant experiences) to follow friendly behaviors (Blalock and Wilken 1979).

Dissonance Reduction

While we usually think of attitudes as affecting behavior, the causal effect sometimes goes in the other direction. There is a great deal of evidence that when people are induced to change their behavior, they often will change their attitudes in the same direction (Sherwood et al. 1969). This change in attitude may occur because their new behavior leads to new perceptions and new experiences. Thus, a prejudiced white person who is forced to work with blacks may discover new, more positive information about blacks that changes his perceptions. Also, his new experiences in the interracial situation may be so pleasant that his feelings toward blacks become more positive (through the process of "affect conditioning" discussed above).

There is another important reason why changes in behavior may be followed by changes in attitudes. People tend to feel uncomfortable—to experience "cognitive dissonance"—when their behaviors are not consistent with their attitudes (Brehm and Cohen 1962). If a person who has a negative attitude towards Hispanics finds himself working alongside Hispanics, his behaviors and his attitude are inconsistent. He could leave his job or ask to be transferred but such actions might risk serious inconvenience or penalty. To reduce cognitive dissonance in this situation, he may change his attitude towards Hispanics, emphasizing their positive characteristics to himself.

However, when a person's behavior and attitudes are inconsistent, he may not be motivated to reduce the dissonance if he believes that his behavior has been forced on him, rather than chosen freely (Sherwood et al. 1969). James Wiggins and his colleagues explain: "For individuals to experience dissonance, they must believe they acted *voluntarily*; otherwise, they will not feel responsible for the outcomes of their decisions. . . . If, in contrast, they are compelled to act contrary to their beliefs, they can avoid dissonance by reasoning, 'I was forced to do this; I really did not have any choice'" (1994: 254).

Thus, attitude change is more likely in a contact situation in which people participate voluntarily, at least to some extent. However, the dissonance-reduction effect is still relevant where people are induced (rather than forced) to interact with members of other ethnic groups. For example, people may be induced to attend a "magnet" interracial school by offering an attractive curriculum and people may be induced to live in an interracial neighborhood by providing affordable attractive houses. In such circumstances, they are apt to want to make their attitudes consistent with their behavior.

Societal Context of Intergroup Contact

The contact that may occur between members of different ethnic groups in particular settings—neighborhood, school, work, and so on—cannot be isolated from the broader economic, social, and political relationships among groups in the larger society. First, the opportunity for contact is affected by social institutions and practices. For example, racial segregation of neighborhoods limits opportunities for contact between whites and blacks in their neighborhoods and neighborhood schools; discrimination against Latinos by some companies in their hiring reduces opportunities for contact between Latinos and Anglos in the work place.

When contact between those from different ethnic groups does occur, the conditions of contact are shaped by broader society. For example, if educational, economic, and legal institutions permit, or even promote, the placement of minority groups in lower-skill jobs, while the white majority generally have higher-skill jobs, interethnic contact in the work place will be between those of unequal status and power.

The outcomes of interpersonal contact between members of different ethnic groups also may be affected by political conflicts between these groups in the larger society. For example, the positive impact of contact between Catholics and Protestants in Northern Ireland have been limited by a societal context of political struggle between the two religious groups (Trew 1986). Similarly, in Israel the potential benefits of contact between Jews and Arabs have been reduced by the ongoing conflict between these groups for control of territory. Similar political, economic, and social divisions in other societies (Horowitz 1985) may limit the effects of interpersonal contacts between members of different groups.

Thomas Pettigrew comments: "The use of intergroup contact as a means of alleviating conflict is largely dependent on the societal structure that patterns relations between groups" (1986: 191). We will be considering some of the broader societal factors affecting intergroup relations in later chapters.

Summary

Let us now summarize the evidence and conclusions presented in this chapter.

Contact between members of different ethnic groups, in schools, neighborhood, work places, and other settings, sometimes leads to more positive attitudes and more friendly behaviors. However, contact sometimes has little effect or even negative effects on intergroup relations.

The impact of contact on attitudes and behavior toward another group depends on the nature of the contact, the contact situation, and the characteristics of those involved in the contact. In order to have any substantial positive effects, contact between those from different groups must be relatively frequent and non-superficial; that is, the contact must be intimate enough so that people can get to know each other as individuals.

Contact has more positive effects on relations among groups in certain types of situations than in others. In general, situations that facilitate the most favorable outcomes include the following features: 1) common rewards for the groups and/or a task that requires working together; 2) equal status for each group; 3) support by authorities for friendly relations, and nondiscriminatory treatment of groups by the authorities; and 4) support for friendly relations by face-to-face groups, such as family and peers.

Positive outcomes as a result of intergroup contact also become more likely, in general, as members of each group are more similar to each other, especially in attitudes, values, and behavioral styles. More favorable initial attitudes toward the other group are also conducive to more positive change following contact.

There are a number of specific effects that contact between persons from different ethnic groups may have. Contact between groups may affect the ways in which people categorize those with a different ethnic background (as "they" or part of "we"); the content of perceptions they have of another ethnic group; their feelings toward another group; their expectations about the rewards and costs of alternative behaviors; and their behavior towards members of that group.

Following interaction with members of another ethnic group, people may see the others less in terms of their ethnic identity and more as individuals. While intergroup contact may reduce the rigidity of boundaries between groups, usually members of both ethnic groups will continue to recognize some distinction between the groups. However, intergroup contact may change the type of image that members of each group have of the other group. Changes in perceptions of the other group may occur because contact results in new information about the other group. Such information, if it is mainly positive, may counter previously-held stereotypes.

However, new information does not always bring changed stereotypes. Images are resistant to change. A variety of cognitive mechanisms act to reduce changes in existing images. In order to break through the cognitive barriers that resist change, the contact situation must bring those who have negative views of another group into repeated contact with "typical" members of the other group who clearly do not fit their negative stereotypes.

In addition to its effects on people's perceptions of another ethnic group, contact between groups may affect their feelings toward members of the other group. Feelings may change as a result of changed perceptions; for example, seeing the others as less harmful than before may reduce anger. In addition, feelings toward another group may change because members of this group have been associated with pleasant or unpleasant experiences in the contact situation.

Changes in perceptions of, and feelings toward, members of another group that occur as a result of contact generally result in changes in behavior. In addition, intergroup contact may affect behavior towards those of another group more directly. It may do so by requiring or encouraging certain behaviors and/or by rewarding or punishing such behaviors when they occur.

Friendly actions toward another group are more likely to occur initially and to be repeated when they are encouraged and rewarded by those in authority and by members of the groups (e.g., schoolmates, co-workers) to which people belong. When people are induced to act in more positive ways toward those in another ethnic group, they then may change their attitudes (perceptions and feelings) toward that group in order to reduce inconsistencies between their behavior and their attitudes. However, it is better to induce, rather than to compel, positive behavior since people are less likely to change their attitudes to match their behavior when they feel their behavior is not voluntary.

The nature and outcomes of contact between people in a particular setting (a classroom, an office, a military unit, a neighborhood, and so on) are affected strongly by the broader institutions within which these contacts occur. These various social institutions and patterns affect important aspects of the contact situation and they affect important characteristics of those coming into contact. In the next chapter, we turn our attention to these larger institutions of society.

5) Population Composition and the Economy

The preceding chapters have discussed the feelings of identification that individuals have with their own ethnic groups, the attitudes that they have towards other ethnic groups, and what happens when members of different groups come into contact with one another. These reactions by individuals and these contacts between individuals take place in the context of a larger social setting and are shaped by that larger setting. For example, the attitudes of individuals toward their own and other ethnic groups are likely to be influenced by the relative size of those groups in the larger society, by the kinds of occupations that those in each group tend to have, and by depictions of that group in the media.

Thus, to understand the types of relationships that exist between members of different ethnic groups—high or low contact, friendly or unfriendly, equal or unequal—examining the position of the groups within the larger society is important. Among the significant features of any society that shape the relationships between ethnic groups are the following:

1. The size and composition (e.g., the age distribution) of different groups.
2. The place of each ethnic group in the economy.
3. The school system.
4. The political and legal systems.
5. Other societal institutions, such as churches and the media.

In this chapter we will discuss the first two of these aspects of society. The other aspects will be discussed in the following chapters.

Size of Ethnic Groups

One basic circumstance that affects the relationships among different ethnic groups is the size, especially the relative size, of these groups. The size of the groups affects a number of relevant outcomes, including opportunity for contact, perceptions of threat, avoidance, and neighborhood segregation.

Opportunity for Contact

In his formal theory of population structure, Peter Blau (1994) points out the "mathematical truism" that the smaller the proportion of a total population represented by a particular group, the more contact by sheer chance it is likely to have with members of other groups. Some of these contacts will lead to friendly relationships, and even to marriage. Thus, the smaller a religious group's proportion of the total population, the higher is its rate of marriage outside its own group (Bealer et al. 1963; Burma 1963; Blau and Schwartz 1984). For example, the smaller the proportion of Jews in a given area, the larger the proportion of Jews who marry non-Jews.

As Blau points out, the smaller the proportion that a given group represents in a population, the greater also is the probability that its members will have negative or unfriendly contacts with members of other groups. Because they are more likely to come into contact with other groups, members of smaller groups are more likely than those from larger groups to be involved in intergroup conflict. Supporting evidence comes from several studies of the victims and perpetrators of crimes. One study looked at those involved in four crimes (rape, robbery, assault, and larceny). Although victims and assailants tended to belong to the same racial group, the smaller a group's proportion of the population in a neighborhood, the more likely its members were to be victimized by a member of another race (Sampson 1984). Another study found that an increase in the proportion of blacks increases the chances that both blacks and whites will be robbed by blacks (Messner and South 1986). Studies of interracial rape also conclude that the likelihood that either the victim or the perpetrator will be black or white is strongly related to the relative size of the black and white populations (O'Brien 1987; South and Felson 1990).

Perceived Threat

As another ethnic group becomes larger in size relative to their own group, those in a majority or dominant group tend to become more uncomfortable about and often to feel threatened by the other group. For example, a

114

study in twelve countries of western Europe found that as the percentage of immigrants from outside western Europe (e.g., Turkey or North Africa) increased, hostility towards the minorities also increased (Quillian 1995). Dominant group members may see two primary types of threats—economic and political—as the relative size of the minority group increases. (They also may see a threat to their cultural dominance or "way of life.")

With regard to perceived economic threat, Blalock writes: "The larger the relative size of the minority, however, the more minority individuals there should be in direct or potential competition with a given individual in the dominant group. As the minority percentage increases, therefore, we would expect to find increasing discriminatory behavior" (1967: 148). Consistent with this reasoning (though there may be additional reasons for the association), as the proportion of blacks in given regions of the United States (counties, metropolitan areas, and states) increased, racial differences in income also increased (Blalock 1967; Brown and Fuguitt 1972). A recent study using the fifty states as units reported that inequality in income between African Americans and white Americans is correlated .52 with the percentage of African Americans in the state—that is, states with the largest proportions of blacks have the largest gaps between black and white median household incomes (Jenness and Grattet 1993). However, as the proportion of African Americans in an area increases, disparities in occupational level between the races do not increase in the same way that income differences do. A relatively large non-white population may "spill over" into white-collar and professional occupations and may support a larger number of professionals and business people in its own community (Brown and Fuguitt 1972).

As the relative size of a minority group increases, it tends to be seen by members of a dominant group as posing a greater threat to their political control, as well as to their economic position. Blalock (1967) argues that the perceived threat, and the mobilization of the dominant group to maintain their political dominance, accelerates as the proportion of the minority group increases:

> In a very real sense, the curve [relating discrimination to minority percentage] can therefore be conceived as a "threat curve." The power threat, in the form of a need for increasing degree of mobilization, may be relatively slight with small minority percentages, but the theory predicts that the threat should increase with [minority percentage] at a continually accelerating rate. The need for a high degree of mobilization of resources to maintain dominance becomes extremely great as the minority becomes larger. (1967: 154)

Use of discrimination by dominant ethnic groups in order to maintain control over large ethnic minorities has been seen in a variety of nations—including South Africa, where a numerically inferior white group excluded blacks from political participation for many decades; in Northern Ireland, where Protestants maintained political control over a large Catholic minority; and in Israel, where Jews withheld political rights from a relatively large Arab population in the occupied West Bank territory.

In the United States as well, political behavior by members of the dominant white group sometimes has been related to racial proportions. Before federal legislation enforced voting rights for African Americans, blacks were much less likely to be permitted to vote in counties with large proportions of blacks (Matthews and Prothro 1963). The proportion of whites voting for racist political candidates also has been greater as the proportion of blacks in a region is greater (Pettigrew and Cramer 1959; Wright 1977). On the other hand, as a minority group becomes larger, it becomes more attractive as a potential coalition partner. For example, in the American south in the 1890s, the Populist Party attempted for a time to ally itself with the large black constituency in order to win political control, although it eventually changed its strategy and turned against blacks (Woodward 1957).

Avoidance and Segregation

As the relative size of a minority ethnic group increases, discomfort among members of the dominant group about contact with the minority tends to increase. This seems to be especially true when the minority group in question generally occupies lower-status positions in society.

Studies of reactions of white Americans to the prospect of contact with members of racial minorities—especially with blacks—indicate that willingness for such contact declines as the proportion of the minority increases. Following the U.S. Supreme Court ruling that segregation of public schools was unconstitutional, resistance to school desegregation was highest where the proportion of blacks in an area was highest (Pettigrew and Cramer 1959; Van Fossen 1968; Giles 1977). In more recent years, when white Americans were asked whether they had "any objection to sending your children to a school where a few of the children are black" respondents were almost unanimous (95 percent in 1983) in saying they had no objection. But the proportion having no objection dropped to 76 percent "where half of the children are black" and to only 37 percent where "more than half of the children are black" (Schuman et al. 1985).

Neighborhood Segregation

Ethnic group proportions also affect the extent of segregation in neighborhoods. Farley and Frey (1994) studied residential segregation in all American metropolitan areas with substantial black populations, as shown in census data for 1990. They found that as the relative size of the black population of an area increased, segregation scores for that city also increased. Furthermore, the higher the percentage of blacks that whites were exposed to in their neighborhoods in 1980, the less the decline in residential segregation during the decade of 1980 to 1990. The metropolitan areas that had the largest reductions in segregation during the decade were those that had the lowest percentage of blacks.

For Hispanic Americans, while residential segregation is much less pronounced than is the case for African Americans, contact with other Americans also is affected by the relative size of the minority population in an area. A study of the fifty largest American metropolitan areas, plus ten others that have large numbers of Hispanics, found that as the percentage of Hispanics in an area increased, the degree of exposure or potential contact between Hispanics and majority group members (Anglos) decreased significantly (Massey and Denton 1988).

One major reason that increasing proportions of a minority leads to greater segregation is the reactions of members of the majority or dominant group. When asked about their reactions if African Americans "with the same income and education as you have moved into your block," a large majority of Americans say that it would *not* make a difference to them. However, when asked about their preferences for living in neighborhoods of varying racial compositions, only a minority express a preference for living in a neighborhood that is not mostly white (Schuman et al. 1985: 106–108).

Studies of attitudes toward neighborhood racial composition among people in the Detroit metropolitan area (Farley et al. 1994) help to clarify the processes that lead to residential segregation of neighborhoods. When shown a diagram of a hypothetical neighborhood in which one black family (out of fifteen) lived close to them, 84 percent of whites said they would feel comfortable in such a neighborhood, 87 percent said they would be willing to move into such a neighborhood, and only 4 percent said they would try to move out. But as the proportion of blacks rose, more negative reactions among whites increased. If five black families (out of fifteen) lived in the immediate neighborhood, only 56 percent of the whites would feel comfortable, 29 percent would try to move out, and only 42 percent would be willing to move in. If blacks were a slight majority in the neighborhood (eight of fifteen families), a large majority of whites would feel uncomfortable, most would try to move out, and few would be willing to move in. In contrast, African Americans interviewed in the Detroit area study were almost unanimous in being willing to move into a neighborhood that was almost half white, and a large majority (87 percent) said they would be willing to move into a neighborhood that was predominantly white.

In an earlier study of the racial attitudes of Detroit area residents, Farley and his colleagues plotted the percentage of whites who would be willing to move into, and the percentage who would want to move out of, neighborhoods of varying racial compositions. They found that when the proportion of blacks in a neighborhood reached about 30 percent, a greater proportion of whites would move out of than would move into such a neighborhood. Thus, above this percentage, the neighborhood would be expected eventually to become black—to "turn over" (1978).

Tipping Points

The minority proportion at which an ethnic "turn over" may occur has been referred to as a "tipping point." The idea of a racial tipping point originated in a study of the Chicago Housing Authority, in which the researchers found that if "there were more than about one-third Negroes in an otherwise white project, the whites would leave until the project eventually became all Negro" (Myerson and Banfield 1955).

Some studies of school racial composition have found a similar phenomenon. A study in fifteen northern cities found that, once the student body of a school became about half black, it rapidly became all black (U.S. Commission on Civil Rights 1967). A study of school districts in Florida found that parents of white students became more likely to switch their children to private schools when the proportion of black students in a school rose above 30 percent. However, this study also found that after an initial jump in leaving by white families following the crossing of the 30 percent minority "threshold," the exodus of white families declined in following years (Giles et al. 1977).

While a number of studies indicate that members of a majority group may withdraw when the proportion of a minority group rises above about one-third, there is no magical "tipping-point" proportion. Blalock (1982) asserts that the exact tipping-point in a given situation will depend on such things as the racial distinctness of the minority, the extent to which the norms of the groups differ, the extent to which the minority group segregates itself, the amount of competition between the groups, and the stereotypes of the minority held by members of the members of the majority. The Detroit area studies of whites' attitudes concerning the racial composition of their neighborhoods (Farley et al. 1978; 1994) illustrate that majority reactions to particular racial proportions may change over time, thereby moving the exact tipping-point at which larger numbers of whites will move out of, than will move into, a given type of environment.

Self-Sufficiency

When an ethnic group is very small, it must depend on the larger group for many things it needs—jobs, customers, stores from which to buy essential goods, services of various kinds (from doctors, dentists, lawyers, plumbers, auto mechanics, and so on), and even for social contacts. In this context, the *absolute* size of the minority group, rather than its relative size, is the more important feature. But as the size of an ethnic group increases, it may become more self-sufficient. Members of the group may, if they wish (or if they are excluded by the larger group), meet many of their needs within their own ethnic group (Blalock and Wilken 1979). Many may be employed by members of their own ethnic group, as has been the case for large numbers of Chinese immigrants. In such cases, people may be able to shop in stores run by fellow ethnic-group members, eat in ethnic restaurants, be served by doctors, dentists, and others from their own ethnic community. Ethnic institutions—a black-owned bank in Harlem, revolving credit associations among Chinese Americans, Hispanic-centered churches, and so on—may be created to serve the needs of the ethnic community.

One consequence of the greater self-sufficiency that is likely to accompany larger group size is that the group may become more isolated from other groups. As an ethnic group develops a larger network of services, organizations, and institutions to serve its own members, individuals in that group have less need to venture beyond the warmth and acceptance of their own group. Contacts with those outside their own ethnic group may sometimes be unpleasant or cause them to feel uncomfortable. Thus, as Blalock and Wilken observe: ". . . with increased self-sufficiency combined with non-negligible costs, any level of [outside] contact may appear to be more costly than it is worth" (1979: 574).

There is evidence that as the absolute size of the non-white population (primarily blacks) increases, residential segregation in American cities also increases (Marshall and Jiobu 1975). Marshall and Jiobu suggest that a large black population is more likely to be residentially segregated at least partly because specialized black-oriented institutions and facilities can emerge and be supported in these circumstances. They also note, however, that a large minority population may increase majority-group feelings of threat, as well as discrimination by the majority group, and that such reactions by the majority may help to explain greater residential segregation when the minority is large.

Change in Ethnic Proportions

In addition to effects of the absolute and relative sizes of a minority group on its relations with a majority group, the relative growth of the minority population may be important. If one group is growing in size more rapidly than another, especially if it already is large, it is more likely to be seen as a threat by the larger dominant group. For example, Protestants in Northern Ireland have seen the higher birth rate of Catholics in their area, compared to Protestants, as a threat to their political and cultural control of Northern Ireland. Similarly, many Israeli Jews have been concerned that a high birth rate among

Arabs could eventually result in a much larger Arab minority, or even an Arab majority, and thus change the nature of the presently-Jewish state.

In the United States, large and rapid influxes of minorities—such as Puerto Ricans into New York after World War II, of Cubans into Miami starting in the 1960s, and of blacks from the south into northern cities—led to perceptions among many residents that their cities were being "taken over" by alien persons (e.g., see Burkholz 1980). Morrill asserts that during the 1940s and 1950s the "influx [of blacks] was far greater than the cities could absorb without prejudice," resulting in the ghettoization of blacks (1965: 340).

Evidence that rapid growth in a minority population (black, Hispanic, or Asian) is associated with greater residential segregation in American cities is weak (Marshall and Jiobu 1975; Massey and Denton 1988; Farley and Frey 1994). However, this relationship is likely to be complex for several reasons, including the fact that rapid growth in a minority population may lead to temporarily mixed neighborhoods as the transition from one ethnic group to another occurs.

Overall, while the evidence is based primarily on qualitative historical cases, one can reasonably expect that rapid growth of a minority, especially where it threatens the control of a dominant group, will trigger alarm and resentment.

Other Population Characteristics

In addition to the fundamental aspect of size, several other aspects of ethnic group populations may affect the relationships between groups. These include the dispersion of groups, their gender ratios, and their age compositions.

Dispersion

If members of a particular ethnic group are concentrated in a certain area, this tends to maintain some social and cultural separation from other ethnic groups. Greater dispersion among group members tends to promote cultural and social assimilation. For example, Japanese Americans who were concentrated geographically and occupationally (in agriculture and related businesses) in California before World War II tended to remain separate from the larger community. After becoming more geographically and occupationally dispersed after World War II, Japanese Americans became more culturally and socially assimilated, as indicated, for example, by a high rate of intermarriage (Kitano and Daniels 1988).

Concentration of any ethnic group in a limited area may increase the political power of that group. The fact that American Jews are disproportionately concentrated in certain areas, especially in the New York City area, gives that group greater power in city, state, and national politics than it would have if Jews were dispersed more uniformly around the country. The same is true for Puerto Ricans in New York City, Cubans in Miami, Hispanics in the southwest, and blacks in the central sections of many American cities. Thus, many American cities with large proportions of black residents have elected blacks as mayors and as other city officials. (On the other hand, the concentration of low-income blacks in inner-cities has contributed to a variety of problems, including a low tax base for education and other city services.)

Age Composition

The distribution of different age categories within given ethnic groups may affect their relationships with other groups. Younger people differ in their interests, behaviors, and economic position from older people. For example, crimes of violence are most often committed by teenagers and young adults (20s to 30s) who are likely to earn less than those in their middle years.

Age distributions are different for some major ethnic groups in the United States. Hispanics have the youngest population, non-Hispanic whites the oldest, and blacks are intermediate. In 1990, about 35 percent of Hispanic Americans, 32 percent of blacks, and only 23 percent of non-Hispanic whites were under eighteen years of age. The proportion of Hispanics and blacks who were young adults (eighteen to thirty-four years old) was higher than that for non-Hispanic whites. Non-Hispanic whites were somewhat more likely than members of the other groups to be thirty-five to fifty-four years of age and especially more likely to be fifty-five years or older. Twenty-four percent of non-Hispanic whites, 15 percent of blacks, and only 11 percent of Hispanics were in the oldest age categories (Murdock 1995).

These age differences among the major ethnic groups contribute to a number of other behavioral and economic differences. The relatively youthful composition of Hispanics and of blacks contributes, for example, to higher rates of violent crime and lower incomes in these groups. The different age distributions also creates a situation in which generally poorer Hispanic and black workers pay taxes for social security and Medicare that go disproportionately to relatively affluent older non-Hispanic whites.

Gender Ratios

The proportions of males and females in a given group may affect its position in society or its relation to other groups. Historically, some immigrant (or conquering) ethnic groups have been predominately male. This was true, for example, of the Spanish who first came as conquerors to Mexico. Such gender disparity has led to the rapid biological mixing of the ethnic groups, as men who lacked women from their own ethnic group found sexual partners (in or outside marriage) from other groups.

In the United States today, various ethnic groups generally have a balance of men and women. However, such gender balance is not entirely present among African Americans. Because of high rates of imprisonment and of death (usually from violence) among young African American males, there is a greater number of marriage-age women than men among African Americans. Combined with problems that many black males have in getting well-paying jobs, this results in a shortage of marriage partners for black women (Raley 1996). This shortage undoubtedly contributes to high rates of birth outside of marriage and reduces income in black families, since one-parent families earn less than two-parent families. These social and economic effects, in turn, contribute to the separation of many black Americans from white Americans.

Socioeconomic Distribution

Another important characteristic of population groups is their distribution among various socioeconomic levels, especially of occupation and income. This aspect of ethnic groups will be discussed later in this chapter in the section on economic inequality, as part of a broader consideration of economic relationships among ethnic groups.

Population Composition of the United States

U.S. Bureau of the Census counts showed the total population of the United States to be almost 249 million in 1990. Of these, almost 76 percent were non-Hispanic whites, almost 12 percent were non-Hispanic blacks, 9 percent were Hispanics, and almost 4 percent were of other races (Asians, Native Americans, Pacific Islanders, Eskimos, and Aleuts).

In the last few decades, there has been a steady decrease in the proportion of Americans who are non-Hispanic whites, from about 83 percent in 1970, to 80 percent in 1980, to 76 percent in 1990. During this same period, the proportions of blacks rose slightly, from 11.1 percent in 1970 to 11.7 percent in 1990. Larger and steady increases have occurred in the proportions of Hispanics and of Asians. In 1970, 4.6 percent of Americans were of Hispanic origin. This proportion increased to 6.5 percent in 1980 and to 9.0 percent in 1990. During the same period, the proportion of those of other races (primarily Asians) rose from 1.2 percent in 1970 to 2.3 percent in 1980 and 3.6 percent in 1990 (Murdock 1995).

The recent substantial increase in the proportion of Americans who are of Hispanic and Asian origin is primarily the result of changed patterns of immigration. Up until the mid-1960s, most immigrants to the United States came from Europe, although substantial numbers came from elsewhere in our own hemisphere (Canada and Latin America). Following changes in American immigration law in 1965, most immigrants have come from Latin America and Asia. In the decade of the 1980s, 47 percent of all immigrants came from Latin America, 37 percent from Asia, only 10 percent from Europe, and the rest from other parts of the world (Murdock 1995).

The growth of the Hispanic population in the United States also has been abetted by a high fertility rate among Hispanics. In 1990, the average Hispanic woman had 3 children, as compared to an average of 2.5 children for non-Hispanic black women and 2.0 for non-Hispanic white women. (However, while most Hispanic groups are relatively high in fertility, some—notably Cubans—are below the national average.)

Large family size does not appear to be a major contributor to the rapidly growing size of the Asian American population. The average Asian American woman was estimated in 1992 to have 2.3 children. Birth rates were relatively high for Native American women, who were estimated in 1992 to average 2.9 children (Gill et al. 1992; Murdock 1995).

Future Changes

What changes in the composition of the American population may we expect in the future? The U.S. Bureau of the Census has made alternative projections about the possible ethnic composition of the population in coming decades. These projections are based on varying assumptions about the fertility rates and the death rates for each group and about the size and composition of immigrant flows into the country. Projections based on mid-level assumptions show (see table 5.1) that by the year 2030, the non-Hispanic white portion of the American population would decline to about 60 percent (down from 76 percent in 1990); almost 14 percent of the population would be non-Hispanic blacks (up from under 12 percent in 1990); Hispanics at more than 17 percent would have become the largest minority (up from 9 percent in 1990) and others, primarily Asians, would have more than doubled their population proportion (to almost 9 percent, up from 3.6 percent in 1990).

Projections using the same mid-level assumptions about birth and death rates and about immigration show that by the middle of the next century (2050) non-Hispanic whites would constitute only a little more than half of the American

TABLE 5.1

Projections of the Population in the United States By Race/Ethnicity From
1990 to 2050 (Percentages)*

| | Race/Ethnicity | | | |
Year	White Non-Hispanic	Black Non-Hispanic	Hispanic	Asians and Others
1990	75.6	11.7	9.0	3.6
2010	67.6	12.8	13.2	6.3
2030	60.2	13.8	17.2	8.9
2050	52.7	15.0	21.1	11.2

* Projections made by U.S. Bureau of the Census, using middle assumptions about fertility, mortality, and immigration.
Source: Computed from U.S. Census data presented in Murdock (1995), 34. Rows may not add to exactly 100 percent due to rounding.

population; 15 percent would be non-Hispanic blacks, 21 percent would be Hispanics and 11 percent would be of other, primarily Asian, origin (Murdock 1995).

Three factors account for the projected rapid growth of minority populations: younger age distributions, higher birth rates (especially for Hispanics), and higher levels of immigration for most of those groups compared to the non-Hispanic white population. Of course, exact projections cannot be made, since relevant factors such as birth rates and immigration flows are difficult to predict. But whatever the specific assumptions made, the direction of change in the ethnic composition of the United States seems clear. After discussing alternative population projections made using different assumptions, Steve Murdock states:

> However examined, it is apparent that population growth in America in the coming decades will largely occur as a result of minority population growth. These patterns demonstrate that the minority populations of the United States will come to form increasingly larger proportions of the total population. . . . The percentage of the total population that is Anglo would decrease under each scenario and the percentage of the population that would be in each minority group increases. . . . Increasing diversity is likely under virtually all scenarios. (1995: 46, 52)

The trend towards increasing ethnic diversity is likely to have many effects on American society. Undoubtedly, some of these effects will be positive. The nation will benefit from the diversity of cultures and talents represented by different ethnic groups. It will benefit also from the energy and dynamism of many of the new immigrants and their children.

Our review of the effects of population sizes earlier in this chapter also suggests a number of other effects, some of them positive, some of them more problematic. As the non-Hispanic white population becomes relatively smaller in size, its members will tend to have more contact with those from other racial and ethnic groups. Some of these more frequent contacts will be positive, resulting sometimes in friendships and even in marriage. However, non-Hispanic whites also are likely to have more negative contacts, ones that involve conflict, with blacks, Hispanics, and Asians.

We have also seen that as the proportion of a minority increases, particularly when it grows rapidly in size, a sense of threat and resulting hostility increases among the majority group. Thus, we may expect that many non-Hispanic whites will feel a threat to their cultural, economic, and political dominance and will be resentful of the increasingly numerous minority groups. Such hostility already has been evident in places such as California and Florida, where rapid increases in minority population, especially through immigration, have occurred.

Other consequences may occur as a result of the shifting ethnic composition of the American population. For example, minority groups are likely to exert greater political influence at local, state, and national levels. There may be greater competition between the major minority groups (blacks, Hispanics, Native Americans, and Asians) with respect to jobs, political influence, distribution of government benefits, and other advantages. On the other hand, the larger relative size of minority groups will provide greater incentive for them to form coalitions that may be able to win great political influence or even control.

We will consider at greater length the issues that are raised by the increasing diversity of American society in later chapters (see especially chapters 9 and 10).

Economic Position of Groups

One of the central determinants of the relations among ethnic groups is their economic position within a society. Economic position refers especially to the kinds of occupations filled by those in particular groups. Some theorists have seen the distribution of occupations—the "division of labor"—in a society as crucial to the relations among its members (Lenski 1966; Hechter 1975). Also relevant are the income of each group, the extent to which each owns economic assets, such as land and business enterprises, and the extent to which it enjoys opportunities for economic advancement.

The impact of groups' economic positions on the relationships among them may be seen in countries all around the world. For example, in the Philippines, people of Chinese ancestry came to dominate trading in agricultural goods and frequently provided credit to farmers. Their prosperity and economic power was resented by many indigenous Filipinos and this hostility boiled over into violence

125

against the Chinese on many occasions (Hunt and Walker 1974). In Quebec, Canada, several decades ago, those of English background and language dominated the higher-status positions in major industries, while French speakers were found disproportionately in lower-level working-class positions (Porter 1965). These differences of occupational status fueled anti-English sentiment and a movement for political autonomy among the French speakers. In South Africa, where up until recently blacks did not have equal political rights, whites long monopolized skilled and high-paying occupations and lived a life of material affluence, while most blacks had jobs of low skill, such as common laborers and miners. These economic differences contributed to attitudes of superiority but also of fear among the smaller population of whites and to agitation for changes among the blacks (South African Institute of Race Relations 1992).

Occupations in the United States

In the United States, historically there was considerable variation in the types of occupations filled by those in various racial and ethnic groups. Differences between blacks and whites were especially great. For example, in 1940, only 6 percent of African Americans, as compared to 35 percent of whites, were employed in white-collar occupations. There were, for example, few black managers, physicians, engineers, or pharmacists. Thirty-four percent of African Americans were employed as service workers (many as cleaners or maids), whereas only 10 percent of whites were in service occupations (Marger 1994).

In more recent decades differences in the occupational distribution of ethnic groups in the United States have decreased considerably. However, there still are substantial occupational differences among these groups.

Table 5.2 shows the occupations of employed persons in the United States in the early 1990s, separately for those of different sex, race, and Hispanic origin. Compared to whites, a much smaller proportion of blacks, especially of black males, were in managerial and professional jobs. Blacks were much more likely than whites to be in service jobs (e.g., household workers, food service, janitors, and cleaners) or to be in the "operators, fabricators, and laborers" category (e.g., machine operators, moving equipment operators, and laborers).

The occupational profile of Hispanic-origin workers generally was similar to that of black workers. Compared to blacks, Hispanics were somewhat less likely to have managerial or professional jobs and, especially among men, had a higher proportion in the "farming, forestry, and fishing" category, mostly as farm laborers.

Asian Americans (who are grouped in the census with a much smaller number of Pacific Islanders) have a very different occupational distribution than other racial minorities. A higher proportion of Asian American men than of white men were in each of the two highest-status occupational categories (managerial and

TABLE 5.2

Occupation of Employed Persons, By Sex, Race, and Hispanic Origin
(Percentages) in Early 1990s

Occupation, Males	White	Black	Hispanic Origin[a]	Asian/Pacific Islander
Managerial and Professional	27.5	14.7	11.4	33.2
Technical, sales, and administrative support	20.6	17.6	16.3	26.3
Service	9.8	20.0	17.7	15.6
Farming, forestry, and fishing	4.3	2.0	7.8	2.3
Precision production, craft, and repair	18.5	15.0	19.4	9.9
Operators, fabricators, and laborers	19.3	30.7	27.5	12.7
Occupation, Females	White	Black	Hispanic Origin[a]	Asian/Pacific Islander
Managerial and Professional	29.9	20.1	16.4	26.4
Technical, sales, and administrative support	43.2	39.4	39.6	42.8
Service	16.8	26.9	24.8	16.7
Farming, forestry, and fishing	1.2	0.2	1.7	0.5
Precision production, craft, and repair	2.1	2.5	2.9	4.2
Operators, fabricators, and laborers	6.8	10.8	14.6	9.4

a. Those of Hispanic origin may be of any race.
Source: Data on whites and blacks are from U.S. Bureau of the Census, *Black Population in the United States, March 1993 and 1994* (1995); Data on Hispanics from U.S. Bureau of the Census, *Hispanic Population in the United States, March 1992* (1992); Data on Asian Americans and Pacific Islanders from U.S. Bureau of the Census, *The Asian and Pacific Islander Population in the United States, March 1991 and 1990* (1992b).

professional; and technical, sales, and administrative support). Asian American men were less likely than white men, and much less likely than African American or Hispanic American men, to be found in most blue-collar jobs (e.g., as production workers and operators).

Asian American women were only slightly less likely than white women to have managerial or professional jobs. Also, like white women, they were less

127

likely than African American or Hispanic American women to be in service jobs. Overall, the generally high-status occupational profiles of Asian Americans are more similar to those of whites than to the other major ethnic minorities.

Data on the occupational distribution of people in various ethnic groups do not take into account the number of people in each group who are not employed at all. In 1993, 5 percent of whites were unemployed, while almost twice as high a percentage of blacks (9.8 percent) and Hispanics (9.0 percent) were not employed (U.S. Bureau of Census 1994a). Figures on unemployment include only those persons who are actively looking for a job. Many other people have stopped looking for work and basically have dropped out of the labor force. Participation in the labor force has declined in recent decades, especially among non-white (primarily black) men. Between 1960 and 1979, the proportion of non-white men in the labor force fell from 83 percent to 71.9 percent. The proportion of white men in the labor force fell much less, from 83.4 to 78.6 percent during the same period. By 1994, the proportion of black men in the civilian labor force had declined further to 66.5 percent, compared to 74.6 percent of non-Hispanic white men in the labor force (U.S. Bureau of the Census 1995). The greater decline in labor force participation among black men appears to reflect declining job opportunities while the smaller decline among white men probably is primarily due to their retiring at a younger age than in earlier years (Shulman 1989). Overall, the evidence indicates that blacks, and especially black men, are much more likely than whites to have no job at all.

Black women do not differ much from white women in their labor force participation; in 1994, about 59 percent of women in each racial group were in the civilian labor force. (However, as for black men, unemployment among black women in the labor force consistently has been much higher than unemployment among whites.)

Among Hispanics, participation by men in the labor force has been relatively high. In 1993, 79.2 percent of Hispanic males sixteen years of age or older were in the labor force, a higher percentage than for non-Hispanic white males. However, labor force participation by Hispanic women was lower than that for nonHispanic white women (U.S. Bureau of the Census 1994b).

Asian Americans (sixteen years or older), particularly Asian American men, have been slightly less likely than white Americans to be in the labor force; in 1991, 72 percent of Asian and Pacific Islander males aged sixteen years and over were in the labor force compared to 76 percent of white men (U.S. Bureau of the Census 1992b). This small difference may be due primarily to the larger percentage of young Asians who were in school.

Family Income in the United States

Another important aspect of various ethnic groups' economic positions in America is their income. Table 5.3 shows the money income of fam-

128

TABLE 5.3

Money Income of Families by Income Level, Race, and Hispanic Origin, 1992
(Percentages)

Income	White	Black	Hispanic Origin[a]	Asian/Pacific Islander
Under $15,000	13.1	36.9	28.7	14.9
$15,000–24,999	15.1	18.9	21.4	11.3
$25,000–34,999	15.3	13.3	17.0	12.2
$35,000–49,999	20.1	14.6	15.5	19.9
$50,000–74,999	21.0	10.9	11.8	21.3
$75,000 and over	15.2	5.4	5.6	20.3
Median Income	$39,320	$21,761	$24,926	$43,418

a. Those of Hispanic origin may be of any race.
Source: U.S. Bureau of the Census (1994a). Columns may not add to exactly 100 percent due to rounding.

ilies in 1992 for each of four ethnic categories: whites, blacks, Hispanics, and Asians and Pacific Islanders. The highest median family incomes were received by Asian Americans. A larger percentage of Asian Americans (20.3 percent) than of any other group were in the highest income category ($75,000 and over), although a slightly larger proportion of Asians than of whites were in the lowest income bracket (under $15,000). Next in income level are whites, whose median family income was a little below that of Asian Americans and who had a smaller proportion receiving the highest incomes than was true of Asians. Far below both Asians and whites were Hispanics and blacks, whose median family incomes were 63 percent and 55 percent that of whites, respectively.

Income of Workers in the United States

Since the income of families depends in part on the number of people in each family who are working, as well as the combined income of both men and women, it is useful also to look at the income of individual workers. Table 5.4 shows the median incomes in 1992 of white, black, and Hispanic workers, separately for men and for women. Among men, whites had the highest median income; black men were considerably lower (72 percent of the income of whites) and Hispanic men had the lowest median incomes (64 percent that of white men).

129

TABLE 5.4

Median Income of Year-Round Full-Time Workers, by Sex, Race, and
Ethnicity, 1992.

Male Workers	Median Income	Percent of White Income
White	31,737	—
Black	22,942	72
Hispanic[a]	20,312	64

Female Workers	Median Income	Percent of White Income
White	22,423	—
Black	20,299	91
Hispanic[a]	17,743	79

a. Persons of Hispanic origin may be of any race.
Source: U.S. Bureau of the Census (1994a).

Black and Hispanic women did better relative to white women than was the
case for the men in their ethnic groups. The median earnings of black women
were 91 percent those of white women while Hispanic women earned 79 percent
as much as the whites.

The reasons why black and Hispanic workers earn less than whites include
(along with other possible reasons) average differences in amount of education,
in work experience (related to age), in number of hours worked, in the knowledge
of English (especially in the case of Hispanic immigrants), in area of residence
(blacks are over-represented in the south, where wages tend to be lower), and
sometimes discrimination based on race or ethnicity (Farley 1984; Chavez 1991;
Cancio et al. 1996). (See section on "Causes of Inequality" later in this chapter.)

Differences in the earnings of Asian workers and white workers have been
relatively small. The income of Asian male workers has been somewhat lower
than those of white male workers; in 1990, Asian and Pacific Islander males who
worked full-time year-round had median earnings which were 93 percent of the
earnings of comparable white men. These differences in earnings are explained,
at least in part, by the tendency of white men to be in specific occupations that
pay more than the specific occupations of Asian men of comparable education.
For example, among college-educated men, whites were more likely than Asians
to be in executive and managerial positions. However, Asian and Pacific Islander
women who worked full-time year-round earned a little more (106%) than white
women and there were no appreciable differences in earnings between Asian and
white women with similar educations (U.S. Bureau of the Census 1992b).

Assets in the United States

Another important indicator of people's economic position is the value of the assets they have accumulated (homes, motor vehicles, stocks, bonds, and so on) The advantage of whites compared to blacks and Hispanics is much greater with respect to their assets than it is for yearly income. In 1991, white households had a median net worth of $44,408, while the median assets of blacks totaled only $4,604 and that of Hispanics was only $5,345 (U.S. Bureau of the Census 1994a). Thus, whites not only earn more than African Americans and Hispanic Americans but have accumulated much more wealth as well.

Effects of Inequality

Many social scientists see inequality between ethnic groups as the root cause of conflicts between them. In his classic work on inequality, Lenski (1966) sees inequality (which he views as inevitable) as leading inevitably to conflicts between social groups, including ethnic groups, which differ in social (especially economic) position. A number of scholars concerned with the causes of nationalism also point to economic inequalities between ethnic groups as crucial. They see economic inequalities, unequal economic development, and/or economic discrimination among regions as leading to an upsurge of ethnic or nationalistic feeling among those disadvantaged. Thwarted economic expectations lead to frustration and to resentment of the more advantaged ethnic regions (Hechter 1975; Gellner 1983; Kellas 1991). (However, Gurr [1993] finds that separatist movements may occur also among ethnic groups that are not economically disadvantaged.) In addition to inequalities of wealth, inequalities in the distribution of economic and educational opportunities have been seen as leading to resentments and tensions between ethnic groups (Horowitz 1985).

Conflicts between ethnic groups may also arise from inconsistencies between their relative economic status and their other (especially political) statuses (Rokkan and Urwin 1983; Kellas 1991). For example, people of Malay origin in Malaysia have enjoyed political dominance but have been resentful that this political control has not extended to the economic domain, where those of Chinese origin have been dominant (Horowitz 1985).

Besides often leading to resentments by low-status groups (often ethnic minorities) and to conflicts over the distribution of wealth, economic inequalities tend to create social barriers of interaction between groups. The more inequality there is within a society, the more opportunity people have to come into contact by chance with people whose socioeconomic position is different than their own (Blau 1994). Thus, the more that ethnic groups differ in socioeconomic position (occupation, education, and income), the more opportunity there is for interethnic contacts to occur between people who also differ in socioeconomic status.

But people tend to avoid contact with those whose social positions are lower than their own. There is considerable evidence that whites in the United States are more likely to stay apart from minority groups in neighborhoods and schools as socioeconomic differences between the races increase. The extent of desegregation in southern and border states was found to be highly correlated with the social class of the non-white population of each state: higher social class, more desegregation (Van Fossen 1968). Residential segregation of minorities (blacks, Hispanics, and to a lesser extent, Asians) from Anglos has been found to decrease as members of the minority groups are higher in occupation and in income (Massey and Denton 1993). Also, residential segregation of blacks from whites decreases as socioeconomic differences between the two racial groups decrease (Marshall and Jiobu 1975; Farley and Frey 1994). Segregation—in neighborhoods or elsewhere—can be seen as the outcome of attempts to maintain social distance from those who are different than oneself in socioeconomic status, as well as in culture (Massey 1985).

The extent of socioeconomic differences between racial groups also affects rates of intermarriage. A study of all large metropolitan areas in the United States showed that as socioeconomic inequalities among races increased, rates of racial intermarriage decreased (Blau and Schwartz 1984; Blau 1994). Blau comments, "The great socioeconomic differences between races reinforced the prejudice and group pressure against racial intermarriage" (1994: 64).

Causes of Inequality

Why does inequality between ethnic groups occur? Some scholars have pointed out that, in their competition for resources, some ethnic groups have cultural advantages—that is, possess certain knowledge, skills, and social organization—which gives them an advantage over other ethnic groups (Shibutani and Kwan 1965). For example, the rapid economic success of Jews who came to the United States in the late nineteenth and early twentieth centuries probably was helped by the emphasis on learning and by the business skills that were possessed by many in this group (Glazer 1958).

In addition, inequality between ethnic groups has often been the result of one group exercising its greater power to establish dominance over one or more other ethnic groups (Noel 1968; Marger 1994). Such power may be physical, as where one ethnic group conquers another by force (e.g., Germans dominating Poles during World War II). Power also may be exercised by control over economic resources (land, jobs, and so on) and control over political bodies, such as those that enact laws.

An ethnic group that enjoys a power advantage over another group or groups may use its power to control and exploit members of other groups. In South

Africa, for example, a powerful white minority dominated and exploited the labor of black South Africans for many decades (Thompson and Prior 1982). Some have seen the relations between white Americans and minority-group Americans in a similar light. They have described these relations as "internal colonialism," and argued that white outsiders (the police, school teachers, social workers, store owners, landlords, and so on) control and exploit minorities in inner-city "ghettos" (Blauner 1969).

Discrimination and Dominant Group Benefit

Some have maintained that discrimination against minority workers occurs because those in the majority benefit by such discrimination (or at least think they do). However, there are different views about who in the dominant white society may benefit from discrimination against minority workers. Some analysts, particularly those with a Marxist perspective, have seen ethnic inequalities as being a result of the functioning of capitalism. In this view, employers attempt to create a pool of low-wage workers (and also to split the working class) by separating and stigmatizing minority group workers (McAll 1990). They point, for example, to employers' past use of blacks (whom they usually denied employment) as strikebreakers against striking white workers.

Others have argued that it is white workers, rather than employers, who are primarily responsible for discrimination against minority workers. In this view, white workers try to keep their own wages high by promoting a "split labor market" in which whites occupy jobs that are denied to minority workers and for which they are paid more than minorities. These theorists point, for example, to the exclusion of blacks from many labor unions, especially craft unions (such as plumbers, electricians, and carpenters), both before and after World War II (Bonacich 1972).

Do whites in the United States actually benefit economically from keeping blacks and other minorities down? The evidence on this issue is mixed. Several studies have found that the larger the percentage of blacks in an urban area (especially within the south), the higher was the occupational position of whites (e.g., more whites were in managerial jobs), the greater the proportion of whites who had high incomes, and the lower the amount of unemployment among whites; however, the average income of whites did not necessarily rise as the proportion of blacks in an area increased (Glenn 1963; Dowdall 1974). A study of one hundred counties in North Carolina, using 1980 data, found that as the percentage of blacks in a county increased, poverty among whites decreased, but only when a white elite did not control most of the land (Tomaskovic-Devey and Roscigno 1996).

Several other studies have found that greater income inequality between whites and blacks is associated with lower income for white workers. One study examined the ratio of black to white earnings for male workers in each U.S. state in 1970 and related that ratio to the earnings of whites. The lower the black-white earnings ratio (that is, the greater the inequality in earnings), the lower were the earnings of whites were (Szymanski 1976). Another study, based on the fifty most populous metropolitan areas in the United States in 1960 and 1970 also found that as inequality in wages between white workers and black workers increased, the wages of white workers decreased; in other words, white workers appeared to suffer, rather than benefit, from racial inequality (Reich 1981).

However, a number of studies suggest that while most whites may not benefit in income as a result of racial inequality, some higher-status whites may benefit. As inequality in wages between whites and blacks increase, there may be greater inequality in income among whites. Moreover, greater racial inequality appears to be associated with higher incomes for upper-status whites and with higher profits for employers (Dowdall 1974; Szymanski 1976; Reich 1981). Dowdall comments: "Racist practices against blacks do not benefit all whites equally; as with other issues in American inequality, gains accrue more rapidly to those at the top" (Dowdall 1974: 182).

Szymanski (1976) and Reich (1981) assert that the average white worker is affected negatively by racial inequality, while elite whites benefit, because inequality divides workers and prevents them from forming coalitions (especially unions) against employers. Reich also maintains that inequality between the races is associated with more inequality in schooling among whites, since low funding of schools in a central city affects poor whites as well as blacks. (The conclusions that racial inequality hurts the average white worker, raises the profits of employers, and hurts union membership, are not supported by one study of changes in these variables over time [Beck 1980]; however, these results may have been distorted by other economic changes over time, such as the types of jobs available [J. Farley 1995].)

When considering whether whites benefit from racial inequality, it is relevant to consider not only the possible economic benefits to particular groups but also the impact of racial inequality on the total society. If blacks, Hispanics, Native Americans, and others have low skills and low incomes, this will reduce national productivity, competitiveness, and economic growth. Furthermore, inequality contributes to social problems—crime, drugs, broken families, and so on—that tend to destabilize society and raise the taxes that people have to pay to fund programs to deal with such social problems. Thus, while some whites may benefit at times, the bulk of the evidence indicates that most whites do not derive advantage from, and in fact tend to be disadvantaged by, racial inequality.

This is not to say, however, that discrimination against minorities—with respect to hiring and promotion, and in other ways—does not continue to occur.

Such discrimination has become less common than in the past, due to changes in white attitudes (Schuman et al. 1985) and to changes in the law that make discrimination on the job illegal (see section on political and legal institutions in chapter 6). However, recent studies have found that some employers continue to discriminate against black job applicants, especially black males. Such discrimination is often based on negative beliefs about black workers—for example, as being unreliable and not willing to work hard (Beers and Hembree 1987; Kirschenman and Neckerman 1991). Also, Latino job applicants are sometimes the targets of discrimination (Bendick et al. 1993). Some of the strongest evidence of continuing discrimination comes from "tester" studies in which a white job applicant and a black or Latino job applicant who are matched in relevant characteristics, such as education and work experience, are sent out to apply for the same job. For example, in one study in the Washington, D.C. area from 1990 to 1992, blacks were treated worse than equally qualified whites in 25 percent of the cases and Latinos were treated worse than whites 22 percent of the time (Edley 1996).

There is disagreement among researchers about the extent to which the income of minorities continues to be affected by racial discrimination. Research has focused on determinants of differences in earnings between blacks and whites. Using data on the hourly wages of young workers (ages twenty-eight to thirty-three) in 1985, Cancio and her colleagues (1996) concluded that while 74 percent of the gap in hourly wages between blacks and whites could be attributed to nonracial factors (such as education and geographical region), 26 percent of the gap was due to discrimination. However, in a study of the hourly wages of young (ages twenty-six to thirty-three) workers in 1991, Farkas and Vicknair (1996) found that, when cognitive skills as well as other nonracial factors (such as education) were taken into account, the black-white gap in wages was explained; they concluded that racial discrimination was not an important cause of the wage gap. Commenting on the Farkas-Vicknair study, Cancio and her colleagues (Cancio et al. 1996) argue that using the tests of cognitive skills used by Farkas and Vicknair to help explain differences in wages between blacks and whites is inappropriate. They maintain that the tests used (on word knowledge, paragraph comprehension, arithmetic reasoning, and mathematical knowledge) also test for "exposure to the values and experiences and the white middle class" (1996: 561).

Available Jobs

As the preceding section suggests, the disadvantaged economic position of certain minorities—especially of segments of the African American and Hispanic American populations—is not due entirely, or even primarily, to discrimination based on their race or ethnicity. In 1978, the eminent black sociologist William J. Wilson addressed the economic and social situation of black Americans in his book, *The Declining Significance of Race* (Wilson 1978). He argued that the changed structure of the American economy had become more important than racial discrimination in accounting for blacks' economic position.

Others have questioned the extent to which discrimination against minorities, especially against African Americans, has been reduced (Schulman 1989). There is little doubt, however, that many of the economic problems experienced by minorities are due to general changes in the American (and world) economy. The past few decades have seen dramatic reductions in the number of jobs in a variety of manufacturing industries—steel, automotive, rubber, and so on (Johnson and Oliver 1990). These industries had provided large numbers of relatively well-paying, usually unionized, jobs to people with relatively little education. A substantial proportion of such workers in industrial cities around the country—Detroit, Chicago, Cleveland, Toledo, St. Louis, and so on—have been minorities, especially black men. Many of their jobs have disappeared, due to a combination of factors, including greater automation, movement of factories owned by American firms outside the United States, and stronger competition from foreign producers. For example, one study found that between 1979 and 1984 about half of the black male production workers in five Great Lakes cities lost their jobs (Hill and Negrey 1985).

The loss of jobs available to minority workers has been magnified by the movement of many business firms, in retail and wholesale trades as well as in manufacturing, away from the central city (where most minority people live) to more outlying areas, in the suburbs of cities and in smaller cities and towns. This geographical out-movement is the result of a number of economic forces: the need by modern enterprises for large tracts of land, the improvement of truck transport, and the concentration of more affluent people (and thus customers) in the suburbs (Kasarda 1989; Schneider and Phelan 1990).

The movement of businesses and jobs outside the inner cities is not due entirely to impersonal economic forces; it is also influenced by racial attitudes. The concentration of affluent whites in the suburbs that attracts retail stores there stems in part from the desire of most whites to live away from the racial minorities and the social problems found in the inner city. In addition, fear of crime in minority areas and complaints about minority employees contribute to the decisions of some business managers to leave or avoid the inner city (Kirschenman 1990).

Changes in the types of jobs in the economy—fewer low-skill manufacturing and clerical jobs and more high-skill professional and technical jobs—has meant that the educational qualifications required for jobs, and especially for good-paying jobs, have generally risen. Increasingly, those with less than a college education are unlikely to find jobs that pay enough to support a family. Sometimes the educational qualifications required by employers have been inflated beyond those actually necessary for the job and have been designed, instead, to get employees whose life styles and social skills will help them to "fit in" (Berg 1971). Unnecessary educational requirements were made illegal by the 1991 Civil Rights Act but may continue in some instances.

Those members of minority groups who have been able to obtain the types of education and skills required in the changing economy have done well. There is a growing African American and Hispanic middle class composed of managers, professionals (doctors, lawyers, accountants, teachers, and so on) and other skilled workers. At the same time, there has been a growing proportion of blacks and Hispanics who are poor, having either extremely low-paying jobs or no jobs at all.

Some writers have described these trends as being part of a general tendency toward "polarization" of the work force (applying to majority whites as well as to minorities). More Americans are at the top of the income ladder, more are at the bottom, and fewer are in the middle than was true several decades ago (Jaret 1995). While this may be a general tendency cutting across racial and ethnic lines, minorities remain disproportionately represented at the bottom end of the ladder.

Family Composition

Another reason for the low income of minority families compared to that of white families is the high proportion of families headed by a single parent (usually female) in minority groups. This is due to a high proportion of children being born to unmarried women as well as to high rates of separation and divorce (Jaynes and Williams 1989).

The average income of families having only one parent present, especially if that parent is a woman, is much lower than that of married couples. In 1992, American families (of all races) with only the mother present had a median income of $18,587. This was only 44 percent of the median income ($42,140) of families with a married couple present (U.S. Bureau of Census 1994a).

African Americans are much less likely than most other Americans to have families in which both a wife and a husband are present. Table 5.5 shows the composition of white, black, and Hispanic families in the United States in the early 1990s. Among whites, 82 percent of all families included a married couple. Married couples were present in 69.3 percent of Hispanic families, but in only 46.5 percent of black families. Almost half of all black families had only a woman with no spouse present, compared to 14 percent of whites and 23.8 percent of Hispanics. Since the average income of families with only a woman present is much lower than that of married couples, the greater frequency of this

TABLE 5.5

Family Composition by Race and Ethnicity in the Early 1990s
(Percentages)

Type of Household	White	Black	Hispanic
Married Couple	82.0	46.5	69.3
Female Householder, no spouse present	14.0	47.9	23.8
Male Householder, no spouse present	4.0	5.6	6.9

Sources: Data for non-Hispanic whites and for blacks are from U.S. Bureau of the Census, *Black Population in the United States* (1995). Data for Hispanics are from U.S. Bureau of the Census, *Hispanic Population in the United States, March 1991* (1991).

type of family among African Americans pulls down their family income considerably.

The proportion of black families with only a mother present has risen sharply in recent decades. In 1970, 28 percent of African American families had a woman, with no spouse, present. In 1994, this was true of close to half (48 percent) of African Americans families. A parallel increase in mother-only families has occurred among whites during the same time period, though the absolute proportions for whites are much lower (from 9 percent to 14 percent). This suggests that some general changes in our society have affected all segments of the population.

One society-wide change is the restructuring of the economy, just discussed, that has eliminated large numbers of well-paying jobs previously held by men with relatively little education. This restructuring has meant that many young men, and especially young black men, are not able to support a family in the same way they were in an earlier era. Thus, they are less attracted to marriage and less desirable as marriage partners.

There has also been a general weakening of the social norms that traditionally dictated that women should not have children outside of marriage. While the causes of this change are complex, the general emancipation of women, especially their fuller entry into the work force, as well as related changes in standards of sexual behavior, appear to have played a role. Many analysts have also suggested that the social welfare policies of government—especially the rules and availability of "welfare" (Aid to Families of Dependent Children)—have facilitated the formation of single-parent families by making their economic survival possible and by providing an incentive not to have a man present.

Whatever the causes of increased single-parent families, once they become more common, they become more socially acceptable. This is true among all ethnic groups, but is especially true among some African Americans because single-parent families among this group are common in certain areas (Jaynes and Williams 1989).

Competition Between Ethnic Groups

Conflicts between racial or ethnic groups sometimes occur not primarily because of inequality but rather because of competition and conflict of economic interests between the groups. The initial stages of contact between ethnic groups have often been characterized by competition for resources and economic advantage—with respect to territory, raw materials, jobs, customers, and so on (Shibutani and Kwan 1965; Noel 1968; Brass 1991).

Disputes between ethnic groups over possession of land are widespread. To take just a few examples: fighting between Native Americans and whites in North America that continued sporadically over two centuries was fundamentally over efforts of Europeans to seize more and more of the land occupied by Native Americans (McLemore 1994). Hostility and flare-ups of overt conflict between Mexican Americans and Anglos in the American southwest during the nineteenth century and beyond were based in part on efforts by Anglos to gain control of land that had been possessed by the Mexicans (Nostrand 1992). Similarly, armed conflict among ethnic Serbs, Croats, and Muslims in Bosnia has centered on the desire of each group for control over particular territory (Bjork and Goodman 1993).

Competition for jobs between those of different ethnic groups also may lead to mutual hostility. For example, in some Asian and African countries, educated people from different ethnic groups compete for government positions that are a major source of employment for the well-educated (Horowitz 1985). In the United States, competition between workers of different ethnic groups has been central to ethnic conflict at times. Hostility and discrimination by white workers toward blacks, especially during the late nineteenth and early twentieth centuries, resulted from perceived and sometimes actual competition from blacks (sometimes used by employers as strikebreakers) for jobs held by white workers (Rudwick 1964; Olzak 1992). In California in the mid-nineteenth century, white miners attacked and attempted to expel Chinese laborers whom they saw as "unfair" competitors (Lyman 1974).

Working-class people usually are most likely to experience and/or perceive job competition, as well as other types of competition, such as that for housing. The perception of threats to their welfare from such competition may contribute to the relatively high level of ethnic prejudice found among lower-income whites (Schuman et al 1985; Dyer et al. 1989).

Competition between business people of different ethnic backgrounds also may cause intergroup friction. For example, Chinese laundrymen in Montana and Chinese fishermen in California, who competed with whites, were the targets of hostility and discrimination during the nineteenth century (Lyman 1974). More recently, during the 1970s, Vietnamese fisherman along the Texas coast were attacked by whites angered at their "intrusion" into nearby fishing grounds (J. Farley 1995).

139

One of the most important and widespread types of conflict of economic interests involves groups that have been called "middleman minorities" (Bonacich 1973). These are groups that are in the middle of the socioeconomic ladder, being neither at the top nor at the bottom. They also play a middleman role in the economy, engaging in activities (such as storekeeper, broker, lender, or sales agent) that mediate between those at the top (such as large landowners and industrialists) and those at the bottom (consumers, renters, and so on).

Throughout much of European history, Jews were the prototype middleman minority, since they tended to fill the roles of merchant, lender, and agent (e.g., tax collector) for the dominant elite. Chinese in Southeast Asian countries, such as the Philippines and Indonesia, and Asians (Indians and Pakistanis) in some African countries, such as Kenya and Uganda, are other examples of ethnic groups that have tended to fit the description of "middleman minority." In the United States, middleman groups, mostly in the role of small businessmen, have included Jews, Chinese, such as those in the Mississippi delta (Loewen 1971), and Koreans, such as those who own grocery or liquor stores in the central part of some cities (Bonacich 1995). However, one should note that while certain middleman occupations, such as store owner, may be occupied disproportionately by members of particular groups (Jews, Chinese, Koreans, and so on), most of those in each group actually are in other occupations.

Groups that tend to be economic middlemen in a society consistently have been the targets of hostility and often violence. They tend to be looked down on by the elite of society, who have dominant economic and political control. Those at the bottom of the economic ladder—who come into contact with the middlemen as the owners of stores where they shop or as landlords of the residences they rent, or as the boss in a small business where they work—commonly feel resentful towards the middleman. Often, they feel they have been cheated and exploited (by high prices, high rent, low wages, and so on). They are often resentful that "outsiders" are making money from people of their own ethnic group. They resent the relative economic success of the middleman group and often feel that it is "not right" for outsiders to be doing well (probably by using unfair methods, in their view), while their group remains disadvantaged.

The resentment, envy, and anger that large numbers of low-status people may feel against middleman minorities frequently has led to violent action against them all over the world. Jews were attacked, killed, and expelled from many European countries including England, Spain, Germany, and Russia. A similar fate was met by Chinese in the Philippines, Indonesia, Vietnam, and other Southeast Asian countries. In the 1950s, Asians were expelled from the African nation of Uganda. In the United States, blacks burned down stores owned by Koreans during the Los Angeles riot of 1992, and have boycotted and had other violent confrontations with Korean merchants in other American cities. Jews and Asian business owners who deal with the African American community have been labeled as "bloodsuckers" by Louis Farrakhan, the leader of the Nation of Islam, a black religious group (*New*

York Times 1995b). The generally high level of anti-Jewish sentiment among African Americans undoubtedly is based in part on the perceived conflicts of interest that some blacks see between themselves and Jews with whom they do business.

Causes of Competition

What leads to heightened competition between members of different ethnic groups?

Barriers Falling

One important cause is that barriers (of education, job openings, credit availability, and so on) that previously tended to segregate groups into different occupations may break down, at least to some extent. Thus, members of different ethnic groups increasingly compete for the same school openings, the same jobs, the same customers, the same housing. For example, during the late nineteenth and early twentieth centuries the caste-type relations between blacks and whites in the United States began to break down. Blacks began to compete with whites much more than before, especially for jobs and for housing, as large numbers of blacks moved from southern farms into cities of the south, northeast, and midwest. The "competitive race relations" during this era (J. Farley 1995) often led to hostility from whites. Such hostility from whites, who felt threats to their jobs and housing, erupted in a number of major riots by whites against blacks, including riots in East St. Louis (1917), Chicago (1919), and Detroit (1943).

Susan Olzak has hypothesized that "ethnic conflict surges when barriers to ethnic group contact and competition begin to break down" (1992: 209). To test this idea, she analyzed data on 262 ethnic and racial conflicts and protests that occurred in the seventy-seven largest American cities from 1877 to 1914. Consistent with Olzak's hypothesis, as minority groups moved more into occupations previously reserved for whites, intergroup conflict increased. Olzak also suggests that a greater number of confrontations between those from different ethnic groups increases the salience of ethnic boundaries and thus leads to an escalation of conflict.

Increasing Minority Numbers

Another factor that contributes to greater competition is an increase in the numbers of one or more groups (Shibutani and Kwan 1965; Olzak 1992). Hostility by whites against blacks from World War I to World War II and beyond was heightened by the fact that millions of blacks flowed out of the rural south to northern and midwestern cities. The rapid increase of the black population in northern cities—Detroit, Chicago, Cleveland, Philadelphia, and so on—alarmed many whites who felt a threat to their jobs, neighborhoods, schools, or control over local government.

Large-scale immigration of ethnic groups that are different from the dominant population also often provokes alarm and hostility. During the nineteenth and early twentieth century, the arrival in America of large numbers of Catholics and Jews from eastern and southern Europe (Russia, Poland, Hungary, Italy, and so on) led to an upsurge of antiforeign, anti-Jewish, and anti-Catholic sentiment in the United States. Many white Protestant Americans felt that the "avalanche" of foreigners posed a threat to their jobs and their way of life (Higham 1955). Attacks on immigrants from Europe increased as the flow of immigration rose (Olzak 1992). Similarly, immigration of Chinese on the west coast and Mexicans into the southwest led to concerns among many whites about economic competition from these groups. Anglo labor unions opposed large-scale immigration, arguing that immigration depressed wages and increased unemployment (Foner 1947; Olzak 1992). More recently, in western European countries such as Great Britain, France, and Germany, increased immigration from other parts of the world (such as Asia and North Africa) have been a major contributor to antiminority sentiments (e.g., see Richmond 1986).

Economic Hardship

Hostility resulting from perceived competition from members of other ethnic groups is likely to increase during periods when economic problems are increasing. Susan Olzak's study of ethnic conflict in the United States from 1877 to 1914 found that increases in the rate of business failures were associated with increases in the number of conflicts between white and black Americans. The same study found that years in which there was an economic depression had significantly higher numbers of conflicts involving white immigrants (though not racial minorities) than nondepression years. In addition, urban racial violence against minorities increased at times of economic slumps (1992). Recent evidence from western European countries (Quillian 1995) also shows that hostility towards minorities is higher when economic conditions are poorer. (A combination of poor economic conditions and a large proportion of immigrants resulted in the greatest hostility.) Apparently, economic hard times make many people more concerned and angry about real or perceived competition from native minority groups and/or from immigrants.

On the other hand, members of the majority group may feel less threatened by competition from other groups during times of economic prosperity. Wilson (1978) argues that this was true in the United States after World War II, especially during the 1950s and 1960s. This was a time of prosperity, generally with a plentiful supply of jobs and with rapid increases in income for most Americans. In this favorable economic climate, whites generally were receptive to greater equality for blacks and other minorities.

The fact that competition between members of different racial or ethnic groups often results in conflict between them leads Olzak (1992) to draw what may seem to some a surprising conclusion. The conclusion is that desegregation, rather than segregation, causes ethnic conflict. While Olzak's conclusion derives mainly from her research on intergroup conflict in an earlier era, she believes that the general principle is relevant to more contemporary race relations. Thus, she suggests that rapid desegregation of schools and housing in recent decades has tended to produce higher levels of racial conflict.

Different Effects of Occupational Differences

Overall, greater status differences between racial and ethnic groups—especially their location in different types of occupations—are likely to have separate and opposite effects on intergroup conflict. On the one hand, if different groups have distinctive types of occupations, especially occupations of differing skill and status, this leads to greater inequality. Inequality produces resentments that result in more ethnic conflict. On the other hand, if members of different ethnic groups do different kinds of work, they are less likely to compete with each other, thus tending to reduce intergroup conflict.

Certainly, few people would prescribe rigid status and occupational separations of racial and ethnic groups in order to reduce conflict between them. We should be aware, however, that changing the established order of things makes many of those who previously enjoyed advantages feel threatened and angry. It is likely that anger and hostility provoked by greater competition from other groups will wane as people become accustomed to a changed situation. Eventually, competitors may be seen less in terms of their ethnic identities and more in terms of other relevant characteristics.

Exchange and Interdependence

While inequality and competition have been advanced as explanations of conflict between ethnic groups, another type of economic relationship has been seen by some as an integrative force, tying ethnic groups together. This is functional integration, whereby each group provides necessary resources or services to the other (Eitzen and Zinn 1994). Blalock and Wilken (1979) proposed a model for understanding continuing exchange relations between ethnic groups. Such factors as the dependence of groups on each other and an absence of alternative exchange partners leads to the maintenance of reciprocal exchange relations. For example, Israelis and Palestinians in territories occupied by Israel (the "West Bank" and the Gaza Strip) have had a mutual economic dependence. Israelis needed the labor of the Palestinians and the Palestinians needed the jobs (Lustick 1994). Though this economic interdependence was not enough to prevent hostility and conflict between these groups, it did provide incentives to both sides to try to resolve their conflicts.

Of course, exchanges may involve a great inequality in relative outcomes for different groups. Nevertheless, both parties may see the continuation of exchange as their best alternative in a given situation. This contributes to stability in their interactions.

A Model of Economic Effects

Figure 5.1 summarizes the possible relationships among economic variables that may have an impact on ethnic group relations. The right side of the figure indicates two central aspects of ethnic group relations—the amount of contact and the amount of conflict between groups. These aspects of ethnic group relations are affected by three central economic variables: the extent of inequality, the amount of competition, and the amount of interdependence between members of the ethnic groups. Contact between groups decreases as groups become more unequal but increases as they become more interdependent. Intergroup conflict increases both as inequality increases and as competition increases.

Inequality between ethnic groups is determined primarily by the division of labor between the groups (i.e., the occupations filled by each group). The division of labor, in turn, is affected by 1) cultural differences that may make those in a particular group more interested in or more successful at certain occupations; 2) power differences that enable one group to claim the best positions for its own members; 3) the economic system (e.g., the types of jobs available); and 4) the educational system, which provides job-relevant skills to those in various ethnic groups.

In addition to these four factors that may affect inequality indirectly, inequality between groups may be affected *directly* by power differences between groups; for example, the demonstrated power of blacks in the United States to cause social disruption in the 1960s may have led the federal and local governments to try to placate poor blacks with more available and more generous welfare payments (Piven and Cloward 1971).

Figure 5.1 also shows the factors that may affect the amount of competition between groups. Competition is affected first by the division of labor between groups; in general, the less the division of labor between groups (i.e., the more similar their occupations), the greater the competition between them. Competition among groups also may be affected by the relative size of the groups (the

144

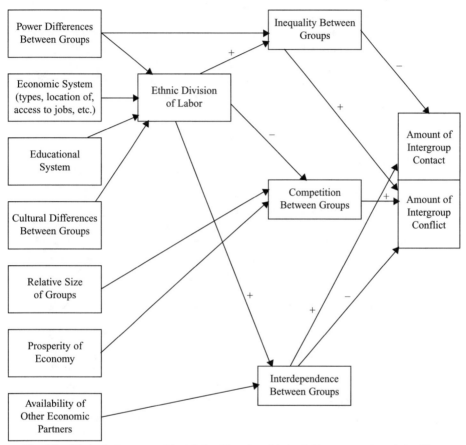

FIGURE 5.1

Economic and Related Factors Affecting Relations Between
Ethnic Groups

Note: Plus (+) sign indicates a positive relationship; minus (–) sign indicates a negative relationship.

more equal their sizes, the more potential competition) and by the overall vitality of the economy (competition is greatest in a weak economy).

Figure 5.1 also shows that interdependence among ethnic groups is (like inequality and competition) affected by the ethnic division of labor; greater division of labor makes members of each group more dependent on each other. The extent of interdependence also may be influenced by the availability of alternative exchange partners—for example, other ethnic groups within the society or possibly people of ones' own ethnic group in other countries.

It should be noted again that the division of labor among ethnic groups, which plays a fairly central role in this conceptual scheme, has several different

145

effects that ultimately influence interethnic relations in *opposite* ways. Greater division of labor between ethnic groups produces greater economic inequalities between groups that may lead to conflicts (due to resentments) and to less contact between groups (e.g., because of residential separation of groups with unequal incomes). But greater division of labor also should result in lesser competition among members of different ethnic groups and therefore in less conflict. Also, greater division of labor may lead to more interdependence and thus to more contact and stronger ties between members of different ethnic groups. Which of these different and opposing effects will be the stronger is an empirical question. Their relative effects will differ in different circumstances.

Summary

Relations among racial and ethnic groups are influenced by the absolute and relative size of these groups. In general, the larger a minority group is and the more rapidly it grows in size, the more it is apt to be seen as a threat (economically, politically, and socially) by members of the dominant group. As the size of a minority increases, relative to the size of the dominant group, segregation of and discrimination against the minority tends to increase. In the United States, this has been true especially with respect to reactions of whites toward blacks.

In the United States today, the size of racial and ethnic minorities, relative to the non-Hispanic white population, is increasing rapidly. The Hispanic American and Asian American populations are growing at an especially fast rate. While about three out of every four Americans were non-Hispanic whites in 1990, it is estimated that by 2050 only slightly more than one in two Americans will be a non-Hispanic white; the rest will be primarily Hispanics, blacks, and Asians.

The rapid changes in the racial and ethnic makeup of America are likely to provide a challenge to good intergroup relations. Already many whites feel threatened by the growing numbers of people who look different from themselves, who may not speak English well, and who may have different customs from themselves. Some whites feel alarmed (especially in areas like Los Angeles and Miami, which have large immigrant populations) at what they see as a change in the traditional character of their communities. Frictions also have arisen, and undoubtedly will continue to arise, between different minority groups—for example, between blacks and Hispanics over jobs and between blacks and Asian merchants. Reducing the anxieties of those who fear change, reducing frictions that arise among ethnic groups, and building unity out of growing ethnic diversity is a major challenge for Americans. (We will discuss issues of diversity and unity in greater detail in chapter 10.)

Relations between racial and ethnic groups are not shaped only by the relative size of the groups. Much depends also on the economic position of each group. First, the extent of inequality between groups is important. The more the economic inequality (in occupational position, income, and so on) between ethnic groups, the more segregation and the more conflict between groups there is likely to be. In the United States, while minorities have made economic gains, African Americans and Hispanic Americans still lag considerably behind whites (as well as behind Asian Americans) in occupational level, levels of employment, income, and accumulated wealth. Inequality of income between the ethnic groups does not appear to benefit the earnings of most whites. But discrimination against minorities by some employers continues.

In addition to discrimination, there are important other causes of continuing economic inequality. One is a reduced availability of good-paying jobs to people with relatively little education. Thus, to try to reduce existing economic inequalities between groups, it is necessary not only to try to prevent discrimination in hiring and promotion but also to provide more education and training for minorities to quality them for good jobs in the changing economy (see chapter 7 for a discussion of education). Another reason for the lower income of minorities, especially among blacks, is their high proportion of single-parent families. With only one parent (usually a mother) present, family income usually is much lower than in families where both parents are present.

Economic competition may be another source of conflict between racial and ethnic groups. One type of group that often has real or perceived conflicts of economic interest with others is the "middleman minority"—typically traders or shopkeepers belonging to a particular ethnic group (e.g., Jews or Chinese). Resentment by low-status groups directed at middleman groups frequently has erupted in interethnic violence. Racial or ethnic groups also may come into competition over jobs, housing, and other advantages. Competition between the dominant majority and members of minority groups is likely to increase as barriers to the progress and equality of the minorities weaken. Intergroup competition also is increased as the relative size of minority groups increases and as available resources (such as jobs) decrease, as happens in economic hard times.

As occupational differences between racial and ethnic groups decrease, there may be some increase in intergroup hostility based on increased competition (for jobs, customers, housing, and so on). Hopefully, any negative effects will decrease as members of the dominant group become more accustomed to competition from minority-group members. Also, any negative effects from increased competition should be counterbalanced by the positive effects of greater equality that have been noted: fewer resentments and greater acceptance of those of similar status.

6) Political-Legal Systems and Religion

In the previous chapter we considered the ways in which the attitudes and behaviors of people from different racial and ethnic groups are influenced by the make-up of the total population (especially the size of the different groups) and by their positions in the economy. In this chapter we examine how the relations among groups are affected by several other major institutions of society and by the role of various racial and ethnic groups in these institutions. Specifically, we will consider how the relations among groups are influenced by the political-legal systems of the society, including its criminal justice system, and by organized religion. We conclude the chapter by discussing how the relations among ethnic groups may be affected by differences or similarities in their other social characteristics.

The Political and Legal Systems

The relations between ethnic groups in a society are affected in a number of ways by the political and legal arrangements of that society—that is, by who makes the laws and what the laws are. First, the economic positions of various groups within society are affected by political-legal decisions. For example, government actions may help or impede opportunities for members of various groups to obtain skills, jobs, and wealth (Carnoy 1994) and they may increase or decrease residential segregation (Massey and Denton 1993). Because the relations among ethnic groups depend on their positions within the larger society (on their relative status, conflicts of interest, and so on), government actions that affect groups' positions are likely to influence intergroup relations as well.

Secondly, relations among ethnic groups are affected by the roles that they play (or do not play) in the making of government policy and law. Each ethnic group naturally wants to have the greatest influence possible on government decisions. A group whose members feel that they have been shut out of political decision-making, or who feel that their influence is less than deserved, are likely to be resentful towards more influential groups. Thus, harmony among ethnic groups depends in part on political arrangements that are satisfactory to all major groups.

In the next section we will discuss first some of the ways in which government and the legal system affect various ethnic groups and relations among groups. Then we will consider the arrangements by which ethnic groups may exert political influence and the ways in which political competition and cooperation may affect intergroup relations.

Actions Supporting Discrimination

In some nations, both in the past and in the present, governments have supported or actively promoted discrimination against certain ethnic groups. For example: in nineteenth-century Russia, the Czarist government decreed that Jews could not own land or even live in certain regions. In South Africa after World War II, the government passed laws to establish a system of rigid racial segregation, which restricted the rights of black citizens (and mixed-race and Asian people as well) to live in certain areas, to vote and to go to schools with whites (Hunt and Walker 1974). In the Canadian province of Quebec, the provincial government controlled by French speaking people (themselves the target of much past discrimination) has restricted the rights of English speakers to display store signs in English and to send their children to English speaking schools.

Blacks in the United States

In the United States, for most of the nation's history until and even after World War II, the actions of government and the legal system usually supported and sometimes required discrimination against blacks and other racial minorities. The U.S. Constitution counted each black slave as three-fifths of a person for purposes of apportionment of Congress and the federal government recognized the legality of slavery in the southern states. All states were required to return fugitive slaves to their owners. After the Civil War, the federal government briefly attempted to promote equal rights for blacks in the defeated south, but soon lapsed into acceptance of segregation and the treatment of blacks as second-class citizens in the south and elsewhere (Franklin 1969; W. Wilson 1973; J. Farley 1995).

Throughout the later part of the nineteenth century and the first half of the twentieth century, state and federal governments and courts required or encouraged the separation of African Americans from other Americans. In the south and states bordering the south (e.g., Maryland and Kentucky), an elaborate system of segregation of blacks from whites was put into place. Seventeen states and the nation's capital (the District of Columbia) had laws at one time requiring that schools be racially segregated. Public facilities such as parks, swimming pools, and buses also were segregated by these laws. Moreover, operators of private facilities open to the public often were required to bar blacks or to separate the races. For example, Florida, Tennessee, and other southern states mandated segregation of railroad passenger trains. Even marriage between whites and blacks was prohibited by law.

Outside the south, in the northern and western parts of the United States, separation of the races usually was not required by law. However, laws calling for racial segregation were adopted in some northern states and localities (e.g., in Indiana) and marriage between blacks and whites was illegal in some northern states. Even where racial segregation was not mandated by law, the actions of local governments in the north—for example, in locating schools and housing projects in all-black or all-white neighborhoods—were often intended to promote separation of the races.

The authority and influence of the federal government and courts also generally supported racial segregation during this earlier period. In 1896, the U.S. Supreme Court upheld the doctrine that public facilities that were "separate but equal" were acceptable and other court decisions upheld racial segregation. Also, the actions of the federal government supported and promoted segregation in housing. For example, from 1935 to 1950 the Federal Housing Administration advised banks to provide mortgage loans to prospective homemakers in a way that preserved racially separate neighborhoods.

In addition to actions that required or encouraged separation of blacks from whites, state governments in the south, and sometimes elsewhere as well, discriminated against African Americans in other ways. Especially important were laws and administrative actions of southern states that effectively prevented most blacks from voting. A variety of devices—including a "poll tax," restriction of Democratic Party primary elections to whites, and the decisions of voting registrars—were used to deny the ballot to black people. The "whites only" primary election was upheld by the Supreme Court for decades on the ground that these elections were concerns of the political parties involved, until the court reversed itself in 1944 and finally ruled this practice illegal. Poll taxes remained in effect in five states until 1964, when they were outlawed by the Twenty-fourth Amendment to the Constitution.

151

Other Minorities in the United States

Early American government policy towards other racial minorities usually reflected hostility (Nash 1974; Snipp 1989; McLemore 1994). Actions of the government towards Native Americans reflected the attitudes of most whites that these people should be chased away (or killed if necessary) so that they did not stand in the way of whites acquiring their land and establishing European American communities upon it. Throughout the eighteenth and nineteenth centuries, whites displaced more and more Native American tribes, pushing them farther and farther westward , breaking treaty after treaty in the process. To take just one example, in 1830 President Andrew Jackson proposed the Indian Removal Act, designed to force all the Indians in the southeastern states to move west of the Mississippi River. Their forced exodus, along the "Trail of Tears" under the surveillance of the U.S. Army, resulted in the death of thousands from starvation and disease (Spicer 1980).

Asians also were the targets of discrimination by government (Peterson 1971; Lyman 1974). A federal law passed in 1790 provided that only "free white persons" would be allowed to become U.S. citizens. While people of African or Mexican descent later were allowed to become citizens, a revision of the naturalization law in 1870 barred Chinese, Japanese, or other Asian immigrants (except Filipinos after 1906 in rare cases) from U.S. citizenship. During the latter part of the nineteenth century, California passed a series of laws intended to end immigration of Chinese. For example, in 1858 the California legislature passed "An Act to Prevent the Further Immigration of Chinese or Mongolians to This State." After most of such California laws were struck down by the federal courts, Congress, beginning in 1882, passed a series of laws to restrict Chinese immigration. Such legislation endured until the repeal of the Chinese Exclusion Act in 1943. Immigration of Japanese was also restricted by legislation and by agreement with Japan around the turn of the century. The Oriental Exclusion Act of 1924 barred all subsequent immigration of Japanese.

In addition to discriminatory laws against immigration of Asians, state and local governments on the American west coast passed laws discriminating against Asians in other ways, including their activities in certain occupations and their residency in certain areas. For example, special taxes levied in California were directed at foreign (mostly Chinese) miners and Chinese fishermen. Charges of "unfair" Chinese competition in the laundry industry in the western United States led to a series of harassing laws regulating such matters as the times that ironing could be done. When fire destroyed Chinatowns in smaller cities, as occurred in Monterey, California in 1906, officials refused to permit the Chinese to return and rebuild. In larger cities, the Chinese were confined within isolated ghettos and not permitted to settle elsewhere (Lyman 1974).

In the first decade of the twentieth century, the San Francisco Board of Education ordered that Chinese, Japanese, and Korean pupils attend a separate school for "Orientals." Concern among some whites about acquisition of land by efficient Japanese farmers led the California legislature to pass the Alien Farm Law in 1913. This law prohibited ownership of land by people who were "ineligible for citizenship." The law clearly was aimed at Japanese, who were at that time ineligible to become naturalized American citizens. When Japan attacked the United States in 1941, all people of Japanese ancestry on the West Coast, whether aliens or U.S. citizens, were forced to leave their homes and be transported to "relocation centers." During 1942, more than 110,000 people of Japanese ancestry (most of whom were American citizens) were placed in these camps by the U.S. government and were confined there during the war years.

Federal and state actions directed at Hispanics were generally fewer than those targeting blacks, Native Americans, or Asians. But, when officials did act, their actions were usually harmful to Hispanics (Moore and Pachon 1970). Some public schools in Texas segregated children of Mexican descent from Anglo pupils. During periods such as World War I when agricultural labor was in demand, the federal government permitted large numbers of Mexican farm laborers to enter the United States to work for low wages under conditions of unsanitary housing and inadequate schooling and health care. During the Great Depression of the 1930s, in an effort to reduce job competition and welfare costs, federal and local authorities cooperated to round up tens of thousands of people of Mexican descent, including some naturalized U.S. citizens, and deported them to Mexico.

Overall, the actions of the government and the courts toward racial minorities up until World War II, and to some extent beyond that period, generally had the effects of increasing the physical separation of these groups (in neighborhoods, schools, occupations, and so on) from the dominant majority of Americans of European descent. Government actions during this time also tended to degrade the economic status of people in these minority groups, further increasing their separation from most other Americans.

Actions Supporting Equality

During and after World War II, a combination of changes in American society and in the U.S. position in the world combined to change the attitudes of government officials and the courts regarding equality for minorities. Need for support from blacks and other minorities in World War II, the increased presence of black voters in northern cities, pressures from primarily-black labor groups (especially railroad workers), revulsion at the racial supremacy ideology of the Nazis, and America's new leadership position in a multiracial world were among the forces that led government leaders to try to reduce discrimination against blacks and other minorities.

Among the early actions taken against racial discrimination were the following:

- In 1941, President Roosevelt issued a presidential order barring racial discrimination in defense plants and government agencies.
- In 1946, federal courts ruled that segregation in interstate travel (such as trains and buses) was illegal.
- In 1948, President Truman ordered racial integration of American armed forces.
- In 1948, the Supreme Court ruled that agreements that racially restrict the sale of houses were not legally enforceable.

These administrative and court decisions, though not always well-enforced, initiated modest progress towards equal treatment of Americans of different races. More significant court and legal actions followed. Especially important were actions that affected segregation of public schools, segregation of housing, voting opportunities, and discrimination in jobs, businesses, and public accommodations. We will briefly outline the government and court actions taken in each of these areas and discuss the extent to which these actions have accomplished their goals.

School Desegregation

In a landmark decision, the Supreme Court ruled unanimously in 1954 that legally-mandated segregation of public schools according to race was unconstitutional. Separate schools were inherently unequal, the Court said, because segregation of minority children "generates a feeling of inferiority as to their status in the community that may affect their hearts and minds in a way unlikely ever to be undone."

The Court order to desegregate public schools was initially met with great resistance throughout the south. Stratagems such as giving students "tuition grants" to attend segregated schools were widely used to circumvent the Court's ruling. When the governor of Arkansas tried to block desegregation of schools in Little Rock in 1957, President Eisenhower had to use U.S. Army troops and a federalized National Guard to enforce the law. Gradually, legal segregation of schools came to an end in the border states in the late 1950s and early 1960s, and soon after in the deep south. The percentage of black students throughout the country attending schools having more than 90 percent minority enrollment fell from 65 percent in 1965 to about 40 percent in 1974 and to 32 percent in 1988 (Armor 1992). These nationwide declines in school segregation primarily reflected changes in southern and border states. (Racial segregation in schools is now actually lower in the south than elsewhere in the country.)

While legal segregation of public schools has ended, separation of black children from white children in schools has not ended. Because neighborhoods generally are highly segregated (Massey and Denton 1993) and because most school assignments are based on neighborhood residence, many schools continue to be substantially segregated by race. In many large cities such as Cleveland, Chicago, and Washington, D.C., a majority of blacks attend schools that have few white students. Students from other minority groups, including Hispanics and Native Americans, also often attend schools that have predominately minority student bodies. As the proportion of Hispanics in the American population has risen, Hispanic students have increasingly become separated from whites. The percentage of Hispanic students attending schools that had over half minority student bodies rose from 54 percent in 1968 to 73 percent in 1992 (J. E. Farley 1995).

To try to overcome school segregation stemming from racially segregated neighborhoods, as well as from other historical causes, busing of students outside their neighborhoods sometimes has been ordered by courts, or occasionally adopted voluntarily by school authorities. Such programs to bus students to achieve racial mixing of schools have often generated anger among parents, especially among white parents, and controversy in communities (Feagin 1980).

Overall, then, court and government efforts have resulted in substantial reduction of racial segregation in schools, especially in the south. But much separation of black students (as well as of Latino students) from white students, based primarily on racial segregation of neighborhoods, continues.

When desegregation of schools has occurred, its impacts on the quality of education of minority students and on the racial attitudes of white and minority students have been mixed. We will consider the effects of racial mixing in schools in chapter 7.

Equal Job Opportunities

Following several years of protest demonstrations and civil disobedience by black activists and their white supporters, led especially by Martin Luther King Jr., Congress passed the Civil Rights Act in 1964. Among the provisions of this act were those that outlawed racial discrimination in public accommodations and facilities, such as hotels and restaurants, and gave the federal government new powers to enforce school desegregation. The new law was largely effective in opening public places to people of all races and also was helpful in the desegregation of public schools in the south.

Perhaps the most important section of the 1964 Civil Rights Act was the one that prohibited discrimination by employers and by labor unions on the basis of race, color, national origin, religion, or gender. (The law was extended in 1972 to cover employees of state and local governments, as well as federal employees.) This portion of the law has a great potential effect on the economic status of minorities, but it has also been the most difficult portion to enforce and has generated the most controversy.

Laws requiring equal job opportunities are inherently difficult to enforce because determining whether or not discrimination has occurred is often difficult. Employers may require certain educational attainments, scores on tests, particular skills, or previous experience. Often, knowing when such requirements are, in fact, necessary for effective performance of the job and when they may be unnecessary or even intended to screen out minorities is also difficult. In addition, the federal agency most directly concerned with job discrimination, the Equal Employment Opportunities Commission (EEOC), was given little authority in the 1964 law to act against violators. Although the law was amended in 1972 to try to make it more effective, large backlogs of individual complaints about discrimination continued to pile up without action. Moreover, the government generally filed relatively few "class action" suits on behalf of groups of people who may have been discriminated against by employers, and the numbers of such suits declined during the 1980s (Rose 1994).

Federal agencies have taken some actions to try to enforce equal job opportunity. In the late 1960s and early 1970s, the Labor Department required bidders on federally-assisted construction contracts to set numerical goals and timetables for employing blacks and other minorities in the building trades. During the 1970s, the EEOC and the Justice Department succeeded in forcing greater employment opportunities for minorities and for women in a number of larger companies and industries, including the American Telephone and Telegraph Company (AT&T), the steel industry, and the trucking industry. In addition, the federal government obliged a large number of other businesses and organizations, such as universities, that had government contracts, to adopt goals and timetables to hire more minorities and women (Lundberg 1994; Rose 1994).

As these examples indicate, government agencies have often used numerical measures of employment outcomes to assess whether or not job opportunities for minority groups are equal to those of white males. This results-based approach— part of an "affirmative action" policy—has been used because of the difficulties in assessing whether or not employers are practicing discrimination and because minority group members often tend to score lower than others on tests used to select employees (Lundberg 1994).

Opposition to policies that make decisions on the basis of race led government officials in the 1980s (during the Reagan administration) to move away from the policy of setting goals for hiring and promotion of minorities (Rose 1994). More recent administrations also have been more hesitant than those in the 1970s to pressure employers to set goals and timetables for employment of minority groups.

The question of what criteria and procedures employers should use when deciding who to hire or promote was addressed again in the Civil Rights Act of 1991. This act makes it illegal to use racial, ethnic or gender quotas—that is, to set aside a specific percentage of jobs or promotions—to be filled by members of a given group. It also prohibits the "race-norming" of job test scores, that is, scoring candidates within their own racial group rather than scoring all candidates together. While these provisions restrict actions that have been used to help minority people, two other provisions make discrimination against minorities less likely. One states that tests, credentials, or other requirements may not be used in hiring or promotion decisions unless the employer can prove that they are "job related for the position in question and consistent with business necessity." Also, the new law increased the potential monetary damages that employers must pay if found guilty of discrimination, thus providing a greater deterrent.

How much of an impact on the economic status of African Americans have laws requiring equal employment opportunity had? First, it is important to recognize that overall improvements in employment and wages for African Americans have not necessarily coincided with government efforts to combat job discrimination. Donahue and Heckman comment: "The federal effort appears weak during the period [before 1975] in which black breakthroughs in employment and wages took place. As enforcement budgets grew, black relative gains fell off and actually receded in some places" (1994: 199). Other social changes not under government control—including a large migration of African Americans out of the south until about 1970 (which raised their wages) and reductions in demand for lesser-skilled workers (which reduced black employment and wages)—have also been important influences. Nevertheless, the overall government effort to promote civil rights, including school desegregation and nondiscrimination in jobs, appears to have had some success in improving employment among African Americans, especially in the south (Donahue and Heckman 1994).

Open Housing Efforts

Racial segregation of neighborhoods—primarily of blacks from whites—is pervasive in the United States. It developed from, and has been maintained by, the racial attitudes of the dominant white population and by systematic discrimination by realtors and banks. Prior to the 1960s especially, federal and local governments and courts also contributed to segregation of the races in neighborhoods throughout the country. They did this by enforcing restrictive covenants that made the sale of property to certain groups illegal, by encouraging banks to discriminate against minorities (especially blacks), by putting housing projects in racially segregated areas, and in a variety of other ways (Tobin 1987; Massey and Denton 1993).

From about the early 1960s, however, the weight of government, especially the federal government, swung in the direction of opposing discrimination in housing. Already, in 1948, the Supreme Court had ruled that restrictive covenants intended to prevent sale of housing to any group were not legally enforceable. In 1962, President Kennedy issued an executive order directing federal agencies to "take all necessary and appropriate action to prevent discrimination" in federally-supported housing. However, this order was not enforced effectively and had little practical impact. Efforts to end discrimination in housing then shifted to

Congress, which, in 1968, passed the Fair Housing Act. This act made the refusal to sell a home or to rent housing to any person because of race unlawful.

However, the opponents of this act had succeeded in seriously weakening its enforcement provisions. The Department of Housing and Urban Development was empowered only to investigate complaints of housing discrimination and to engage in "conference, conciliation, and persuasion" to deal with the problem. Complainants could sue alleged discriminators themselves, but penalties under the act were small and individuals who sued were liable for all court costs and attorneys' fees. Reviewing the weakness of the 1968 Housing Act, Massey and Denton comment: "In practice, therefore, the 1968 Fair Housing Act allowed a few victims to gain redress, but it permitted a larger system of institutionalized discrimination to remain in place" (1993: 198).

The 1968 Fair Housing Act also directed the secretary of housing and urban development (HUD) to promote racial desegregation in federally supported housing. However, HUD moved slowly and hesitantly in response to this mandate. When, in 1972, HUD did issue regulations concerning desegregation of new federally-assisted housing projects, local authorities stopped construction of such projects. Racial segregation of existing public housing projects continued at a high level.

To try to correct weaknesses in open housing efforts, Congress passed some additional legislation during the 1970s including a prohibition on discrimination in home-lending by banks. However, during most of the 1980s the Reagan administration was hostile to enforcement of open housing laws and the Justice Department virtually stopped initiating any fair housing cases. By 1988, concern about lack of progress towards desegregation in housing led Congress to pass some significant amendments to strengthen the 1968 Fair Housing Act. The new provisions included stiffer penalties for violators and required the U.S. Attorney General to take action against violators in many circumstances (Schwemm 1990). These changes in the law may help to somewhat reduce blatant discrimination. Racial segregation of neighborhoods in the United States did decline slightly during the 1980s but neighborhood segregation of African Americans from other Americans remains at a high level (Farley and Frey 1994).

The Criminal Justice System

Another aspect of the political and legal institutions of a society that affect the well-being of ethnic groups, and relationships among them, is the operation of the criminal justice system. Every modern society must define illegal behavior and deal with such behavior when it occurs. Sometimes, however, the legal system discriminates against some ethnic groups. For example, during the 1980s the government of Bulgaria decreed that people of Turkish ethnicity had to take Slavic names on their mandatory identity cards and fined them if they spoke Turkish in public (Gurr 1993). In Sri Lanka, people of Sinhalese ethnicity have been allowed to occupy state land, while those of Tamil ethnicity have been evicted when they attempted to establish similar settlements (Arasaratnam 1987). Such discriminatory treatment by legal authorities (the police, the courts, and so on) is often harmful to the welfare of minority groups (including their personal liberty) and may cause them to feel hostile towards and alienated from dominant groups and the larger society.

In the United States, racial minorities and especially African Americans historically have been treated more harshly than white Americans by criminal justice systems. During slavery, black slaves basically had no legal rights and could be punished at the whim of their white masters. After the end of slavery, widespread discrimination against African Americans continued, most blatantly in the south but elsewhere as well. African Americans were more likely to be arrested for some activities (e.g., "disrespect" for a white man), less likely to get adequate legal representation or bail, less likely to have their cases dismissed, less likely to serve on juries, and more likely to receive harsh sentences (Spohn 1995).

The extent to which African Americans and other minorities continue to suffer discrimination by the criminal justice system has been the subject of a considerable amount of research. Let us see what that research has shown.

Arrests

The frequencies with which people of different racial groups are arrested depends on the type of crime. For much "white collar" and corporate crime—what has been called "respectable crime" (Sheley 1995), that includes such offenses as embezzlement and bribery—whites are more likely than minorities to be arrested. Blacks are much more likely than whites to be arrested for "street crimes" included in the FBI's Uniform Crime Reports (U.S. Department of Justice 1994). These crimes include those of violence (murder, forcible rape, robbery, and aggravated assault) and property crimes (burglary, larceny-theft, and motor vehicle theft). Arrests of blacks for these crimes consistently have been much higher than those of whites. For example, in 1992 African Americans comprised 12.4 percent of the American population, but constituted 44.8 percent of arrests for crimes of violence and 31.8 percent of arrests for property crimes (Conklin 1995: 112).

The FBI data show arrests only for those of different race—whites, blacks, American Indians, and Asians or Pacific Islanders. The FBI data do not show arrests separately for Hispanics or for other more specific ethnic groups. However, there is some evidence that crime rates for Mexican Americans are higher than those for the total population, while those for Chinese Americans and Japanese Americans are lower than average (Conklin 1995).

Some scholars and other observers have suggested that more frequent arrests of African Americans for street crimes may reflect discrimination against blacks by police. There is, in fact, evidence that police sometimes stop and investigate minorities on the street more than they stop whites and may be more quick to arrest members of minority groups (Elliott and Ageton 1980; Smith and Visher 1981; Barlow 1993). However, surveys of the victims of crime (who themselves are disproportionately black) confirm that blacks are greatly over-represented among the perpetrators of street crime (as identified by the victims). The data from the victimization surveys are in close agreement with official arrest statistics. Sheley comments: "Victims seem to be telling us that arrests are probably a reasonable indicator of real race-based differences in the commission of street crimes" (1995: 120).

Bail, Charges, and Jury Selection

Differences in treatment of various ethnic groups may occur after individuals are arrested. Historically, minority group members, most of whom were poor, were unlikely to be able to afford lawyers to represent them. However, Supreme Court decisions since the 1960s have stated that a person is entitled to legal counsel for any alleged offense for which he or she may be imprisoned. Thus, though the quality of their legal representation is not always high, minority poor people, like other poor, at least have the help of a lawyer.

After a person is arrested and charged with a crime, a judge may release him on bail, that is, release him until trial upon deposit of a specified amount of money with the court. Of course, poor people are less likely to be able to pay bail and minorities are more likely to be poor. As to whether race itself affects judges' bail decisions, research results are somewhat contradictory. Some researchers conclude that bail decisions are affected primarily by the accused person's prior record and the seriousness of the offense and that race in itself has no effect. However, several studies found that a defendant's race (African American or Native American versus white) directly affects bail decisions or that race has an impact on these decisions under certain conditions. One study found that, among those who had a prior conviction for a felony, bail decisions for African Americans were more harsh than for whites (Spohn 1995).

When a person is arrested, the specific crime (if any) with which he is charged, whether he is brought to trial, and if so, whether he is convicted, depends on decisions of prosecutors and of juries. Evidence concerning possible racial discrimination by prosecutors is not consistent. Some studies have found that prosecutors are less likely to dismiss charges against African Americans and Hispanics than against whites. However, other studies have found no evidence of racial bias by prosecutors (1995). Whether prosecutors generally treat minorities more severely appears to vary from one locality to another. Also, actions of prosecutors may depend on both the race of the accused and the race of the victim. One study (LaFree 1980) found that African Americans arrested for raping white women were more likely to be charged with felonies than those in other rape cases. Another study (Radelet 1981) found that suspects arrested for murdering whites were more likely to be charged with first-degree murder than were those arrested for murdering blacks.

If a person accused of a crime is brought to trial, much depends on the composition of the jury that decides on his guilt or innocence. Although black people are more frequently chosen for jurors today than was true in the past, there is some evidence that the jury selection process still sometimes is racially biased. Prosecutors may use preemptory challenges to exclude African American jurors from cases with African American defendants (Turner et al. 1986).

Sentencing

If a person is found guilty of a crime, does her ethnicity affect the severity of her sentence? Since equity in sentencing is a central element of a fair justice system, this issue has been the subject of considerable research. This body of research indicates that ethnicity does not consistently influence sentencing. At times, African Americans and Hispanics even seem to be treated more leniently than Anglo whites for some offenses. For example, one study (Peterson and Hagan 1984) found that blacks convicted of homicide received shorter sentences than either whites or Hispanics and another study (Unneever and Hembroff 1988) found that Hispanics were treated more leniently than Anglo whites for assaults and homicides. However, a substantial number of studies have found that African Americans, Hispanics, and Native Americans received more severe sentences than Anglo whites (Kramer and Ulmer 1996). The extent of discrimination against minorities in sentencing appears to depend in part on the type of crime and on the race of both offender and victim. Race appears to matter in sexual assault and murder cases. The longest sentences were imposed on African Americans convicted of sexually assaulting or murdering whites. Similarly, the death penalty is more likely to be imposed on African Americans who murder whites than on African Americans who murder other African Americans or on whites who murder someone of either race (Spohn 1995).

The Justice System and Intergroup Relations

The operation of the criminal justice system and the involvement of different ethnic groups with that system has important effects on intergroup relations.

High rates of street crime among blacks lead many whites to stereotype blacks as criminals and to avoid contact with blacks, especially by not living in racially mixed neighborhoods (Taub et al. 1984). Such white reactions lead the great majority of law-abiding black people, in turn, to be resentful at being negatively stereotyped by whites.

Many African Americans also continue to see the criminal justice system as biased against them. For example, they point out that sellers or users of crack-cocaine (who are often poor blacks spotted by police in public places) are more likely to be arrested than sellers or users of powder-cocaine (often more affluent whites who are able to carry out their illegal activities in private) (Blumstein 1993). The perception among many blacks (and Hispanics and Native Americans as well) that they are discriminated against by police, prosecutors, juries, and judges leads many minority group people to feel alienated from the larger society. Of particular importance is the anger that many feel towards the police. African Americans, Hispanics, and Native Americans are much more likely than Americans of European descent to complain about police brutality (Parisi et al. 1979). During the 1960s, rioting by blacks broke out in about two hundred American cities, including large-scale riots in Los Angeles, Detroit, Newark, and Chicago. From 1964 to 1972, these riots resulted in an estimated two hundred fifty deaths, ten thousand serious injuries, sixty thousand arrests, and billions of dollars of property damage and expenditures by police, national guard, and even U.S. Army troops called in to quell some of the riots (Marable 1984). While there were many underlying causes of these riots (Spilerman 1970), the incidents that triggered many, if not most, of them involved such police actions as arrests, shootings, or searching of blacks in ghetto neighborhoods. Neighborhood residents were angered by what they perceived as unnecessary use of force by the police officers (U.S. National Advisory Commission on Civil Disorders 1968).

While the number of riots by blacks has declined since the peak of the 1960s, some riots have continued to occur. Incidents that have precipitated such riots continue to involve interaction between the police and blacks. In 1980, a major riot by blacks in Miami was touched off by the alleged killing of a black man by police. In 1992, the acquittal of police officers charged with the beating of a black motorist led to a major riot in Los Angeles, resulting in over fifty deaths and property damage of over one billion dollars (Jaret 1995). An example of the much smaller but more frequent disruptions caused by conflict between minorities and police occurred in Indianapolis in 1995—a crowd of blacks threw bottles and rocks at police officers to protest the alleged beating of a handcuffed man during a drug arrest (*Lafayette Journal and Courier* 1995).

Research concerning whether or not police actually exhibit racial discrimination in their actions has produced mixed evidence. There is some evidence that discriminatory actions by police against blacks occur in some localities; for example, police in Memphis, Tennessee, were more likely to shoot blacks than whites even if involvement in criminal activity was taken into account (Fyfe 1981). But studies in other cities, including Boston, Chicago, and Washington, D.C., have found no evidence of discriminatory behavior by police or even found a greater likelihood of police using force against whites than against blacks (Reiss 1971; Geller and Karales 1981). Whatever the actual extent of discriminatory treatment by police towards blacks and other minorities, mutual feelings of mistrust, hostility, and fear are much greater between inner-city minority groups, especially blacks, and the police than between whites and police. A large proportion of black people see police as a kind of occupation force in their neighborhoods, representing the white establishment, having little respect or concern for them, and offering them inadequate protection. Police officers, in turn, sometimes act harshly in reaction to what they see as disrespect and lack of assistance by minority residents, out of fear of losing control in a hostile environment, and out of fear for their own personal safety.

One of the factors that has often contributed to poor relationships between the police and minority residents is the low representation of minority group members on police forces. In recent years the representation of blacks and Hispanics on police forces has increased significantly, but in most cases they still are below the minority's percentage of the total population (U.S. Department of Justice 1990).

Political Influence of Ethnic Groups

Variations Among Nations

As we have just seen in previous sections, political and legal institutions can have important influences on people's lives—where they go to school, where they live, their chances of getting a job, how they are treated by the police, and so forth. Partly for such practical reasons, different ethnic groups within societies around the world try to maximize their influence in government.

In addition to the practical advantages that may stem from political influence, ethnic groups often compete for political power because members of each group see political control as indicating that the nation or community is "theirs," rather than belonging to another group. For reasons of prestige and self-esteem, as well as for practical reasons, they are often loathe to be "dominated" by some other group.

Competition for control of the government and the legal system often leads to hostility and sometimes to armed conflict among ethnic groups. Those ethnic groups that see themselves as permanently dominated politically by another group are likely to feel alienated from the larger society and are potentially rebellious (Horowitz 1985; Gurr 1993).

The competition among ethnic groups for political influence may be handled within a society in a number of ways. In many nations (e.g., in Guyana, Nigeria, and Sri Lanka), political parties have represented different ethnic groups. Where political parties are based on ethnicity, conflict among ethnic groups tends to be

163

increased. The social division based on ethnicity overshadows all other social divisions, such as those based on region or on income. When, on the other hand, political parties are not formed along ethnic lines (e.g., in Great Britain and New Zealand), ethnic conflict tends to be moderated. Each party tries to pull voters from more than one ethnic group with programs that are attractive across ethnic lines. However, where ethnic divisions are sharp, it is difficult to maintain multiethnic parties or even coalitions between separate ethnic parties (Horowitz 1985).

Some nations (such as Belgium, Switzerland, Canada, and Nigeria) have devised arrangements for ensuring that minority ethnic groups have some voice in political decisions. Among the institutional arrangements that have been used (alone or in combination) are the following: 1) a government composed of representatives of all major ethnic groups; 2) representation of each ethnic group proportional to its fraction of the total population; 3) a mutual veto system, whereby each ethnic group can veto government decisions in matters of vital concern to it; 4) autonomy for each ethnic group, through local governments or through separate institutions, such as schools.

Political systems with such features that ensure sharing of power among different ethnic groups have been termed "consociational systems" (Lijphart 1977; Kellas 1991). The nations that have adopted such power-sharing systems have done so in the hope that they would permit people of different ethnic groups to live together without serious conflict in one unified society. But, despite some success (most notably, it has been largely successful in Switzerland), power-sharing arrangements often have not succeeded in avoiding serious ethnic conflict (e.g., in Lebanon and Sri Lanka). Nevertheless, Kellas comments: "As the present state system is a rigid one which opposes the secession of nations from multinational states, the consociational model and its variants give such nationalisms some hope of satisfying their nations without resorting to civil war" (1991: 145). (See chapter 10 for further discussion of these and other political arrangements used in various countries to give greater influence to ethnic minorities.)

United States

In the United States, minority groups have had little influence on political decisions throughout most of the nation's history. Following the Civil War, the Fifteenth Amendment to the Constitution prohibited the denial of the right to vote to any person on account of his or her race. However, southern states soon devised ways to circumvent this provision through such means as a tax on voting (the "poll tax"), restricting voting in party primaries to whites, requiring literacy tests, and giving voting registrars wide discretion in deciding whom to allow to register to vote. These practices prevented most blacks in the south from voting. They also prevented many Hispanics (especially in Texas) and Native Americans from voting. While obstacles to voting were greatest in the south, literacy tests in other parts of the county also reduced voting among blacks, Hispanics, and Native Americans, who often had little education and (in the case of Hispanics) often spoke little English (J. E. Farley 1995).

Efforts to End Voting Discrimination

To try to end discrimination against minorities in voting, Congress passed the Voting Rights Act in 1965. Among its provisions, this act suspended the use of literacy tests, established federal review of any changes in state voting laws, provided for the assignment of federal registration examiners and election observers, and directed the U.S. Attorney General to bring lawsuits against states that retained the poll tax (Grofman et al. 1992).

The 1965 Act was intended to protect the right to vote of blacks in the south. In 1975, the act was amended to extend its protection to language minorities. Voting registration and elections could no longer be conducted only in English in those states or localities where more than five percent of the voting-age population belonged to one of the following language minorities: Alaskan natives, Native Americans, Asian Americans, or people of Spanish heritage. These changes to the Voting Rights Act not only expanded its coverage to other minorities, in addition to African Americans; it also expanded its coverage beyond the south to many other areas of the United States, including many areas of the west (including California), the southwest, and Alaska (1992).

Effects of Voting Rights Act

The Voting Rights Act of 1965 had a quick and dramatic effect on political participation by blacks in the south. Surveys indicated that the proportion of blacks registered to vote rose sharply in seven affected southern states, from an average of 29 percent before passage of the act to over 52 percent only two years later. Since then, registration rates among blacks have moved upward much more slowly but the disparity between black and white registration rates now is small.

The proportion of blacks actually voting in southern states also rose substantially following passage of the Voting Rights Act. For example, while only about one in three blacks in Alabama and Mississippi voted in the presidential election of 1964, more than one in two blacks in those states voted in the next presidential election of 1968. (Among Americans throughout the nation, about 60 percent of those of voting age actually voted in those elections.) After reviewing the dramatic increases in political participation among blacks in this period, Grofman et al. comment: "The suspension of literacy tests and the use of federal examiners unquestionably played a major part in clearing the path to the polling places for nearly three-fifths of the voting-age blacks in this area of the south. In this respect, the Voting Rights Act—and the civil rights movement of which it was an outgrowth—was an unquestioned success" (1992: 22–23).

Political Representation and Influence

At the same time that minorities in the United States have been more free to exercise the right to vote, an increased number of minority-group members have been elected to political office in recent decades. More African Americans have been elected to political offices at all levels—in city, state, and federal governments. For example, between 1970 and the late 1980s the number of African Americans elected to office (local, state, and federal) increased from about thirteen hundred to over seven thousand. The number of Hispanic American office-holders has also increased greatly and more Asian Americans also have been elected or appointed to political offices.

The election of minority group persons to high office has been especially noteworthy in American cities. Thus, blacks have been elected as mayors in a large number of cities, including Chicago, Atlanta, Cleveland, Philadelphia, Minneapolis, New York, and Los Angeles. Hispanic Americans have been elected as mayors of a number of other cities, including San Antonio and Denver. However, despite recent increases, elected officials (especially at the state and federal levels) who are African American, Hispanic American, or members of other minorities still are far below the proportions of these groups in the total population (Grofman et al. 1992; Jaret 1995).

The underrepresentation of minority group members among elected officials stems from a number of causes. Some European Americans are reluctant to vote for minority candidates (although some black and Hispanic officials, such as the governor of Virginia and the mayors of Los Angeles, Minneapolis, and Denver have been elected in heavily-white areas). In addition to prejudice among white voters, certain electoral devices have sometimes reduced minority representation. One such arrangement is the "at-large" district, in which a number of representatives (say, to a city council) are chosen by the entire electorate, rather than having each of a number of smaller districts (some of them with mostly minority residents) choose its own representative. Another device that sometimes has reduced minority representation is the "gerrymandering" of election districts—that is, drawing district boundaries in a way that results in few or no districts in which African Americans or Hispanics are a numerical majority. An example of the "dilution" of the votes of minorities is found in a suit brought against the city of Los Angeles in the 1980s in which it was pointed out that although Latinos constituted a third of that city's population and a quarter of the population of the county, there were no Latinos on their governing boards (Cain 1992).

Congress tried to deal with the problem of underrepresentation of minorities in amendments to the Voting Rights Act in 1982. Voting practices or procedures were said to be illegal if they resulted in members of a minority group having "less opportunity than other members of the electorate to participate in the political process and to elect representatives of their choice" (Grofman et al. 1992: 39). The extent to which members of a "protected class" (i.e., certain minority groups) have been elected to office was said to be one circumstance to be considered in judging violations of the law, although any right to proportional representation was explicitly ruled out.

The issue of how to draw election districts fairly has reached the Supreme Court on a number of occasions. In a major decision on the subject in 1986, the Court upheld lower court decisions requiring North Carolina to redraw some state legislative districts in a way that gave more influence to black voters. The Supreme Court stated then that three circumstances were necessary to uphold a minority group's claim of vote dilution: 1) the minority group must be sufficiently large and geographically compact to constitute a majority of a single-member district; 2) the minority group must be politically cohesive; and 3) the white majority votes sufficiently as a bloc to enable it usually to defeat the minority's preferred candidate.

The combined effect of changes in the Voting Rights Act, court decisions, and pressure from the U.S. Justice Department led some states and localities to re-draw election districts to try to provide for election of more minority group representatives. The underrepresentation of minority groups in elected offices, and the efforts to remedy this situation, has also led to a debate on the subject of proportional representation and that of special political rights for minorities more generally. Some have argued that mechanisms to provide proportional representation to minorities such as African Americans and Hispanics are necessary if members of these groups are to have an adequate voice in political decisions. Otherwise, they say, members of these groups will not only lack a fair role in decision making but will also feel alienated from the larger society. They point to arrangements to share power among ethnic groups that have been adopted by other nations (e.g., see Guinier 1992).

Others (e.g., see Thernstrom 1987) see dangers in a system of proportional representation. They see such a system as increasing racial divisions and separations and increasing racial tensions. Further, they allege that it gives certain minority groups rights that others do not have and is contrary to the American tradition of individual, rather than group, rights.

Whatever the merits of the idea of proportional representation, it seems unlikely to be adopted fully in the United Stated in the foreseeable future. In 1995, the Supreme Court decided that using race as a basis for drawing election districts had gone too far. In a case concerning lines for a Georgia Congressional district, it ruled that using race as a "predominant factor" in drawing district lines is unconstitutional (*New York Times* 1995a); the Supreme Court confirmed this doctrine the following year by ruling that congressional district boundaries in North Carolina and Texas, drawn so as to create majority-black districts, were also unconstitutional (*New York Times* 1996).

Increasing Influence of Minorities

How can an ethnic group that is a numerical minority exercise significant influence in the American political system? Increasing the number or elected representatives from that group may help. But even if African Americans, Hispanic Americans, Asian Americans, or Native Americans were represented at all government levels in proportion to their percentage of the total population (which, as we have noted, is unlikely to occur), each of these groups could still be outvoted by other Americans.

Ever since the presidency of Franklin Roosevelt in the 1930s, most black voters have supported the Democratic Party, seeing it as championing equal rights and help for their race. In recent presidential elections, between eight and nine out of every ten African American voters have supported the Democrats (Huckfeldt and Kohfeld 1989; *Wall Street Journal* 1996). Democratic Party guidelines adopted in 1972 provide for substantial minority (as well as female) representation at presidential nominating conventions. But while blacks are influential within the Democratic Party, many observers see the Democrats as becoming a minority party in the country. A majority of whites (especially white men) now do not vote for the Democratic Party at the national level; in some recent elections, only about one in three whites voted for the Democratic presidential candidate (Huckfeldt and Kohfeld 1989).

What seems to have happened, in part, is that the Democrats became identified in the public mind as the party that represents the interests of blacks, while Republicans have been seen as defending the interests of whites. This perception has alienated particularly the lower-income whites who were an important part of Democratic Party support from the 1930s to the 1960s. Commenting on stronger positions the Democratic Party has taken on race issues, Huckfeldt and Kohfeld state: "This new Democratic Party . . . is the natural culmination of the need to secure heavy support among blacks, which in turn compromises the party's ability to secure the support of lower-class whites. Thus the Democratic Party has frequently become the party of blacks and liberal whites, two groups that do not constitute a winning coalition in American politics" (1989: 15). This is especially true in the south, which used to vote solidly Democratic but which now has swung behind the Republicans, especially in national elections. A Democrat, Bill Clinton did win the presidency in 1992, but with only 43 percent of the total vote and a minority of the white vote (in a three-person contest). President Clinton won again in 1996, but again with a minority of the white vote and after taking a moderate stand on race-related issues, such as affirmative action programs and welfare reform, that left many minority leaders dissatisfied. (For a discussion of President Clinton's attempt at "blunting the wedge issue of race," see Omi and Winant 1994).

In addition to pushing its agenda within a major political party, a minority group may consider other possible ways to exert greater political influence. Some

African Americans, frustrated with the outcomes of the essentially two-party system in the United States, have talked of creating a separate black political party. It may be that such a party, if created, might succeed in exercising some influence—for example, by being the balance of power in a closely split legislature. But a separate minority party might simply ensure the election of conservative candidates. Moreover, ethnically-separate parties in other nations have tended to increase ethnic conflict (Horowitz 1985).

Another strategy is to try to form winning coalitions with other groups. There is considerable evidence that minorities sometimes are able to exercise significant political influence or be elected to office by forming such coalitions. Sometimes a coalition with other minority groups increases the influence of a particular minority. For example, the political clout of a particular Asian American group, such as Chinese or Japanese, has been greatly increased by the coalitions that these groups have formed (Espiritu 1992). Coalitions between minority groups and some whites also have been successful in some cases. For example, alliances between blacks and liberal whites in Chicago and Philadelphia and between Hispanics and some whites in San Antonio and Denver were successful in electing minority group members as mayors (Browning et al. 1990).

To become part of a winning coalition a minority group must find common interests and common positions with those in other groups. Although together, blacks, Hispanics, and Asians constitute a large and growing proportion of the American population, there are differences in political views among and within these general groupings. There are, of course, differences also among each of these groups and "Anglo" whites. But there also are commonalties of interests and of views that cut across ethnic lines—for example, common interests of those with a similar income. As the work of Sniderman and Piazza (1993) emphasizes, the line-up on a given political issue relevant to race—such as government help for disadvantaged people—is rather fluid and varies with how an issue is framed. Thus, the building of winning coalitions based on common interests, especially economic interests, appears to be a possible way for numerical minorities to enjoy significant political influence.

Religion

While people from different racial and ethnic groups may relate to each other on the basis of their economic and political interests, they may relate also on the basis of their religious affiliations.

Religion is important in the life of many people. In the United States, almost half of the population attend religious services nearly every week or more often (Roof and McKinney 1987). Those who participate in the religious life of a congregation generally also participate in its social life. They socialize with and make friends with fellow congregants, with whom they have common beliefs and a common religious identity. If people from different racial and ethnic backgrounds attend the same churches, they have an opportunity to interact in positive circumstances (shared beliefs, equal status, shared goals for the church). If they do not attend the same churches, such opportunities are lacking.

Religious Identifications

We may consider first the extent to which Americans from different ethnic groups follow the same religions. A large-scale study of Americans' religious identifications that was conducted in 1990 (Kosmin and Lachman 1993) provides relevant information. The religious identifications that were found among various ethnic groups are shown in table 6.1. Among nonHispanic whites, well over half said they were Protestant and a little over one in four identified themselves as Catholic. An even larger majority of African Americans than of whites identified themselves as Protestant and about 9 percent said they were Catholic. Among Hispanic Americans, about two out of three people identified themselves as Catholics and one in four as Protestant. Protestants are most likely to be found among Hispanics whose origins are in Puerto Rico or Central America while Mexican Americans overwhelmingly identify as Catholic. Among Asian Americans, about one-third said they are Protestants, about one-fourth said they are Catholics, and the rest said they had no religion or professed other religions. Overall, then, most Americans of every major ethnic group share a personal identification with Christianity. Moreover, although the proportions of Protestants and of Catholic vary among ethnic groups, those who are either Protestant or Catholic have many co-religionists in all of the other groups.

Relations Based on Religion

While Americans of various ethnic groups share some general religious identifications and beliefs, this does not necessary mean that religion brings them much closer together. We may consider the situation first with respect to relationships between Hispanic Catholics and other Catholics and then with respect to relationships between African Americans and whites.

Hispanic Catholics

There is no tradition of a Hispanic Catholic Church that is separate or autonomous from the Catholic Church as a whole. However, Kosmin and

170

TABLE 6.1

Religious Identifications of Americans, by Racial/Ethnic Background
(Percentages)

Religious Identification	White, Non-Hispanic	Black Non-Hispanic	Hispanic	Asian
Protestant	57.9	81.8	24.6	33.6
Catholic	26.5	9.2	65.8	27.1
Other Religion[a]	5.3	1.9	2.5	16.2
No Religion	8.5	5.9	6.3	19.6
No Answer	1.8	1.2	0.8	3.5

a. Main other religions are: 1) for whites: Jewish, 2.2 percent, Mormon, 1.7 percent;
2) for blacks: Muslim, 0.9 percent; 3) for Hispanics: Mormon, 0.8 percent; 4) for
Asians: Buddhist, 4.1 percent, Hindu, 3.7 percent, Muslim, 3.0 percent.
Source: National Survey of Religions Identification in 1990 (Kosmin and Lachman 1993).

Lachman assert that "the Catholic Church feels foreign to many [Hispanic] immigrants" (1993: 139). They note that feelings of estrangement may stem in part from the small number of Hispanic priests; only 4 percent of Catholic priests in 1990 were Hispanic, while Hispanics comprised 14% of all U.S. Catholics. Other writers estimate that less than one-fourth of Hispanic Catholics actually practice their religion (Gonzales and LaVelle 1988).

The Catholic Church in the United States has long been dominated by men of Irish extraction, while Hispanic Americans have often felt left out of church decision making. They have wanted a greater role in deciding church policy and wanted the Church to give greater recognition to Hispanic culture. Recently, the Church hierarchy has been moving to accommodate the special needs of Hispanic Americans and to bring about a fuller participation by Hispanics (Weyr 1988; Kosmin and Lachman 1993). The Catholic Church is an institution that has the potential to bring together both Hispanic and non-Hispanic Americans in common belief and purpose; but this potential has not yet been fully realized.

Blacks

While Hispanics are formally part of the same churches (usually Catholic) as non-Hispanics, the same is not true for African Americans. African Americans originally were members of the same Protestant denominations to

171

which whites belonged. But, in the early 1800s, separate black Methodist denominations were organized and in the later 1800s separate black Baptist and black Presbyterian associations were formed. Black churches were centers of black community life and were the one institution over which blacks had control. But, as Roof and McKinney comment: "The price of autonomy and self-control was a separate and segregated church, one effectively cut off socially and religiously from white America" (1987: 140). Of course, blacks, especially in the south, had no option other than to be in separate churches, at least at the level of the local congregation.

One study estimates that about 86 percent of blacks who are church members belong to separate black denominations (such as the National Baptist Convention, U.S.A., and the African Methodist Episcopal church). About 8 percent of blacks were estimated to be Roman Catholics and about 5 percent to be members of predominately white Protestant groups (Lincoln and Mamiya 1990).

Those religious denominations that are predominately white generally have only a small percentage of African Americans among their members. About 5 percent of Roman Catholics are black. Most Protestant denominations have only a tiny proportion of blacks (most have two percent or less). A few Protestant denominations have a fairly substantial percentage of blacks; Pentecostals/Holiness (16 percent) and Jehovah's Witness (18 percent) are the outstanding instances. Even where blacks are fairly well-represented in a denomination, they often belong to congregations that are all or predominately black (Roof and McKinney 1987; Lincoln and Mamiya 1990).

In one survey, more than one-third of the whites said that they attended religious services with blacks (Roof and McKinney 1987). Such experiences are more likely to be mentioned by Catholics than by Protestants. But given the small proportions of blacks in predominately white churches, most whites are likely to attend services with only a very small number of upwardly mobile blacks. As Roof and McKinney comment: "The extent of racial integration in the 'white' churches, whether Protestant or Catholic, is still small, hardly enough to refute the charge that the church remains among the most segregated major institutions in the society" (1987: 143). As Winter has said, "eleven o'clock on Sunday morning is the most segregated hour of the week" (1962), and this statement is still largely true.

Religion does have the potential to unite people on the basis of common values. Kosmin and Lachman state: "Religion will continue to be a binding element in our diverse society . . . it asserts that the Scriptures can speak directly to hugely diverse groups of people irrespective of their origins and that these texts transcend time, place, and background in their ability to uphold our common values" (1993: 156). Evidence on the small amount of interaction between ethnic groups, especially between blacks and whites, suggests, however, that the "binding" force of religion largely remains an unrealized potential.

Overlapping and Cross-Cutting Differences

In this chapter and in previous chapters, we have discussed a number of ways that members of different ethnic groups may be different or similar other than in ethnicity—in social class, language, political affiliation, religion, and so on. Sometimes, people of different ethnicity differ considerably with respect to other socially important characteristics. For example, in Northern Ireland those of British (English or Scottish) ancestry differ from those of Celtic (Irish) descent and generally these two groups also differ from each other with respect to religion and social class (those of British descent being Protestant and, relatively, economically well-off, while those of Celtic descent are Catholic and tend to be lower in occupation and income). When groups that differ in one important social characteristic also differ in other ways these differences may be said to be "overlapping."

An alternative possibility is that while two groups may differ with respect to one salient social characteristic, other important social divisions do not follow the same lines. For example, in Switzerland a major social division is between those who speak German and those who speak French. But another important social division, between Catholics and Protestants, does not coincide with the division based on language. In such a situation the social divisions may be termed "cross-cutting."

Overlapping divisions among groups tend to reduce interaction and to increase conflict between these groups. Suspicions and tensions based on one social division are reinforced and strengthened by those based on other divisions. Thus, for example, British-origin Protestants and Irish-origin Catholics in Northern Ireland harbor grievances toward the other group based on their ethnic-based histories, their religious beliefs, and their socioeconomic positions. Identities and loyalties based on ethnic origin are reinforced by those based on religion and social class.

Where social divisions are cross-cutting (independent of each other), on the other hand, friendly interaction is likely to increase and conflict to decrease. For example, in Switzerland those who speak different languages often share a common religion, which increases the chances that they will interact in a friendly way and decreases the chances that they will experience conflict.

The degree of overlap, or intersection, of social characteristics is central to a formal theory of population structure proposed by Peter Blau (1994). Blau states: "If several social differences that independently influence social life are closely related, the barriers to social intercourse are strengthened and the probability of social relations diminishes" (1994: 34). On the other hand, Blau points out, when important social characteristics intersect (cross-cut) each other, the person often will interact with someone who is different in one characteristic (say, ethnicity) but who is similar in other respects (say, occupation). Thus, cross-cutting social differences will promote positive interaction across group lines. Blau also reasons that cross-cutting cleavages reduce the intensity of intergroup conflicts. In such situations, individuals often are subject to cross pressures from different groups that have opposite views on issues. He explains: "A person's ethnic group and colleagues at work, her union and church, his fraternity and professional association often have different political viewpoints and support opposite candidates in elections. Such situations put individuals under cross-pressure, which may lead some not to take sides at all and undermine others' inflexible convictions" (1994: 41).

In support of his theoretical propositions, Blau finds that the more that racial differences in a U.S. metropolitan area are cross-cut by other social differences, such as occupation and education, the greater the likelihood of marriage between those of a different race (Blau and Schwartz 1984; Blau 1994). The same was found to be true with respect to marriage between those of different national origin (e.g., Polish, Italian, Irish, German). In other words, the chances of marriage between those of different races or of different national origins increased when the racial and national groups differed less with respect to their education, occupations, and other social characteristics.

Because relations among groups are influenced by whether cleavages are overlapping or cross-cutting, it is useful to examine racial and ethnic groups in the United States from this perspective. We have seen that differences in religion do not closely overlap ethnic categories. Large majorities in all ethnic groups are Christians. While Hispanics tend to be Catholic and blacks tend to be Protestant, large proportions of non-Hispanic whites are in each of the major subdivisions of Christianity. One group in which religious difference does overlap a racial difference is the Nation of Islam, whose members are black and Muslim. The combination of these differences tends to separate members of this group, more than other blacks, from the rest of society.

Ethnic groups in the United States also do not generally differ with respect to language. English is the language spoken by a large majority of every ethnic group. Some Hispanic Americans, especially those who are recent immigrants, do speak Spanish primarily but most Hispanic Americans born in the United States adopt English as their first language (e.g., see Keefe and Padilla 1987).

The primary differences that tend to overlap those of ethnicity are those related to social class. African Americans, Hispanic Americans, and Native Americans tend to have less education, less skilled occupations, and lower incomes than non-Hispanic white (and Asian) Americans (see section on "Economic Position of Groups" in chapter 5). This overlap of differences in ethnicity and differences in social class tends to reduce interaction, and to increase conflict, between whites and those in minority groups. However, the overlap between ethnicity and social class has been reduced considerably in recent decades. A growing proportion of blacks and of Hispanics has moved up into the middle class. To the extent that social class differences between racial-ethnic groups decline further, relations between these groups may be expected to improve.

Summary

The positions and relationships of ethnic groups in any society are affected by the law and how the law is administered. In the United States prior to World War II, actions by governments (at all levels) and the courts generally had the effects of maintaining physical segregation of ethnic groups and of keeping the economic status of minorities low. In more recent times, government and the courts generally have tried to reduce racial segregation (e.g., in schools and in neighborhoods) and to increase equality (e.g., in employment and in opportunity for political participation). These legal actions have had varying degrees of success in different areas. Major advances have been made in making it possible for blacks and other minorities to vote everywhere in the United States. Separation of blacks from whites in schools has been reduced but considerable racial separation, and also separation of Hispanics from non-Hispanic whites, continues in American schools. Efforts to reduce job discrimination have had some limited success. Least progress has been made in desegregating neighborhoods, although small declines in residential segregation have occurred.

The operation of the criminal justice system has posed serious problems for relations between racial and ethnic groups. Members of minority groups, especially blacks, are much more likely than non-Hispanic whites to be arrested for "street crimes," including both crimes of violence and property crimes. While minorities sometimes are arrested unfairly, the actual rate of street crime by blacks appears genuinely to be high. This crime rate contributes to fear of and avoidance of blacks by many whites, which then angers many law-abiding blacks who feel unfairly stereotyped as criminals.

Discrimination against minorities within the criminal justice system may take place at a variety of stages—including bringing charges, setting bail, and sentencing convicted persons. Evidence is mixed about whether discrimination against blacks and Hispanics generally occurs within the justice system. However, the harshest sentences appear to be issued against blacks who are convicted of serious crimes (especially rape or murder) against whites. Perceptions by minorities of unfair treatment in the justice system leads them to feel alienated from the larger society.

The hottest "flash points" in the criminal justice system occur where minority groups come into contact with police. Anger among minority residents (especially blacks) at what they perceive to be unnecessary violence against or mistreatment of minority persons has precipitated many riots. Efforts to improve relations between the police and people in minority areas are essential for improving intergroup relations more generally.

Minorities have been underrepresented in political offices in the United States. The federal government and courts have made some efforts to increase minority representation, for example in Congress. Some people have advocated proportional representation for minority groups but others have argued against such procedures and neither Congress nor the Supreme Court have endorsed this idea. Blacks and other minorities generally have supported the Democratic Party and enjoy considerable influence within that party. However, many working-class whites have been alienated by the Democrats' support of minorities and a coalition of minorities and white liberals has not been sufficient to win national elections. An alternative way for minorities to enjoy greater political power is to join coalitions based on economic interests that cut across lines of race and ethnicity.

Relationships between people of different race and ethnicity may be affected also by their religious affiliations. Like most whites in the United States, most members of minority groups are Christian, either Protestant or Catholic. Most Hispanic Americans are Catholic but Hispanics have not been well-represented in the Church hierarchy or in the priesthood and many Hispanics have felt alienated from the Catholic Church. African Americans, like white Americans, are predominately Protestant. But a large majority of African American church members belong to separate black denominations. While some African Americans and some whites have contact with members of the other racial group at religious services, religious institutions remain one of the most racially segregated areas of American life. Since most people from different racial and ethnic groups share religious beliefs, churches have a great largely-unrealized potential for bringing people from different groups together in activities based on common beliefs and common purposes.

When people who differ in ethnicity are similar with respect to other important social characteristics (religion, class, language, politics, and so on), interaction among the ethnic groups is apt to be more positive than when the ethnic difference overlaps other social differences.

7) The Educational System

A crucial institution that affects the position of various racial and ethnic groups is the educational system of the society. This chapter is concerned primarily with examining differences among groups in the amount and quality of education they obtain and reasons for such differences. We will also discuss the use of education to try to reduce intergroup prejudices. Finally, we will look briefly at ways in which the media portray minorities and some effects the media may have on relations between ethnic groups.

Educational Outcomes

The educational system of a society is likely to have important effects on relations among the ethnic groups that compose that society. Some of these effects are direct. For example, where children and young adults from various ethnic groups attend school and how much contact with other groups they have in school is likely to affect their attitudes toward other groups.

Probably the most important impact of the educational system on intergroup relations is an indirect one. How much schooling members of various groups get, and how good that schooling is, have major effects on the kinds of occupations and incomes those in each group attain. Thus, the educational system helps to determine the extent to which different groups are equal or unequal with respect to their socioeconomic status within the society. People tend to be more willing to have close contact with, and to get along well with, those of equal status as compared to those of unequal status (see chapter 4).

In this section we will examine the educational attainments of different ethnic groups in the United States, consider some of the factors that affect school experiences and attainments, and discuss some efforts that have been made to make schools more effective, especially in the education of minorities.

Amount of Education

The amount of education that Americans receive has increased greatly during the course of the twentieth century, especially since World War II. The rise has been marked for people of all races, but especially for minorities. Table 7.1 shows that the median years of education completed by young whites in 1920 was 8.5—that is, the typical white person left after completing elementary school. The amount of education for whites jumped sharply from 1920 to 1950 and then rose more slowly, standing at a median of 13.0 years of education (a little beyond high school) in 1993.

Among blacks and other non-whites, the increase in amount of education during the same time span was even more dramatic. In 1920, the median years of education completed by young non-whites (aged twenty-five to twenty-nine) was 5.4 years—that is, the typical non-white left school after about the fifth grade of elementary school. But the median amount of education received by minorities rose sharply in the following decades, coming close to the level for whites by 1970. In 1993, the median years of education for young non-whites was 12.9 years, about the same as that for whites. However, table 7.1 shows that, while the percentage of young whites and young non-whites who completed high school was about the same in 1993, the proportion of non-whites who had completed four years of college lagged behind that of whites (18.7 percent to 24.7 percent).

In the data presented so far, based on census data, persons of Hispanic origin are not shown separately; they are included, as appropriate, in the "white" or "black and other races" category. Other data on education do separate persons of Hispanic background from non-Hispanic whites and nonHispanic blacks. Table 7.2 shows the amount of education completed by young people (aged twenty-five to twenty-nine) in these three ethnic groups in 1993. These data show that young Hispanics were completing less education than whites or blacks of non-Hispanic background. While over 82 percent of whites and of blacks had graduated from high school, less than 61 percent of young Hispanics had finished high school. And only about 8 percent of Hispanic young people, compared to over 13 percent of blacks and over 27 percent of whites, had completed college. The relatively high proportion of immigrants among Hispanic Americans is one factor reducing educational attainment in this ethnic category (Chavez 1991).

Native Americans have also had less education than whites. Census data for

TABLE 7.1

Years of School Completed by Persons Age Twenty-Five to Twenty-Nine,
By Race, 1920–1993

A. Whites

Years of School Completed (Percent)

Year	Less Than 5 Years	4 Years High School or More	4 Years College or More	Median Years of Education
1920	12.9	22.0	4.5	8.5
1940	3.4	41.2	6.4	10.7
1950	3.3	56.3	8.2	12.2
1960	2.2	63.7	11.8	12.3
1970	0.9	77.8	17.3	12.6
1980	0.8	86.9	23.7	12.9
1993	0.7	88.5	24.7	13.0

B. Blacks and Other Races[a]

Years of School Completed (Percent)

Year	Less Than 5 Years	4 Years High School or More	4 Years College or More	Median Years of Education
1920	44.6	6.3	1.2	5.4
1940	27.0	12.3	1.6	7.1
1950	16.1	23.6	2.8	8.7
1960	7.2	38.6	5.4	10.8
1970	2.2	58.4	10.0	12.2
1980	1.0	77.0	15.2	12.7
1993	0.5	87.0	18.7	12.9

a. Persons of Hispanic origins are included as appropriate, in the "white" or in the "black and other races" category.
Source: National Center for Education Statistics, *Digest of Education Statistics, 1995* (1995).

1990 show that, among those twenty-five years old and over, 65.5 percent of Native Americans had a high school diploma or more; this compares to 77.9 percent of whites, 63.1 percent of blacks, and 49.8 percent of Hispanics who had completed high school. Completion of college or higher was reported by 9.3 percent of Native Americans, compared to 21.5 percent of whites 11.4 percent of blacks, and 9.2 percent of Hispanics (National Center for Education Statistics 1995).

The picture for Asian Americans is, in general, very different. In 1990, while other racial minorities had attained less schooling than whites, 77.5 percent of Asian/Pacific Islanders twenty-five years old and over had completed high school, about the same percentage found among whites. Moreover, 36.6 percent of Asians had a college degree or more, a much higher percentage of college graduates than was found among whites (21.5 percent). In 1992, 5 percent of students enrolled in American institutions of higher education were Asian Americans, about double their proportion in the general population (National Center for Educational Statis-

TABLE 7.2

Level of Education Attained, For Persons Twenty-Five to Twenty-Nine,
By Race/Ethnicity, 1993 (Percentages)

Race/Ethnicity	Did Not Complete High School	High School Graduate	College Graduate or More
White, Non-Hispanic	17.5	82.5	27.2
Black, Non-Hispanic	17.4	82.6	13.3
Hispanic	39.1	60.9	8.3

Source: National Center for Education Statistics, *Digest of Education Statistics, 1995* (1995).

tics 1994). Asian Americans are especially highly represented in the natural sciences, computer-information sciences, and engineering (Miller 1995: 39).

There are differences in educational level among different Asian groups. Those of Asian Indian, Japanese, and Chinese descent have high amounts of education on average, while the relatively small groups of Laotians, Hmong, and Pacific Islanders generally have less education than most other Americans. Overall, however, Asian Americans have attained more education than any other broad category of Americans.

Quality of Education

While the gaps in amount of education between white "Anglo" Americans and most minorities, especially blacks, have narrowed considerably in recent decades, differences in the quality of education have remained great. Table 7.3 shows the percentage of non-Hispanic whites, non-Hispanic blacks, and Hispanics who scored at or above a satisfactory level of reading proficiency for students of their age. Among both nine year olds and thirteen year olds, the proportion of both black students and Hispanic students who read at the specified proficiency level increased slightly from 1975 to 1992 but was still far below the proportion of white students at that proficiency level. Among seventeen year olds, reading scores improved from 1975 to 1992, especially for black students, but the percentages in these groups who read at the satisfactory level of proficiency still were considerably below the comparable proportion of white students. Black students and Hispanic students also score, on the average, considerably below white students on tests of other skills and knowledge, including mathematics, physical sciences, and history (National Center for Education Statistics 1994).

The ethnic differences in test scores described so far are based on tests given to all students attending school. In addition, we may compare the test scores of the subset of students who consider going to college and therefore take the Scholastic Aptitude Test (SAT) required for admission by most colleges.

180

TABLE 7.3

Percent of Students at or Above Selected Reading Proficiency Levels, by
Race/Ethnicity and Age, at Three Different School Years

A. Nine Year Olds (Reading Level I[a])

	1975	1984	1992
Whites (non-Hispanic)	69.0	68.6	69.3
Blacks (non-Hispanic)	31.6	36.6	36.6
Hispanics	34.6	39.6	43.1

B. Thirteen Year Olds (Reading Level II[b])

	1975	1984	1992
Whites (non-Hispanic)	65.5	65.3	68.5
Blacks (non-Hispanic)	24.8	34.6	38.4
Hispanics	32.0	39.0	40.9

C. Seventeen Year Olds (Reading Level II)

	1975	1984	1992
Whites (non-Hispanic)	86.2	88.0	88.0
Blacks (non-Hispanic)	43.0	65.7	61.4
Hispanics	52.9	68.3	69.2

a. Able to understand, combine ideas, and make inferences based on short uncomplicated passages about specific or sequentially related information.
b. Able to search for specific information, interrelate ideas, and make generalizations about literature, science, and social studies materials.
Source: National Center for Education Statistics, *Digest of Education Statistics, 1995* (1995).

Table 7.4 shows the average scores of students from various racial and ethnic backgrounds on two sections of the SAT at three different school years (1975 to 1976, 1983 to 1984, and 1992 to 1993). White ("Anglo") students scored considerably higher than others on the verbal section at all three school years, followed by Asian Americans, Native Americans, Mexican Americans, Puerto Ricans, and African Americans, in that consistent order. Between 1975 to 1976 and 1992 to 1993, the gap in verbal scores between whites and blacks decreased somewhat (though still remaining large) and the difference between whites and Native Americans also became somewhat smaller.

Students of the different racial/ethnic groups also ranked consistently at the three time points with respect to their scores on the mathematics section of the SAT. However, average math scores were highest for Asian American students, followed by whites, Native Americans, Mexican Americans, Puerto Ricans and

TABLE 7.4

Scholastic Aptitude Test Averages, by Race/Ethnicity, at Three Time Periods

A. Verbal Scores (Averages)

	1975–1976	1983–1984	1992–1993
White	451	445	444
Black	332	342	353
Mexican American	371	376	374
Puerto Rican	364	358	367
Asian American	414	398	415
Native American	388	390	400

B. Mathematic Scores (Averages)

	1975–1976	1983–1984	1992–1993
White	493	487	494
Black	354	373	388
Mexican American	410	420	428
Puerto Rican	401	405	409
Asian American	518	519	535
Native American	420	427	447

Source: College Entrance Examination Board, 1994. Reprinted in National Center of Education Statistics, *Digest of Education Statistics, 1995* (1995).

blacks. Average math scores rose from 1975 to 1976 and 1992 to 1993 among all groups except whites. The gap between Asian Americans and whites, at the top, and the other groups, especially blacks, remained large. Also, Asian Americans had pulled further ahead of whites.

Overall, then, among students applying for admission to college, whites and Asian Americans clearly showed stronger basic skills than Latino Americans, Native Americans, or blacks. While the average scores of blacks improved both on verbal and on math tests, their scores remained below those of other groups, especially those of whites and Asian Americans.

How can the average differences in educational attainment among various ethnic groups be explained? Two general kinds of factors seem important: those outside the school system (especially social class) and the school system itself.

Factors Outside the School

Social Class

How well students do in school is related strongly to the education and the income of their families (Miller 1995). High academic achievers are much more likely than others to come from families with a high income and to have parents with a substantial amount of formal education. Students whose achievement in school is low are much more likely to come from low-income families and to have parents with relatively little education. The longer that children live in poverty, the more likely they are to do poorly in school (Kennedy et al. 1986).

Growing up in a family at the lower end of the socioeconomic ladder may reduce the achievement and eventual educational attainment of children for a variety of reasons. Parents with little formal education themselves are less able than better-educated parents to give their children information, cognitive stimulation and help with school subjects. The families of low-income students are more likely than other families to move often, thus disrupting the continuity of their children's education (Miller et al. 1992). Children from low-income families are also at greater risk of exposure to preventable health conditions that may affect their school performance, including low birth-weight; prenatal exposure to tobacco, alcohol, or drugs; lead poisoning; malnutrition; and child abuse and neglect (Newman and Buka 1990). Even those children from low-income families who do well in school may be prevented from going to college by a lack of money.

Students whose families have relatively low education and incomes are more likely than others to come from ethnic groups whose school achievement scores are lowest. The parents of black, Hispanic, and Native American students generally have less formal education than the parents of white and Asian American students. Black, Hispanic, and Native American students also generally come from families whose income is considerably below those of most whites and Asian Americans. Moreover, they are more likely to attend schools that have a high proportion of students who come from very poor families. The combination of students being poor themselves and attending schools where a large proportion of classmates are poor appears to result in especially low achievement scores (Kennedy et al. 1986).

Reviewing data on the relation of student achievement to social class, L. Scott Miller concludes that this material "suggests that a substantial portion of the large educational achievement gaps between whites and Asians, on the one hand, and African Americans, Latinos, and American Indians, on the other hand, is related to differences in parent education and family income levels" (Miller 1995: 125).

Social Influences and Motivation

However, social class differences among ethnic groups do not account for all of the differences in school achievement among these groups. At each level of the parents' education, there remains an average difference in reading scores and mathematics scores among students from different ethnic groups.

National samples of student achievement scores from 1978 to 1990 show that, while the gap between whites and blacks and Hispanics had declined somewhat over that time span, differences still remained among students from these groups whose parents had similar education. For example, the average reading and math scores of white students whose parents had graduated from high school were higher than those of black and Hispanic students whose parents had the same amount of education.

The difference among students from different ethnic groups but similar social class background was most noticeable in the SAT scores of the high school students applying for college. At both extremes of parental education (either the parents have no high school degree or at least one parent has a graduate degree), there were sizable ethnic group differences in both reading and math scores. At either extreme of parental education, white and Asian American students scored much higher than black and Puerto Rican students and considerably above Mexican American and Native American students.

The same pattern was also found when SAT scores of students from different ethnic groups at each extreme of income (less than $10,000 a year and more than $70,000 a year) were compared. Both among students whose parents were the poorest and among those parents were the richest, the same differences in achievement among ethnic groups were found (Miller 1995).

There are some especially troubling aspects of the data. One is that ethnic gaps in achievement scores were often larger among students having at least one parent with a graduate degree than among those whose parents lacked a high school education. Another is how low the average verbal, math, and combined SAT scores were for blacks and Puerto Ricans who had at least one parent with a graduate degree. For example, among students whose parent(s) had a graduate degree, African Americans had a combined average SAT score of 830, compared to average scores for whites and Asian Americans of 1053 and 1018 respectively. The average score for blacks was only 25 and 28 points higher, respectively, than the average score of Asians and whites who had no parent with a high school education. These data, Miller states "makes it clear that as a nation Americans must be concerned with improving the educational prospects not only of those living in poverty and those whose parents have little education but also of those who are middle-class minority students" (Miller 1995: 159).

Why do achievement scores of students in some ethnic groups tend to be low even though they come from middle-class families? While the answer to this question is not completely clear, some clues come from research that shows that, among all ethnic groups, students perform best when their academic achievement is encouraged by both their parents and their friends. This research shows also that white and Asian American students are more likely than African American and Hispanic American students to have the combined support of both parents and peers for high achievement in school (Steinberg et al. 1992).

With respect to parental encouragement of their children's achievement in school, there is evidence that Asian American students especially benefit from a cultural tradition that places high value on education and from family norms that dictate educational "duties" for both parents and children (Ogbu and Matute-Bianchi 1986; Schneider and Lee 1990). However, parents from all ethnic groups appear to provide considerable support of their children's school activities. A national survey in 1988 found that, among all ethnic groups (Asian, Hispanic, black non-Hispanic, white non-Hispanic, and Native American) a large majority of parents reported talking regularly with their children about current school experiences, having rules about doing homework and maintaining a certain grade average, and helping their children with homework (National Center for Education Statistics 1994).

However, the peer groups of black students and of some other racial minorities sometimes give little encouragement to, or even oppose, high achievement in school. Some studies have found that the peer culture of African American students rejected behavior that was seen as "acting white." Behaviors defined as "white" included speaking standard English, studying hard, and getting good grades. Black students who worked hard in school sometimes were harassed by their peers (Fordham and Ogbu 1986; Fordham 1990). Such antiachievement attitudes have been described as part of a broader "opposition orientation" to a white-dominated society. While this oppositional orientation is strongest among the most disadvantaged blacks, it appears also to affect middle-class black children who sometimes are under peer pressure not to "act white" academically (Ogbu 1990; Steele 1992). Similar norms of opposition to school authorities and to academic achievement have been found to be fairly widespread among Mexican American youth (Gibson and Ogbu 1991).

Peer pressure against school achievement also has been described among Native Americans. Academically oriented Navaho and Ute students were discouraged by both school peers and adults from doing well in school; high achievement was considered a "white" attribute. Native American students who aspired to attend college also were seen by some as "acting white." Those who went away to college tended to feel they were rejecting their cultural identity (Deyhle 1992).

Aspirations

While some black students and other minority students are subjected to peer pressures against high academic achievement, the educational and occupational aspirations of minority students are generally high. In a national sample of high school seniors in 1992, the proportion of blacks and of non-Hispanic whites who said they planned to attend college right after high school was almost identical (about three quarters in each group). The proportion of Asians with such plans was a little higher (83 percent) and that of Hispanics a little lower (66 percent).

Asked about the occupations they expected to be in at the age of thirty, a slightly higher proportion of blacks than of whites said they expected to have a professional, business, or managerial position (55 percent to 50 percent); Asian Americans were slightly more likely (61 percent); and Hispanics and Native Americans slightly less likely (47 and 43 percent respectively) than other groups to expect to attain one of these relatively high-level occupations. Other differences in occupational expectations among the ethnic groups were small (National Center for Education Statistics, 1994).

These data indicate that the large differences in achievement scores between students of different ethnic groups, especially those between whites and blacks, are not paralleled by differences in the aspirations and expectations of students in these groups. While having high aspirations is desirable, such aspirations may be difficult to fulfill for students whose grades and achievement scores are low and who do not take a course of study that prepares them for higher education. However, expectations about getting the kind of jobs they want appears to be more highly related to school performance among white students than black students (Patchen 1975). Moreover, black and Hispanic students are more likely than white and Asian American students to be optimistic about getting the kind of jobs they want even if they do poorly in school (Steinberg et al. 1992). Apparently, black and Hispanic students often are not well informed about what needs to be done in order to attain the occupational positions to which they aspire. Miller comments: "Black students may also be unable to assess the level of academic performance it takes to prepare for good jobs or to avoid bad jobs, possibly in part because the students know few adults who have excellent education and the associated desirable jobs" (1995: 210).

Overall, then, the academic performance of minority students appears to be affected adversely by the lower education and income of many minority families, sometimes by a lack of support for achievement by peers, and sometimes by a lack of awareness of the connection between school achievement and occupational success.

Factors Within the School

To what extent are differences among ethnic groups in school performance related to differences in the schools they attended and their experiences within school?

Expenditures and Facilities

There are instances in which a lack of adequate facilities or too few teachers can hurt children's education. However, the evidence is strong that the overall variation in student achievement is related very little to variation in spending on schools or on school facilities. The 1964 Civil Rights Act mandated a national study to assess the quality of education received by children belonging to various racial and ethnic groups in the United States. The results of a massive study in over four thousand schools throughout the nation were contained in what became known as the "Coleman Report" (Coleman et al. 1966). This study showed that characteristics of the school (facilities such as laboratories and libraries, class size, curriculum programs available, quality and availability of textbooks, and so on) generally had little effect on the achievement scores of students. Differences among schools had somewhat stronger effects for blacks, Hispanics, and Native Americans than for non-Hispanic whites and Asian Americans. But the effects of school factors were small for all groups.

The achievement test scores of students were most strongly correlated with their home background, especially with the education and occupations of their parents. What was most important about the schools was not their facilities or even their teachers but the characteristics of the student body, such as average attendance and the proportion planning to attend college.

Other more recent studies also indicate the outcomes for students have little relation to school spending. School expenditures per pupil have been found to have little association with scores on achievement tests or with high school graduation and dropout rates (Kaufman et al. 1992; National Center for Education Statistics 1991). Also, in general, money spent per pupil has not been lower in areas and school districts with a high proportion of racial minorities. Other studies also have found that student achievement scores were not related much to teacher salaries or to student-teacher ratios (Miller 1995). Reviewing some of these results, L. Scott Miller comments: "These findings are important because ever since the Coleman Report of 1966 researchers have found little relationship between differences in academic outcomes of racial/ethnic groups and the variations in school resources available to them" (1995).

Tracking, Ability Grouping, and Curriculum

While differences in outcomes for students of different ethnic groups cannot be accounted for by the amount of money that schools spend, the type of education that students receive does make a difference. Specifically, achievement of students and ultimately the amount of education they receive is affected by ability grouping, tracking, and by the kinds of courses in which they are enrolled. Such variations in how students are educated are, in turn, related to the expectations of school administration and teachers.

In many schools, students are grouped by ability so that, for example, some students are assigned to a "slow" math class, others to a "regular" math class, and still others to an "advanced" math class. In addition, high school students are often placed in different curriculum tracks—for example an "academic" or college preparatory program, a "vocational" or "commercial" program, and a "general" program.

A major study of grouping and tracking practices and their outcomes examined more than one thousand classrooms in thirty-eight elementary and secondary schools throughout the United States (Goodlad 1984; Oakes 1985). This study found that poor and minority students were greatly overrepresented in groups and tracks for low-achievers and underrepresented in groups and tracks for high-achievers. Other research also shows that blacks, Hispanics, and those of lower social class are more likely to be placed in low-ability or low-track groups (Useem 1991; Oakes et al. 1992). Some studies indicate that students from lower socioeconomic backgrounds are more likely than others to be assigned to lower tracks even when their test scores are the same (Hallinan 1992).

Placement of students in particular ability groups or tracks affects the amount of academic demands that are placed on them (Goodlad 1984; Oakes 1985). For example, high school students in college-preparatory courses take more math and science courses. African Americans and Latinos, who are less likely to be in academic programs, are likely to take fewer and less advanced courses in these subjects (Policy Information Center 1989). Among high school students who took the Scholastic Aptitude Test in 1992, much larger proportions of white students and Asian American students than of African Americans, Mexican Americans, or Native Americans had taken twenty or more courses in academic subjects.

Having more and higher-level courses generally raises students' achievement scores. For each of the racial and ethnic groups who took the SAT in 1992, combined scores were much higher for those who had twenty or more academic courses than for all students in that ethnic group who took the test. Possibly those students who had taken more academic courses did well for other reasons in addition to their course preparation (e.g., greater ability). However, there is evidence for both black and white high school students, that being in higher ability groups and in an academic program raises overall achievement scores, even when student ability is held constant (Patchen 1982).

The importance of a school's curriculum for student achievement is shown also by research that compares public high schools and Catholic high schools. Students in Catholic schools were much more likely than those in public schools to be enrolled in an academic program of courses and their overall achievement scores were higher than those in public schools. Black students and Hispanic students in Catholic schools were especially likely to score higher in both verbal and math tests than those of similar race or ethnicity in the public schools (Coleman et al. 1982).

In addition to affecting students' achievement, placement in a particular track is likely to affect how far students go in school. Those who are placed in nonacademic tracks may lack the preparation for college-level work and they may discard aspirations for further education.

Placing students in separate ability groups or tracks may sometimes have educational advantages. For example, students with poor preparation in math may have trouble keeping up in a class where mastery of prior material is necessary. Moreover, for students who do not have the interest, motivation, or ability to go on to higher education, providing good vocational education, as some European countries do, may be beneficial to them.

The main problems appear to be that students may be assigned to an ability group or track inappropriately (perhaps influenced by their social class or ethnicity) and especially that placement of a student in a particular group may become too permanent and inflexible. Where schools employ tracking in a flexible manner, so that students' track assignments may change as their school performance changes, there appears to be a higher overall level of achievement and less inequality in achievement (Fuerst 1981; Gamoran 1992).

Teacher Expectations

Students may be placed in particular ability groups or tracks partly because of teacher expectations about their abilities and, once in a given group, their performance may be affected by teacher expectations. More generally, regardless of whether students are separated into ability groups or tracks, teacher expectations may play a role in learning.

Teachers generally perceive some students as having more academic ability and as likely to perform better than others (Good 1981; Dusek and Joseph 1983). In forming their expectations, teachers tend to rely on students' records (test scores and grades), on the reports of other teachers, and on their own classroom experiences with students (Brophy 1983). However, teacher expectations are sometimes influenced by the students' social class and race; they may have higher expectations of middle- or upper-class than of lower-class students and higher expectations of non-Hispanic whites than of blacks or Hispanics (Leacock 1969; Rist 1970; Harvey and Slatin 1975; Dusek and Joseph 1983).

The expectations that teachers have for students' academic performance affect their behavior towards their students. When they have low expectations of students, they tend, for example, to call on them less often to answer questions; give them the answers to questions more often, rather than helping them to find the right answers themselves; criticize them more often when they fail at a task; praise them less often when they succeed; place fewer academic demands on them; and pay less overall attention to them (Brophy 1983; Knapp and Shields 1990; Means and Knapp 1991). Some studies have found that teachers give black students less positive feedback and more negative feedback than white students, although this has not always been found to be true (Irvine 1990).

The expectations that teachers have of their students do appear to affect the students' academic performance (Miller 1995). Low teacher expectations tend to

189

lower students' performance, while high expectations tend to raise performance. These effects undoubtedly occur in part because of the links between teacher expectations and teacher behavior. Teacher expectations may also have an impact on students' performance because they affect students' own expectations. Students become aware of their teachers expectations for them and they tend to have expectations for themselves that are similar to those of their teachers (Weinstein et al. 1987).

While teacher expectations appear to have some effect on students' performance in school, generally, this effect is not large for individual students. Brophy (1983) concludes from his own research, and that of others, that a typical student's academic performance is lowered or raised 5 to 10 percent as a result of teacher expectations.

However, teacher expectations, and those of school administrators as well, do not affect individual students only. They also may affect the programs and the atmosphere of the entire school. Especially in schools that enroll mostly low-income minority students, the teaching staff and administrators may have basically given up on providing a good education. Their low expectations are reflected in the few demands placed on their students (e.g., see Kozol 1991).

However, some schools in low-income city neighborhoods and with primarily black student bodies have been successful in markedly raising the achievement of their students. They have done this by having high expectations and rigorous academic standards for students and by encouraging and supporting students to meet such standards (Weber 1971; Sowell 1974). The higher achievement of lower class and minority students in Catholic schools, as compared to public schools, also appears to be a result not only of the expectation that all students can do well, but of requiring a demanding curriculum for all (Coleman et al. 1982).

Racial Mixing

The large-scale national study in the 1960s on "Equality of Educational Opportunity" (Coleman et al. 1966) found that achievement scores of black students generally rose as the proportion of whites in their classes increased. Those black students who entered desegregated schools in early grades had slightly higher average grades than those who entered desegregated schools in later grades. However, the racial composition of classes explained little (less than 2 percent) of variations in test performance among black students.

While the results of the "Coleman Report" were provocative, and were used to support plans for greater racial integration of schools, to draw firm conclusions about the effects of racial proportions (or of any other school factor) on the basis of a cross-sectional survey is difficult. Among a number of methodological problems, students in segregated and desegregated classes might have differed in scholastic aptitude, or in some other way, when they entered school.

To provide more evidence about the impact of racial mixing on student achievement and on other outcomes, dozens of additional studies have been carried out. Many of these studies are experiments, sometimes comparing those in desegregated settings to similar students in racially segregated settings.

This body of evidence indicates first that desegregation tends to have negative effects on the aspirations and on the self-esteem of minority students. After reviewing a large number of studies, Nancy St. John concluded that "the effect of school desegregation on the general or academic self-concept of minority group members tends to be negative or mixed more often than positive" (1975: 54). She also found that as the percentage of whites in a school increases, the educational and occupational aspirations of black students tend to go down. Stephan (1978) also concluded that blacks in desegregated schools have lower self-esteem than those in segregated schools. Other reviewers find evidence on this matter inconsistent (Zirkel 1971; Weinberg 1977). The tendency of black self-esteem and aspirations to become lower in predominantly white settings probably is due to the fact that their grades tend to be lower than those of white schoolmates.

With respect to the effect of desegregation on achievement, some reviewers see more consistent benefits than do others. All agree that the impact of desegregation varies; sometimes attending racially mixed classes is associated with higher achievement for minority students and sometimes it is not (St. John 1975; Bradley and Bradley 1977; Weinberg 1977). However, desegregation appears to have positive effects on achievement more often than not. Crain and Mahard (1983) reviewed the relationship between desegregation and achievement test scores in 323 samples of students from ninety-three studies in sixty-seven cities. They found that gains for black students on standardized tests or equivalent measures outnumbered losses 173 to 98. Moreover, positive gains in achievement were more common and were larger in studies that had the most rigorous research designs. Attending racially mixed classes appears generally to have the most positive effects when it occurs in early grades (Coleman et al. 1966; Crain and Mahard 1982) and in settings with a high proportion of whites (Wortman and Bryant 1985).

Wortman and Bryant reviewed the best (most methodologically sound) studies of desegregation and found that, on average, black students in desegregated settings appeared to be about two months ahead in achievement, compared to those in segregated settings. However, this estimated achievement gain from desegregation still left black students, on average, considerably behind whites. Also, the gain from desegregated classes appears to be small compared to other educational interventions, such as improved teaching methods (Bloom 1976).

There has been relatively little research on the effects of desegregation on Hispanic students. The information that is available shows a pattern similar to that for black students. The Coleman Report found that Hispanics had higher achievement scores in schools with more white students (Coleman et al. 1966).

A later national study of high school students found that attending mostly white schools was correlated with higher achievement for Mexican Americans, Puerto Ricans, and Cubans (Mahard and Crain 1980).

More rigorous studies of the effects of desegregation in specific areas have shown more mixed results. A study in Riverside, California, found that attending school with "Anglo" classmates did not improve the achievement of Hispanic children in grade school (Gerard and Miller 1975). Another study of Hispanic children who were desegregated in the third grade found that they initially had lower test scores than Hispanics in segregated schools; however, by the eighth grade they were about one year ahead of the segregated group (Morrison 1972). The school situation for Hispanic children is complicated by the fact that many of them do not speak fluent English. (We will discuss bilingual education in the next section of this chapter.)

Why does attending school with white classmates appear sometimes to have a positive effect, even if a modest one, on the achievement of minority students? The Coleman Report emphasized that the achievement of students was affected substantially by the home backgrounds of their schoolmates. Students performed better when their fellow students came from homes in which parents were well educated and provided stimulating materials such as books and newspapers. The Coleman group and others have suggested that minority students might do better when they attend school with middle-class whites because the whites, being from more advantaged backgrounds, have positive attitudes toward learning and good study habits and that they influence black classmates in these directions. Such helpful influences from white peers might be expected to occur especially when minority students are socially accepted by the whites (Katz 1968; Pettigrew 1969; Spady 1976).

There is some evidence that attending school with white classmates has more positive effects on black students' achievement when the whites are from middle-class rather than working-class backgrounds (McPartland and York 1967; Lewis and St. John 1974). However, some research has failed to find evidence that improved school performance by minority students is due to the influence of white schoolmates (Maruyama and Miller 1979; Patchen 1982). For example, a study in Indianapolis (Patchen 1982) found that the achievement of black students was not related to the characteristics of white peers (their parents' education or the peers' own academic values), to friendship with white peers, or to the combination of these factors. The greater effort shown by black students in mostly white classes may reflect the presence of higher academic standards and expectations of teachers in the predominately white classes, rather than the racial mix of the students.

Whatever its possible advantages, greater racial mixing in the schools has been and continues to be hard to arrange. In fact, despite efforts of the courts, governments, and school boards, segregation of blacks in schools has not decreased since the 1970s and segregation of Hispanics has increased since that time (J. E. Farley 1995). Large concentrations of blacks and Hispanics in central cities, with whites being in the numerical minority, often means that most or all schools in the central city will be predominately black and Hispanic. There have been efforts in some places to institute school integration within an entire metropolitan area. However, legal obstacles and public opposition to busing students in order to accomplish racial integration have blocked this approach in most areas. Voluntary plans that encourage parents to send their children to schools having a student body that is predominately of another race have been used in some areas, but such approaches generally have done little to decrease segregation (1995).

However, even where changing the racial or ethnic composition of schools is not possible, improving the achievement of minority students may be. The quality of the school program (especially the curriculum) and the standards and expectations of administrators and teachers appear to have a greater impact on achievement than does the racial or ethnic composition of the school. Moreover, as we have noted, the racial composition of the student body may be important primarily because teacher standards and expectations tend to be higher in a predominately white school. Changing the racial mix of the student body may be one (indirect) way to change the academic climate of a school. But it also is possible to improve the academic climate in schools that have predominately minority student bodies—for example, with more demanding curriculums, improved discipline, and greater support for students (Brookover 1979). There are many examples of predominately minority schools in which a high level of learning takes place.

Language Problems and Bilingual Education

Many children grow up in homes and neighborhoods where the language that is usually spoken is different from the one used in their local schools. This situation is true in many countries around the world, since the population of most nations include people from several different ethnic origins. For example, many children who speak mainly Arabic at home attend schools in France where the language of instruction is French. In Canada, many children from homes in which French is spoken go to schools where teachers speak to them in English.

In the United States, estimates were made that in the late 1980s over six million school-age children were from language minorities, with a majority of these thought to have limited proficiency in English. Those with limited English proficiency were about 8 percent of the total school enrollment in the United States, but the proportion was much larger in some areas; for example, nearly half of the students in Los Angeles County were not fully proficient in English (Council of Chief State of School Officers 1990).

The largest language minority in the United States are those who speak Spanish. But there are children from many other groups for whom understanding the language spoken in school can be a problem. In addition to those children whose families speak languages other than English, understanding standard English sometimes is difficult for African American children who, in their homes and neighborhoods, speak a dialect known as "Black English."

Obviously, if a child cannot understand well what her teachers are saying, she is not likely to learn well. Children with language difficulties may be seen incorrectly by teachers and classmates as stupid. As a result of experiencing difficulties in school, they may lose motivation to learn.

In an attempt to deal with the problems faced by students with limited proficiency in the language of the school, educators in many countries have implemented programs of "bilingual education." Such programs use two languages in their teaching, one of which usually is the students' first (home) language. Specific bilingual programs vary in a number of ways, including the amount of attention given to each language and the rapidity with which the "home" language is phased out for use in teaching as students progress in school (Siguan and Mackey 1987).

Programs of bilingual education have become a matter of government policy and political debate in the United States. The Bilingual Education Act was signed by President Johnson as part of a broader education act in 1968. The original legislation provided for training teachers and aides to work with students who had limited English skills. By 1974, the act was expanded to require inclusion of a child's home language and culture "to the extent necessary to allow a child to progress efficiently through the educational system."

However, opposition to bilingual programs has arisen from those who charge that such programs sometimes discourage students from learning English and retard assimilation into American life. Among the critics of bilingual education in the Reagan administration were Secretary of Education William Bennett and Reagan himself. Defenders of bilingual education have responded that these programs are essential to enable students to make the transition from their home language to English. Some also argue that students' use of their "native" language, such as Spanish, helps them to preserve their cultural heritage (Meyer and Fienberg 1992).

Research on the effectiveness of bilingual programs has produced somewhat inconsistent results and has been interpreted in different ways. The "immersion" method, in which students hear and use only a language that is not their "native" language, has worked well in promoting learning of French by originally English speaking students in Quebec, Canada (Lambert and Tucker 1972). Studies in the United States have presented differing results. Some have found that students become more proficient in English and in other skills when taught in bilingual programs, rather than "immersed" in English only (e.g., see Valenzuela de la Garza and Medina 1985; Lewis 1991). Other studies indicate that bilingual education is neither better nor worse than English-only instruction (e.g., see Medina et al. 1985).

Baker and de Kanter (1983) reviewed reports from thirty-nine bilingual education programs, comparing them to alternative programs (including immersion) and concluded that there was not sufficient evidence to support mandating bilingual education programs. However, another reviewer of the evidence has suggested that bilingual programs have modest positive effects (Willig 1987). To try to provide definitive answers to these questions, two major government-funded studies of bilingual education were conducted in the 1980s. However, a distinguished panel of experts formed to review these studies concluded that the methods used were too flawed to warrant drawing firm conclusions. The one conclusion for which they found good evidence was that a bilingual program that used both Spanish and English in instruction but taught subject matter content only in English was more successful in teaching reading to students in early grades then either an English immersion (all English) program or a program with more extensive use of Spanish (both Spanish and English used in instruction, including content areas) (Meyer and Fienberg 1992).

Overall, we can conclude now only that language problems can be serious impediments to a good education for many children. Bilingual education programs appear to be useful at times in dealing with language problems but are not always more successful than other methods. More needs to be learned about the specific types of bilingual programs that work best and about the conditions under which they are most likely to be successful. Even when effective bilingual programs can be designed, however, their widespread use may be limited by a severe shortage of teachers who speak two languages, especially both English and Spanish.

Effects of Education on Attitudes

We have focused on the amount and quality of education that members of various groups obtain, primarily because education affects position in society and relations among people. But education may also have more direct effects on people's attitudes and behavior toward members of other groups.

Whites with more education, especially those with some college education, generally have more positive attitudes toward members of minority groups—for example, blacks and Jews (Martire and Clark 1982; Schuman et al. 1985). Some have explained this association by arguing that more education provides people with more information (about minorities and about reasons for differences between groups) and with a more broad-minded, sophisticated view of the world (e.g., see Quinley and Glock 1979). Others have questioned whether more education in itself really reduces prejudice, arguing that college educated people have been influenced instead by such other factors as more liberal social norms on college campuses and a low level of competition with racial minorities (Campbell 1971; A. W. Smith 1981). Another suggestion is that well-educated people simply are more sophisticated in expressing and defending their own group interests (Jackman and Muha 1984).

Interethnic Curricula

Aside from any general effects of education on intergroup attitudes, one may try to reduce prejudice by providing people with information specifically about other groups. Many schools, and other institutions as well (e.g., the military), have devised programs that focus on the history and culture of various racial and ethnic groups and highlight the achievements of members of these groups. Stephan and Stephan (1984) have summarized the results of thirty-nine studies that examined the effects of such curricula on prejudice. The majority of these studies (twenty-four) found a reduction of prejudice, some studies (fourteen) found no effects, and one study found an increase in prejudice (the authors of the last study claim that, because students expressed their racial attitudes anonymously, they were more honest than students in some other studies). One study of fifty-one high schools (Slavin and Madden 1979) found that the use of multiethnic curricula was not associated with the racial attitudes of either whites or blacks. However, a study of education students in a required intergroup relations course found that, after taking the course, students became more supportive of policies to increase opportunities for minorities (Davine and Bills 1992).

One aspect of multiethnic curricula is the use of multiethnic texts. An experimental study (Lichter and Johnson 1969) compared the racial attitudes of white second graders who used a multiethnic reader to those of similar white children who used a reader in which all the characters were white. Children in the group that used the multiethnic reader showed more positive racial attitudes than did children in the other group.

Multiethnic programs that have been most successful are those that are most extensive in content. Such courses used multiethnic readers, along with discussions, speakers, films, and field trips. Generally, longer courses have been more successful than shorter ones and the more actively students were involved in the programs, the more successful they have been (Stephan and Stephan 1984). While some research suggests that prejudice is most easily changed in young children (Rooney-Rebek and Jason 1986), age does not appear to be an important factor in the success of multiethnic education programs (Stephan and Stephan 1984).

Programs that teach people more about those in other cultures have also been used outside formal school situations. Triandis (1975) has developed a technique called the Cultural Assimilator, which aims at teaching people the viewpoints of those in other cultural groups. Assimilators have been developed to teach Americans about foreign cultures, such as those of Saudi Arabia, Thailand, and Greece, and also to teach whites about blacks. A study of whites who had taken such training with respect to blacks suggested that such training may reduce stereotyping and anxiety about interactions but found no reduction in prejudice (Randolph et al. 1977).

The U.S. military also has developed cross-cultural training programs. For example, during Army "alien presence" programs, junior officers and enlisted personnel received sixteen hours of briefing about ways in which American culture and the cultures of other countries differ. The lecture information sometimes was supplemented by contact with material aspects of the foreign culture (such as artwork) and by contact with foreign people. Studies of soldiers who went to Korea showed that those who had received the cross-cultural training had more favorable attitudes towards Koreans that did those without this training. In addition, Korean soldiers reported better relations with Americans who had received more of the cross-cultural training than with those who had less such training (Brislin and Pederson 1976).

Overall, the evidence indicates that programs that teach people more about other racial and ethnic groups tend to have positive effects on attitudes. Although the effects of such programs are not always positive, they are more likely to be positive when they are extensive in content, are not too brief, and involve participants in relevant activities. Regarding the content of such cross-cultural programs, Stephan and Stephan comment: ". . . our review of cross-cultural studies suggest that knowledge and understanding of differences as well as similarities among groups is important in reducing prejudice. . . . To promote intergroup understanding, knowledge regarding group differences must be presented in a way that shows respect for the customs, traditions, and culture of all groups" (1984: 238).

Mass Media

The information, images, and feelings that people have concerning other groups are influenced also by what they read and see in the media—television, movies, radio, newspapers, magazines, and so on. Historically, members of some minority groups in the United States were often portrayed by the media in a stereotyped, frequently negative, way. Mexican Americans were often portrayed in fiction and movies as bandits and prostitutes (Pettit 1980) and in national advertising campaigns as lazy, dirty, and irresponsible (Martinez 1972). Asian males often were depicted as sneaky and cruel or as docile and obsequious while Asian women were portrayed as docile or seductive (Liyama and Kitano 1982). Until the early 1950s, African American characters in books or on radio or television typically were stupid, lazy, and cowardly (Lemmons 1977; MacDonald 1993). Blacks appeared on radio and television primarily as maids, cooks, "mammies," and other servants or as "con artists and deadbeats" (Rhodes 1995; Gray 1995a). Herman Gray comments: "Black characters who populated the television world of the early 1950s were happy-go-lucky social incompetents who knew their place and whose antics served to amuse and comfort culturally sanctioned notions of whiteness, especially white superiority and paternalism" (1995a: 75). A study of children's books found that, from the late 1930s through the 1950s, one or more black characters appeared in about 15 to 20 percent of the children's books; usually the blacks were secondary and subordinate characters who were depicted in a stereotyped way (Pescosolido et al. 1997).

The major changes in the American view of race relations that resulted from the civil rights movement and from widespread riots in American cities during the 1960s were accompanied by changes in media portrayals of minorities, especially of blacks. Stereotypes of blacks as stupid, lazy, or servile largely disappeared from the media. However, during the 1960s there was little attention given to African Americans in television. A few black characters who were culturally indistinguishable from whites were included in drama series (Gray 1995b). During the late 1950s and early 1960s, blacks almost disappeared from children's picture books. Pescosolido and her colleagues suggest that book publishers became uncertain about how to present African Americans. "In the face of controversy and with new rhymes and jingles to replace the stereotyped ones, black characters were eliminated and blond haired farm children . . . were added" (1997: 34).

However, beginning in the 1970s, African Americans began to appear in television programming with increasing frequency. Since then, blacks have appeared prominently in a series of comedy programs (with "The Cosby Show" being the most successful); as hosts of entertainment programs (e.g., Oprah Winfrey and Arsenio Hall); with increasing frequency as characters in daytime dramas (soap operas) and in nighttime dramas; as sports commentators; and as anchors of news programs. Writing about the increased visibility of blacks in television programs, Herman Gray remarks: ". . . the variety and sheer number of stories about blacks proliferated in the 1980s to a degree perhaps unparalleled in the history of television" (1995b: 73). Gray attributes the substantial representation of black characters on television partly to the fact that blacks watch television more than whites do and the consequent desire of the networks to attract the urban black audience with black performers who also have appeal to whites. The increase in the visibility of blacks on television in recent decades is paralleled by increasing representation of blacks in children's picture books. From 1970 to the early 1990s, about 25 to 30 percent of such books included one or more black characters, more than in earlier decades (Pescosolido et al. 1997).

While representation of African Americans in the media has increased substantially, this appears to be less true for other minorities. For example,

Hispanics and Asians rarely appear in daytime television dramas (Lindsey 1995).

With the increased visibility of African Americans in the media, how have they been portrayed? Television programs in recent years often have portrayed blacks in positions of equal or even higher status than most whites—for example, as lawyers and judges. Writing about the portrayal of blacks in daytime soap operas, Karen Lindsey writes: "Often, to show that they do not indulge in racial stereotypes, soaps have used black judges to preside over one of the frequent murder or custody trials. Indeed, if you got your knowledge of American culture from soaps alone, you would be fairly convinced that the large majority of judges were black women" (Lindsey 1995: 333).

Although the portrayals of blacks generally were positive and unstereotyped in recent children's books, they tended to be distant and "safe," as when African Americans' distinct cultural heritage was recognized. Depictions of intimate contacts between blacks and whites and of contemporary African American adults were rare (Pescosolido et al. 1997).

Black characters on television tend to be middle-class people who are thoroughly assimilated into American life. Even when blacks are working class they tend to be good humored and share white middle-class values of family and success. Occasional stories of the struggles of blacks against prejudice and oppression, such as "Roots" and "The Autobiography of Miss Jane Pittman," have emphasized the movement from slavery and oppression to achievement and realization of the American dream.

Some commentators have seen a "disconnect" between the picture of American life in most entertainment programs and the realities of segregation, discrimination, and serious social problems that characterize much of black America. Herman Gray contrasts the view of black America presented in most television entertainment programs with the picture that emerges from news programs. He states that television representations of blacks in entertainment programs tend to be:

> celebrations of black middle-class visibility and achievement. In this context, successful and highly visible stars like Bill Cosby and Michael Jackson confirm the openness and pluralism of American society . . . [in contrast] the black underclass appears as a menace and a source of social disorganization in news accounts of black urban crime, gang violence, drug use, teenage pregnancy, riots, homelessness, and general aimlessness. In news accounts (and in Hollywood films such as *Colors*) poor blacks (and Hispanics) signify a social menace that must be contained. (1995b)

Others also have noted that the minority group members, especially African Americans, who receive coverage on television news tend to be those who are criminals (Johnson 1987).

The discrepancy between the generally positive images of race relations presented in media entertainment and often less pleasant realities is illustrated by the contrast between the final episode of "The Cosby Show" in April 1992, and the major riots that were occurring in Los Angeles on that same day. Gray comments: "Next to the rage that produced pictures of Los Angeles burning, the representation and expressions of African American life and experience on "The Cosby Show" (and so much of contemporary television) seemed little more than soothing symbolic props required to affirm America's latest illusion of feel-good multiculturalism and racial cooperation" (1995b: 82). Such comments by Gray and by other critics of contemporary television suggest the importance of not ignoring real societal tensions and conflicts deriving from ethnic inequalities. At the same time, media programs that show shared values, common humanity, and possible cooperation among different ethnic groups may help to reduce prejudices and discrimination that contribute to real-world conflicts.

Another way in which the media can affect relations among racial and ethnic groups is by helping to shape a common discourse and culture. People who read the same books and newspapers, listen to the same radio programs, and watch the same television programs will tend to see the world in similar ways. However, there is an increasing tendency in the media to aim their messages not at the population as a whole but rather at specific audience segments, based on such characteristics as age, gender, race, and ethnicity (Wilson and Gutierrez 1995). For example, there has been a growth in newspapers serving specific minorities (African Americans, Hispanic Americans, Chinese Americans, and so on), special sections of general newspapers targeted at specific minorities, and of television stations which aim at Spanish speakers, black Americans, particular religious groups, and whites of particular ethnic origins. A marketing textbook notes that "ethnic and racial factors have been effectively used as a basis of segmenting markets" (Cunningham and Cunningham 1981).

Particular racial or ethnic groups may benefit from the greater attention to their specific activities and interests that a communications medium targeted on their own group may give. But, as Wilson and Gutierrez (1995) assert, media segmentation of the population may risk "losing the glue" that binds groups together in a unified society. Writing about the media's increasing tendency to communicate separately to different segments or classes of society, Wilson and Gutierrez state: "Class communication points to a society in which people may be integrated in terms of the products they consume, but do not share a common culture based on the content of the entertainment or news media they use. Class communication means that society will no longer be as strongly bonded together by the media" (1995: 359–360). Thus, the composition of the media has possible effects on the extent to which those in different racial and ethnic groups share a common national identity.

Summary

Although differences among ethnic groups have narrowed, African Americans and especially Hispanic Americans complete fewer years of education than do other Americans. In addition, the quality of education received by most minorities, as measured by standardized achievement tests, generally is poorer than that received by whites and by Asian Americans.

School achievement is related substantially to social class. Therefore, the generally lower achievement scores of minority students may be accounted for in part by the relatively low education and occupations of minority students' families. In addition, peer pressures on middle-class, as well as lower-class, minority students that dictate *not* doing well in school sometimes depresses minority students' performance.

The level of school expenditures and the type of school facilities generally have little effect on students' achievement in school. More important are the use of ability groups and program tracks to channel students into different types of courses and courses with different levels of difficulty. Students from minority groups are more likely than others to be placed in lower ability groups and nonacademic tracks, resulting in their taking fewer basic academic courses and covering less advanced materials. Teachers often have lower expectations of students from minority groups. Low expectations by teachers affect the teachers' behavior toward the students and the students' expectations of their own potential.

Racial mixing of school classes has only small and inconsistent effects on the achievement of minority students. However, the effect of racial integration tends to be positive, especially when it occurs in early grades. Students who come from a non-English speaking background may be helped by bilingual education programs, but research has not yet established the effectiveness of such programs.

The keys to raising school achievement of minority children appear to be high expectations and high standards, including the types of curriculum and courses that will provide the knowledge and skills needed for college and beyond. Such a learning environment sometimes is easier in a racially mixed school, although the type of school program, rather than the racial composition of the student body, appears to be crucial for minority students.

The education obtained by people belonging to different racial and ethnic groups has important indirect effects on their intergroup relations because it affects the positions of these groups in society. Education also may have more direct effects on intergroup relations by giving people more information about groups other than their own. Many programs in schools and elsewhere have provided people with information about the history, culture, and achievements of other racial, ethnic, or national groups. Such programs generally have been successful in reducing prejudice toward out-groups.

People's attitudes toward out-groups are influenced also by what they read, hear, and see in various media. Negative stereotypes of minorities were common in American media in the past; they are less common today. However, generally positive portrayals of African Americans presented in entertainment programs often differ widely from more negative images presented in news programs. There is a current trend for the media to target various racial and ethnic groups separately, with different messages sent to each group. While this media strategy may meet some special needs of minority groups, it may undermine a common culture.

8) A Comprehensive View of Relations Between Ethnic Groups

In previous chapters we have tried to understand the relations between ethnic groups at several "levels"—individual attitudes, interpersonal contacts, and societal institutions. Now we will try to "put it all together"—that is, to show how the explanations of ethnic group relations at the different levels fit together. Just how are the attitudes and behaviors of individuals and their contacts across ethnic group lines affected by societal arrangements and institutions?

In previous chapters, many observations and comments relevant to this general question have been made. For example, we observed that greater economic inequality may increase anger among those in a disadvantaged group towards those in more advantaged groups. To take another example, we noted that the more a criminal justice system discriminates against those in a particular ethnic group, the more alienated from the overall society members of that ethnic group are likely to be.

However, we have not yet tried to connect the situation at a societal level to individual contacts, attitudes, and behaviors in a systematic way. Also, we have not yet shown systematically how various societal factors combine to affect individual reactions toward those from different ethnic groups. In this chapter we attempt to do these things. The goal is to present a comprehensive view of ethnic group relations that shows how interactions between individuals are shaped by the institutions and processes of the larger society.

Theoretical Perspectives

A number of theoretical approaches have been used by various scholars to explain the observed relations between racial and ethnic groups. We now briefly review some of these approaches and then consider how the ideas to be presented in this chapter relate to the work of others.

Some theorists have focused on the ethnic attitudes and behaviors of individuals—especially on ethnic prejudice and discrimination. One group of social psychologists has developed and elaborated "social identity theory," which sees discrimination against those in other groups as stemming from individuals' wishes to enhance their self-esteem, which is tied to their group identity (Hogg and Abrams 1988). Other social psychologists have seen hostile or discriminatory behavior towards members of other ethnic groups as stemming from negative stereotypes and feelings regarding these groups. They have presented theories concerning the ways in which perceptions and feelings (attitudes) toward other groups are formed and concerning the connection between ethnic attitudes and ethnic behavior (Mackie and Hamilton 1993).

Some sociologists have also focused on individual behavior. Some of these sociologists have used decision theory or rational choice theory to account for individual acts of discrimination in terms of the costs and benefits expected by the individual. Such individual choices may be seen as aggregating to a societal pattern of discrimination (Becker 1971; Banton 1995).

While many theorists have focused on the attitudes and behaviors of individuals, others have tried to explain relations among ethnic groups by pointing to aspects of the social situation in which people live and interact—including cultures, social institutions, and demographic realities.

Some writers have emphasized the extent of cultural differences among ethnic groups—that is, differences in values, customs, languages, and religions—as crucial determinants of their relations. Greater similarity in cultures of different groups has been said to promote more positive relations (Gordon 1964; Shibutani and Kwan 1965; A. D. Smith 1987). This emphasis on cultural similarities and differences is supported by the work of social psychologists who have found that similarity of attitudes and values contributes to positive attitudes and behavior among ethnic groups (see chapter 4).

Another, and probably the most influential type of explanation views relations among ethnic groups as fundamentally ones, of conflict among groups differing in power. This "conflict" perspective focuses on the unequal distribution of wealth and power among ethnic groups, based on the greater control by one (or a few) ethnic groups of economic resources and institutions (land, factories, banks, and so on). The dominant ethnic group is viewed as controlling other groups in order to serve its own interests, using incentives and indoctrination when possible, coercion when necessary (Cox 1948; Marx 1964; Noel 1968). Closely related to this classical conflict perspective is the more contemporary "resource competition theory," which sees both economic and political competition in society to be frequently organized along ethnic lines, resulting in high levels of ethnic hostility and conflict (Belanger and Pinard 1991; Olzak 1992; Nagel 1995). Such competition among ethnic groups has been seen to be affected by demographic factors, such as the relative size of their populations (Blalock 1967; Blalock and Wilken 1979).

While many theorists have focused on economic and other conflicts of interest between ethnic groups, another sociological perspective has emphasized the interdependence among various groups composing a society and the necessity for various groups to cooperate in order for the society as a whole to function effectively (Parsons 1951; Eitzen and Zinn 1994). This view—sometimes labeled the "order" perspective—is reinforced by research at the small group level, which shows that interdependence between those belonging to different ethnic groups (shared goals and the necessity to work together to reach these goals) can exert a powerful positive influence on intergroup relations (see chapter 4).

The broad framework for explaining ethnic group relations to be presented in this chapter draws on all of the theoretical perspectives just outlined. It includes variables both at the individual level and at the societal level. When trying to explain the subjective reactions and behaviors of individuals, we especially use the ideas of appraisal-emotion theory and of decision theory (presented in chapter 3). When considering aspects of the objective social situation we include variables that are relevant to the various types of explanations just discussed— that is, to cultural differences, to conflicts of interest, and to interdependence between groups. We attempt to show what the key variables are that affect ethnic group relations and how these variables are related to each other.

Some theoretical approaches focus on particular types of relations among ethnic groups (such as conflict) and highlight particular social arrangements (e.g., inequalities of wealth) as crucial in determining these relations. But such explanations do not help much to explain the wide variations in ethnic group relations that occur (e.g., from intense conflict to strong cooperation). Our approach will be to focus on the range of variations in possible relations among groups and to treat possible determinants (such as inequality and conflict of interest) not as fixed factors but as *variables* that may differ (e.g., from little inequality to great inequality, from high conflict of interest to high correspondence of interest between groups).

This chapter will not present a rigorous theory, complete with a set of interrelated propositions that specify precisely the relations among all the variables. Such a product, while desirable, is beyond the scope of this book. The intended contribution here is more modest: to go beyond explanations that are focused only on particular types of intergroup relations, only on selected determinants of such relations, only on individuals, or only on larger institutions, to present a theoretical framework that is more comprehensive. We will try to bring together in one model a set of key variables that affect ethnic group relations at both the individual and societal levels, and to show the general nature of the relations among these variables. Such a broad theoretical framework may help to give a more comprehensive and adequate picture of how interethnic relations are shaped than do more narrowly focused explanations.

Amount and Nature of Interaction

Before presenting a broad model for explaining relations between ethnic groups, we need to specify more precisely the interaction that we are trying to explain. We will first consider briefly the amount of interaction that occurs and then discuss at greater length the nature or type of the interaction.

Amount of Interaction

The amount of interaction is a fundamental aspect of all social relations (Homans 1974). When applied to intergroup relations, the term refers to the volume of behavior that members of each group direct toward members of the other group that communicates information to the others or affects the others. Interaction may be verbal—as where black and white students in a school talk to each other about their class assignments. It may be nonverbal—as where a Latino person smiles (or frowns) at a white person or where a black member and a white member of a basketball team pass the ball back and forth down the court. Interaction usually is reciprocal, as in most conversations, but it sometimes is one-sided, as where one person greets another but the other does not respond.

Approach behaviors by members of a particular group—for example, a European American inviting an Asian American to a party or a white person going next door to introduce herself to new African American neighbor— increase the amount of interaction. Avoidance behaviors—for example, a Christian staying away from a party because he thinks too many Jews will be present— reduce the amount of interaction.

Type of Interaction

Interaction may be of various types. Two key aspects of all social behaviors are a) their friendliness versus unfriendliness and b) how dominant versus submissive the behaviors are (Longabaugh 1966).

Friendly versus Unfriendly Behavior

Friendly behavior is behavior that is intended to help or reward another person. For example: white women and black women living in the same housing project may baby-sit for each other and may buy groceries for each other when shopping; Anglo and Latino students may help each other with their home-work; white and Asian co-workers may cooperate on a work project and may compliment each other on their efforts. Friendly behavior is likely to be mutual, rather than one sided. If one person's friendly behavior is not reciprocated, he is not likely to continue it very long (Blau 1964).

Unfriendly behavior is behavior that is intended to injure another or impede the other from reaching her goals. A white student who uses racial epithets against black, Asian, or Latino schoolmates; a group of blacks who picket a Korean owned store to try to drive its customers away; a white homeowner who urges his neighbors not to sell to black people; a member of a "skinhead" group who desecrates a Jewish cemetery; these are clear examples of unfriendly actions by members of one ethnic group toward those belonging to another ethnic group. Unfriendly actions often are reciprocated—for example, if a Latino gang raids the territory of a black gang, the latter is apt to reciprocate. Thus, a pattern of continuing mutual, and sometimes escalating, unfriendly actions often occurs (Pruitt and Rubin 1986). If, however, there is a great disparity in the power of the two groups, the target of unfriendly action may be unable to respond effectively. For example, black people rarely were able to retaliate against attacks by the Ku Klux Klan during the late nineteenth century.

Dominant versus Submissive Behavior

"Dominant behavior" is behavior that is intended to control the actions of members of the other group and/or to make decisions that affect out-comes for the other group. Anglo farm owners who established unilateral wage rates, hours, work rules, and living arrangements for Chicano field workers; leg-islators in early twentieth-century California who passed laws that, in effect, pro-hibited Japanese from owning land; social workers (usually white) who tell poor black mothers how they must behave in order to get financial aid; black students in some high schools who force white classmates to walk around them in hall-ways—these are a few examples of dominant behavior.

Submissive behavior is behavior that intentionally follows the demands (orders, requests) of others or that accedes to others making decisions that affect one's own outcomes. The Asian house servant (usually of an earlier era) who did what he or she was told with a smile (though, perhaps, with inward resentment); the black person who moved to the back of the bus when told to do so in the seg-regation-era south; black auto workers who, in 1941, were willing to let Ford company management make unilateral decisions on their wages and working conditions rather than backing a union (Goodwin 1994)—these are examples of submissive behavior.

At any given time, members of a particular ethnic group may wish to change their relationship to some other ethnic groups(s) or their position in society relative to other groups. They may wish to change the distribution of rewards (such as income) and of social positions (e.g., representation in certain occupations or in political positions). Members of a given group may become involved in efforts, often in an organized way, to make such changes. These efforts usually involve behavior that is dominant—that is, attempts to pressure members of other ethnic groups to accept certain social changes. Members of other ethnic groups may either submit (in whole or in part) to the demands of the first group, may resist, or may use dominant behavior of their own (such as threats) to try to change the status quo in ways more favorable to themselves.

The interaction between members of different ethnic groups may usually involve dominant behavior by one group and submissive behavior by the other (e.g., the usual interaction between whites and blacks in South Africa when racial domination was legal there). However, just as in other interpersonal relations (such as those between friends of the same ethnic group), the usual interaction may be one of general equality—where neither group dominates or submits regularly but where, instead, alternatives such as compromise or turn-taking are usual.

Patterns of Interaction

Since interaction between those belonging to different ethnic groups may be generally friendly or generally unfriendly, and also may reflect either dominance by one group or relative equality of control, there are four possible overall types of interaction, as shown in figure 8.1. Next, we describe briefly each of these interaction types. Discussion of the circumstances that produce each type of interaction will be presented later in the chapter.

Paternalism

One group may be dominant and the other group submissive, but their interactions may generally be amicable (or at least not overtly hostile). This type of interaction may be labeled "paternalism." Paternalistic patterns have been described by some writers (Van den Berghe 1978) as characteristic of some agricultural societies—for example, the American south and many South Amer-

FIGURE 8.1

Types of Relationships Between Ethnic Groups

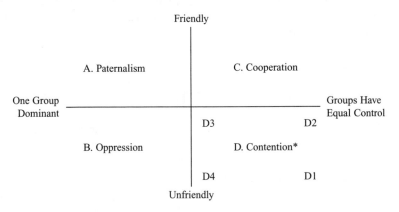

*Points D1 through D4 indicate more specific types of generally contentious relationships

ican countries during the period of slavery. In such societies, there was a rigid caste system based on race, and whites controlled the activities of their slave workers. The system was so firmly established (and backed by overwhelming force) that the dominant group members usually could afford to show a facade of benevolence toward the subordinate group, while the slaves did not dare to show any hostility. Thus, outward conflict was rare. Even after slavery was ended, the same paternalistic pattern of relations often continued between white landowners and black farm laborers or sharecroppers.

Other examples of generally paternalistic relations in more recent times are those between white government officials and Native Americans on reservations, and between Anglo farm owners and Latino farm workers. On many Indian reservations, white government officials, charged with "taking care of" the Indians, have had a dominant voice in decisions about community affairs. In most cases, there has been little visible conflict, in part because the Native Americans have seen little chance to change their situation of dependence and subordination. On some reservations, this situation is changing as Native Americans assert more control over their own affairs (Parman 1994).

Latino farm workers, especially those who come for seasonal work, usually have been under the domination and "care" of Anglo employers, with respect to wage rates, working conditions, and living arrangements. Usually there has not been overt conflict between these groups because the Latino workers usually have felt powerless to challenge the actions of the employers (Taylor 1934; Martin 1992).

Oppression

A second overall pattern of intergroup relations, which may be labeled "oppression," is characterized by the dominance of one group over the other, together with generally unfriendly behaviors by one or both sides. For example, such a pattern was often characteristic of interaction between whites and Native Americans during the nineteenth century. White settlers, backed by soldiers, were often able to control the behavior of Indians, such as by forcing them to move away from land that they had occupied or on which they had hunted. At the same time, the actions of each side toward the other frequently were unfriendly—whites killing and harassing Indians, Indians often retaliating in kind.

Other examples of oppressive interaction are those between Germans and Jews under the Nazi regime (Melson 1992); those between Hutus and Tutsis in Rwanda, where dominant Hutus murdered thousands of Tutsis in 1994 and many Tutsis retaliated (Destexhe 1995); and those in former Yugoslavia where Serbs ethnically "cleansed" Croats and Muslims from certain areas by killing, raping, and forcibly expelling them, while those other ethnic groups fought back as best they could (Cohen 1993).

Oppressive relationships may also be found at a level below that of the entire society, in such places as schools and neighborhoods. For example, if white kids in a neighborhood are dominant over Asian kids with respect to control of a playground space and also "beat up" the Asian kids frequently when they meet, the relationship would fit our definition of oppression.

Contention

A third type of interaction is characterized by the combination of relatively equal degrees of control by each group and generally unfriendly behavior. This pattern may be labeled "contention" or "conflict." Fitting this pattern of interaction to a large extent are what some writers have called competitive race and ethnic relations, especially fluid competitive relations (Wilson 1978). Such a pattern of competitive relations has been described for advanced industrial nations such as the contemporary United States and Great Britain. In these societies, there is at least formal equality among various groups, and no group can completely dominate another, although whites continue to enjoy some dominance informally. There is, however, considerable competition between ethnic groups, especially for economic rewards, and this competition results in a considerable amount of overt conflict (much of it through institutionalized or nonviolent channels, but sometimes erupting into violence).

The pattern of interaction that we have called contention is illustrated by some situations in the contemporary United States. A conflict between white and black firefighters in Memphis, Tennessee, over the distribution of jobs

(Firefighters 1987); the struggle between blacks and whites to win the mayor's office in Chicago (Holli and Green 1984); and the disputes between Asians and whites in California over the number of students in each group admitted to the top colleges in that state (Takagi 1992)—these are examples of such situations of intergroup contention.

Cooperation

A fourth pattern of intergroup relations is characterized by generally equal control and friendly behavior. This pattern, which many people would consider the most desirable, may be labeled "cooperation."

The state of Hawaii seems to fit this pattern fairly well (Grant and Ogawa 1993). The geographic origins of the people there include Europe, Japan, the Philippines, Hawaii, China, and many other parts of the world. Hawaii is not without some measure of racism and discrimination (Haas 1992); however, no one ethnic group is able to dominate the others, politically or economically. And while there are some tensions among groups, there is a great deal of friendly interaction, as indicated especially by high rates of intermarriage; almost half of all marriages in 1991 were interracial.

Other examples of a cooperative pattern of interaction may be found at a level below that of a total society. Many groups and teams in schools and elsewhere (such as those described in chapter 4) are ones in which the members have equal control over decisions and behaviors are predominantly friendly. Unions, such as the United Auto Workers or the International Ladies Garment Workers Union, that include workers from different ethnic groups, are other examples of intergroup relations that generally are ones of cooperation.

The four types of interaction patterns that we have discussed are "pure types." Interaction in the real world is not likely to fit any one of the types exactly. Rarely is one group completely dominant and rarely do both groups have exactly equal control. There are many possible intermediate situations, in which one group may be somewhat more or slightly more dominant than the other. Similarly, there are many possible degrees and mixtures of friendliness and unfriendliness in the behavior of the members of interacting groups.

Thinking of any particular pattern of interaction among groups as fitting anywhere in the two-dimensional space shown in figure 8.1 is more realistic. Thus, for example, within quadrant D of figure 8.1 we may distinguish points D–1, D–2, D–3 and D–4. D–1 indicates a pattern of interaction between two ethnic groups in which both sides have totally equal control and extremely unfriendly behavior—for example, a civil war between two equal power groups. Point D–2 indicates a pattern in which both sides have completely equal power and their relations are only slightly unfriendly—for example, a coalition government in which two ethnic groups have equal representation but disagree often about policy. D–3 indicates a situation in which both sides have some influence but one has more than the other and their relations are mildly unfriendly—for example, a food processing plant in which Anglo supervisors have more power than Latino workers (who have some power because of a union) and where the two groups often bicker over working conditions. D–4 indicates an interaction pattern in which both sides have some, but not completely equal control, and there is a very high level of unfriendliness—for example, a guerrilla war by a somewhat weaker ethnic group against a more dominant ethnic group.

We could distinguish in a similar way (although we have not done so in figure 8.1) between particular points within each of the other three quadrants (A, B, and C). Each point would indicate a particular degree of friendliness or unfriendliness and a particular degree of equality or inequality of control between ethnic groups.

Changes in Patterns of Interaction

The pattern of interaction among any ethnic groups may stay constant for a long time, or it may change, usually gradually but sometimes rather suddenly. For example, the relation between whites and Native Americans generally was one of oppression during the time that whites were seizing control of the great areas occupied by Indians. The relationship became one of paternalism after the Indians were forced onto reservations and became essentially wards of the federal government. Conversely, a relationship can move from paternalism to outright oppression, as where an outwardly peaceful slave system in the south was challenged by a slave rebellion, which was then put down brutally.

The speed with which one pattern of relations may be transformed to another is illustrated by events in Lebanon in the 1970s and Yugoslavia in the 1990s. In Lebanon up until 1975, Christians and Muslims lived together in general harmony under a political system that was carefully tailored to give each group equal power. When new developments (notably an influx of Palestinians) undermined the existing power balance, a long bloody civil war broke out between Muslims and Christians for control of the country (Friedman 1995). In terms of our typology, the groups moved from a general pattern of cooperation to one of strong contention.

A similar story unfolded in what formerly was Yugoslavia. Serbs and Croatians in Croatia, and Serbs and Muslims in Bosnia, had lived together in general harmony and with frequent intermarriage since World War II. But as Yugoslavia split up into its component republics, and ethnic groups vied for political control, their interactions became ones primarily of contention, often carried on with brutality and killing (Cohen 1993).

As we attempt to explain the patterns of interaction that occur among ethnic groups, we will try to explain the ways in which a relationship may move from one pattern to another over time.

General Determinants of Interaction

How can variations in the amount and types of interaction that occur among ethnic groups be explained? Figure 8.2 presents a comprehensive model of factors that affect relations between groups. We will begin this section with an overview of the model, describing in a general way how it may be used to explain the amount and the types of interaction among members of different groups. Then we will discuss how the model may help us to understand each of the four specific patterns of interaction (paternalism, oppression, contention, and cooperation) that we described earlier.

Subjective Reactions

The amount of contact among members of different ethnic groups depends in part on the opportunities for contact that exist. For example, the more that Anglos and Latinos live in the same neighborhood, attend the same schools, and work in close proximity in the same business firms, the more interpersonal contact (talking, and so on) they are likely to have. But, as we saw in chapter 4, people who are in close proximity do not necessarily interact, especially in a nonsuperficial way. They may or may not initiate conversation or joint activities such as a game or work project. They may choose to avoid interaction whenever possible. When they do interact, their actions may be friendly or hostile, dominant or submissive.

Figure 8.2 indicates that—if there is some opportunity for contact—the amount and nature of interaction between two ethnic groups is determined most immediately by a) the perceptions and feelings that members of each group have concerning the other; and b) the expectations of each regarding the probable outcomes of various types of behavior (see chapters 2 and 3 for a detailed discussion of how perceptions, feelings, and expectations affect behavior towards those in another ethnic group).

How much and what types of interaction actually occur depends first on the extent to which members of each ethnic group see those in the other group as outsiders, rather than as members of their own important groups. For example, if black and white students see each other mainly as part of the same team, the same club, or even just as fellow students in the same school, they are more likely to interact than if each group sees the other primarily in terms of belonging to a different racial category; when they do interact, it is more likely to be in a friendly than in a unfriendly way.

How much members of each ethnic group choose to interact with those in another group depends also on their perceptions (appraisals) of and feelings toward the others. If they see those in the other group as doing them harm (or

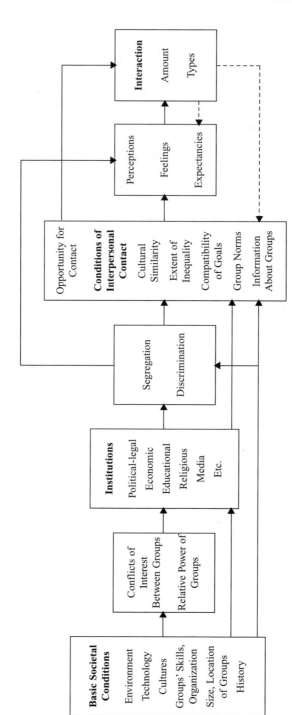

FIGURE 8.2

A Model of Determinants of Interaction Between Ethnic Groups

Basic Societal Conditions

Environment

Technology

Cultures

Groups' Skills, Organization

Size, Location of Groups

History

Institutions

Political-legal

Economic

Educational

Religious

Media

Etc.

Conflicts of Interest Between Groups

Relative Power of Groups

Segregation

Discrimination

Opportunity for Contact

Conditions of Interpersonal Contact

Cultural Similarity

Extent of Inequality

Compatibility of Goals

Group Norms

Information About Groups

Perceptions

Feelings

Expectancies

Interaction

Amount

Types

Key

——— Indicates primary causal connections between sets of variables, going from left to right

- - - - - Indicates secondary, feedback effects of Interaction on some other sets of variables

214

intending to do so), they will be fearful of, and probably angry at, the others, and may try to avoid them; if interaction does take place, it will tend to be antagonistic. For example, a black family that has just moved into a neighborhood in which other blacks have been assaulted is likely to avoid their white neighbors. Similarly, white people walking through a poor black neighborhood may see many of the blacks as potential muggers, and may therefore be fearful of and avoid talking to or even coming close to any blacks they encounter. On the other hand, if those in another group are seen as (potentially) helpful rather than harmful—for example, white students thinking Asian classmates will help them with math assignments—friendly interaction is much more likely.

Other types of appraisals that we have discussed earlier may also affect the amount and types of interaction that occur. The extent to which members of each group see the general behavior of those in the other group as praiseworthy or as reprehensible (i.e., as conforming or not conforming to their standards) may be important. Thus, if a group of Asian American students see their European American classmates as undisciplined, hedonistic, and arrogant and the European Americans see the Asian Americans as too competitive, insincere, and clannish, those in each group are likely to dislike, and even feel revulsion at, the others and thus minimize interaction. To the extent that each group sees the others as having more praiseworthy traits, interaction is likely to be more frequent and more friendly.

Also, when members of one ethnic group perceive that members of another group are receiving undeserved advantages, they are likely to be angry, and such feelings may be shown in protest and in unfriendly behavior.

Based, in part, on their perceptions of, and feelings toward, members of another ethnic group, people also have expectancies that may affect the amount and types of interaction that occurs. The more they believe that friendly approaches to and interaction with members of another group will bring positive outcomes, the more likely they are to seek or to encourage such interaction. For example, if a group of Latino teenagers believes that having a party with Anglo acquaintances would be fun, that the Anglos would respond positively to this idea, and that their own parents would approve, they might try to arrange such an activity. If, on the other hand, they thought it likely that such a party would be unpleasant (e.g., that the Anglos would be condescending), or that the Anglos would rebuff such an idea, or that their own parents would not like such "mingling," they would be unlikely to act on this idea.

The Situation

The interactions between ethnic groups, and the set of perceptions, feelings, and expectations that most directly influence such interactions, are shaped by the objective situation. Figure 8.2 shows that the reactions of individuals toward those from another ethnic group are influenced most directly by the opportunities for and the conditions of interpersonal contact.

Opportunities for Contact

The greater the opportunity for contact (usually a result of physical proximity), the more that people belonging to different groups are likely to interact.

Conditions of Interpersonal Contact

Attitudes and behaviors toward members of other ethnic groups are affected directly by the immediate circumstances in which they have personal contact with each other—such as in neighborhoods, schools, and work places. The conditions of interpersonal contact include the extent of cultural similarity, of compatibility of goals, and of inequality between groups, as well as group norms and information about other groups (see chapters 4 and 5 for a more extensive discussion of how the conditions of intergroup contact affect its outcomes).

- Cultural Similarity—the greater the cultural similarity of members of different ethnic groups (in language, values, customs, activities, and so on), the more likely they are to react towards each other in a friendly way.
- Compatibility of Goals—the more that people from different ethnic groups share common goals or interests (i.e., they will benefit or suffer from similar outcomes), and especially when they are dependent on each other for reaching positive outcomes, the more positive their reactions will be.
- Inequality Between the Groups—inequality with respect to occupation, wealth, prestige, and power has mixed effects. The less the inequality, the less the resentment of one group by another. But greater equality sometimes leads to more competition and contention between groups. Equality is most likely to lead to contention when there is high conflict of interest between groups.
- Group Norms—the more that norms of each group support friendly, egalitarian contacts, the more positive their reactions will be.
- Information—the more positive information people receive about each other, the more positive their reactions to members of the other group will be.

Segregation and Discrimination

The opportunities for contact among members of different ethnic groups and the conditions under which interpersonal contact occur are, as figure 8.2 shows, influenced greatly by two important processes: segregation and discrimination.

Greater segregation of ethnic groups—in their neighborhoods of residence, in their occupations, in schools, in churches, and elsewhere—has the obvious effect of limiting their opportunities for contact. Greater segregation will also increase inequality, since it often keeps low-status groups distant from good jobs and decreases the amount of information they get about jobs and business opportunities.

Discrimination has obvious and important effects on the degree of inequality between groups. The more that members of particular groups are treated differently with respect to their educations, getting jobs, getting promoted, or getting business loans or contracts, the more inequality there will be in economic status and power. Discrimination with respect to the right to vote, and with respect to whom voters will support, will increase differences in political power.

Greater segregation and more discrimination tend to prevent the cultures of different ethnic groups from becoming more similar; these processes may, in fact, result in greater cultural dissimilarity. The more apart groups are from each other (in neighborhoods, work places, churches, and so on), and the more unequal they are (such as in education and occupation) the more they are likely to retain whatever cultural differences they had initially, and the more likely they are to evolve new cultural differences that reflect and help them adapt to their different life conditions and experiences.

Greater discrimination and segregation also tend to increase conflicts of goals between groups. For example, because discrimination and segregation result in ethnic groups having unequal economic positions, these groups are affected differently by many government policies—with respect to taxes, public transportation, welfare, medical care, and so on. These conflicts of interest lead them to support different political parties and different public policies.

However, another type of goal conflict—economic competition—tends to be strongest when discrimination against particular groups is relatively weak, thus permitting members of each group to compete with some effectiveness (see chapter 5 for further discussion of this point). Competition between groups, whether economic or otherwise, is furthered also by some separation between groups. Such separation channels competition along group lines, by encouraging strong group identity and in-group organization.

In addition to their effects on the conditions of contact between groups, the important processes of segregation and discrimination may have some direct effects on people's perceptions and feelings. For example, more actual discrimination by employers, banks, school administrators, and police, in favor of members of one ethnic group over another, is likely to result in anger, often expressed in protests and in unfriendly behavior by those in the disadvantaged group.

Institutions

Figure 8.2 also shows that the central processes of segregation and discrimination are determined primarily by the institutions of society, includ-

217

ing political-legal, economic, educational, and religious institutions, and the media (see chapters 5 through 7 for more detailed discussions of these institutions). Among other institutions that may affect the extent of segregation and discrimination are those of sports, military, medicine, and philanthropy.

The less representation and influence an ethnic group has in political and legal systems, the less likely those systems are to pay attention to the needs and wishes of members of that group; at times, these institutions may also actively discriminate against members of low-influence groups and take actions (e.g., with respect to zoning or to location of schools) that tend to segregate these groups.

Economic institutions may also have major impacts on both segregation and discrimination. Employers may choose only people from a particular racial or ethnic group for certain types of jobs. Unions may refuse membership to persons from particular ethnic groups. Banks may refuse to make loans to members of certain minority groups. Business firms may not hire a subcontractor if that firm is headed by people from certain groups. The more economic institutions act in such (now illegal) ways, the greater the discrimination against and segregation of minority ethnic groups. The amount of such discrimination may vary with features of the economy, such as the supply of labor. An expanding economy that is short of skilled workers may reduce job discrimination against minorities.

Some features of educational institutions affect the extent of group segregation. For example, the places where schools are built, the ways in which school districts are drawn, and provisions for busing some students to school, all affect the ethnic compositions of grade schools and high schools. Within each school, more placement of students into separate ability groups and academic "tracks" tends to separate children from different ethnic groups. On the other hand, more multiethnic activities both in classes (such as project groups) and outside classes (such as intramural teams) tend to reduce segregation.

Religious institutions that are organized formally or informally along racial or ethnic lines (e.g., black church denominations, all white suburban churches) tend to promote or perpetuate ethnic separation.

In addition to their important effects on segregation and discrimination, social institutions may also (as figure 8.2 shows) directly affect the conditions of interpersonal contact between members of different groups. For example, the media, the schools, and religious organizations all affect the information that people receive about other ethnic groups and influence norms about appropriate behavior towards other groups. To take another example, various social institutions, such as the economic system, schools, churches, and the media, may directly affect the extent of cultural similarity between groups. The more these institutions provide similar experiences, inculcate similar beliefs, use the same language, and show the same entertainment to members of different ethnic groups, the more similar in culture these groups will be.

218

Origins of Institutional Arrangements

If institutions have great importance in determining the relations between ethnic groups, how can we explain the emergence of particular types of institutions, and whether they promote ethnic segregation and discrimination? Figure 8.2 indicates that institutions are shaped by the conflicts of interest and the relative power of groups when they come into contact. The degree of conflicting interests between groups as well as their relative power depends on certain basic societal conditions. These conditions include the environment (e.g., the scarcity of land, water, and minerals); technology (e.g., available methods of farming and of manufacturing); the size and location of each group's population; the culture, skills, and organization of each group; and historical events (e.g., wars and conquests).

Groups may have competing interests initially over possession of land, resources, occupations, markets, or even over what kind of religious practices will be favored by the society. The greater power of one group may result from greater numbers, more advanced technical skills (e.g., in military activities), better organization, or some combination of these advantages. These differences in the bases of power often develop from the history and culture of each group. They result in one group being more powerful than the other with respect to its possession of economic sanctions and/or its ability to employ force.

When separate groups, each of which has a sense of ethnic identity, and which also have conflicts of interest, come into contact, the more powerful group is likely to impose its will on the other. This imposition may occur by force or through economic domination. The more powerful ethnic group may then establish institutions that will continue its dominance, by favoring members of its own group and by segregating the groups when it deems necessary.

While a dominant and subordinate group may continue to have different interests, the fact that they are part of the same society makes them interdependent and gives both groups some stake in the effective functioning of the total society. This degree of correspondence of interests is likely to be reflected in some aspects of institutions—for example, in economic arrangements that provide a niche (though not necessarily an equally desirable one) for different ethnic groups.

The basic conditions present in a given society at a given time may have effects on intergroup relations that are separate from those stemming from their effects on conflicts of interest and relative power. First, societal conditions may have direct effects on institutions. For example, the plantation system of the old American south was the result of a particular environment, including land suitable for growing crops such as cotton, a technology that required large numbers of manual laborers, and a world market for harvested crops.

Basic societal conditions may also directly affect the extent of segregation between ethnic groups. For example, the larger and more concentrated a group is initially, the more it is likely to be separated from other groups, irrespective of how much social institutions try to promote segregation. Also, the larger an ethnic group is, and the more dispersed it is geographically, the greater the opportunity for other ethnic groups to have interpersonal contact with members of the group.

Basic societal conditions may even have some direct effects on individuals' perceptions and feelings towards another group. For example, historical events, such as slavery or massacres, tend to make members of one group resentful towards another group, independent of the current conditions of interpersonal contact.

When referring to the basic societal conditions under which ethnic groups come into contact, we must often look at conditions in some historical past. However, to account for ongoing changes in societal institutions, we need to also consider the basic societal conditions at any given time, including the present. For example, there may be recent changes in technology that affect the availability of various skilled and unskilled jobs, and recent changes in the size and location of different ethnic groups. These societal changes may affect conflicts of interest between the groups and their relative power, which, in turn, may affect the amount of discrimination against a minority group; and so on. Relations between ethnic groups continue to change as basic societal conditions change.

In discussing the determinants of ethnic group relations, we have mentioned a large number of variables and a large number (though not all) of the possible connections among these variables (referring again to figure 8.2 to get an overall picture of the causal links among these variables may be helpful).

The causal effects shown in figure 8.2 run mainly from left to right—from basic societal conditions all the way to the amount and types of interaction between members of different ethnic groups. In addition, there may be some "feedback" effects, going in a reverse causal direction. For example, frequent friendly interaction is likely to produce more positive feelings toward the other group; and more friendly interaction and attitudes may modify the immediate conditions of contact (e.g., by leading to more positive group norms). Changes in interaction and in attitudes may also feed back to change social institutions. For example, more negative attitudes and more conflictive interaction between groups may lead a more powerful group to reduce the political rights of another group. Feedback effects may even change some of the basic societal conditions; for example, greater intergroup conflict and the subordination of one ethnic group may cause that group to change its geographical location and to mobilize to defend its interests. Thus, there is a dynamic process by which basic societal conditions, through their effects on various institutions and social processes (such as segregation and discrimination), shape individuals' attitudes and behaviors, which, in turn, may have some effects on these same institutions, processes, and even basic societal conditions. Since the feedback effects generally are secondary, only a few of the most important ones are shown (with dashed lines) in figure 8.2.

Determinants of Particular Types of Interaction

Now that we have outlined a general model for explaining relations among ethnic groups, we will apply this model to the explanation of the four types of interaction that were described earlier: paternalism, oppression, contention, and cooperation. In discussing the determinants of each of these patterns of interaction, we will focus on a few specific cases. First we will describe people's subjective reactions to members of other groups (i.e., their perceptions, feelings, and expectancies, as shown on the right side of figure 8.2). Then we will discuss the objective factors that help to account for these subjective reactions and the resulting behaviors. (The objective factors discussed in each case are some of those shown in figure 8.2 as affecting people's subjective reactions and behavior.)

Paternalism

Subjective Reactions

The pattern of interaction that we have called paternalism (generally friendly but with one group dominating the other) is based first, on perceptions, especially by members of the dominant group, that members of each group belong to rigidly separate categories. Thus, white southerners in the United States during the period of slavery, and later that of segregation, perceived whites and blacks as being very different types of people. Those in a dominant group tend to see members of the lower-status group somewhat as they see children: not a substantial threat, because of their weakness, though potentially dangerous if not kept under control; deserving their disadvantaged status because of their lesser capacities; and possessing some praiseworthy qualities, such as loyalty, although they cannot be expected to live up to "higher" standards of moral behavior.

Such a view of blacks by many white southerners during the segregation era is illustrated by a passage in a cotton trade journal published in Tennessee in 1942: "Anyone who hears Delta Negroes singing at their work, who sees them dancing in the streets, who listens to their rich laughter, knows that the Southern Negro is not mistreated. He has a carefree, childlike mentality and looks to the white man to solve his problems and take care of him" (Goodwin 1994: 313). Such perceptions of blacks by many white southerners often led them to feel a certain amount of affection for those blacks they knew well (such as their house servants) but also to feel some disdain for most blacks, and to fear those who might not "know their place." They expected that acting in a firm way towards blacks with whom they came in contact, being cordial but not allowing

221

any intimacy, would result in a smooth relationship that kept the blacks in their "appropriate" subordinate position.

A rather similar view of Mexican workers is apparent in statements made by Anglo farmers in Texas in the 1930s: "I prefer Mexican labor to other classes of labor. It is more humble and you get more for your money," or "the Mexicans have a sense of duty and loyalty, and the qualities that go to make a good servant," or "no other class we could bring to Texas would take his place. He's a natural farm laborer," or "the Mexicans are ignorant but good laborers" (Taylor 1934: 126).

Such views of Mexican farm workers led the Anglo farmers in Texas and elsewhere generally to treat the Mexican workers in a nonhostile way (and sometimes even with what they thought to be benevolence). At the same time, though the boundaries between Anglos and Mexicans were not quite so strong as the caste lines between whites and blacks, the Anglos were careful to maintain their position of dominance and higher status.

Members of a subordinate group in a situation of paternalism generally see members of the dominant group as being a sharply separate category of people who, by virtue of their ethnicity, enjoy high status and privilege. Members of the dominant group are seen as a continuing threat, since they possess great physical and economic power and often use it in ways that harm those in the subordinate group. If the subordinate group has been subjected to a racist indoctrination by the dominant group (as blacks were), they may believe that members of the dominant group are superior people and that their higher status is part of the natural order of things. They will be fearful of the dominant group and also may feel anger at their treatment but, if so, they generally will suppress and hide the anger. They will expect that expression of discontent, and certainly any overt aggressive behavior, will lead to swift and possibly severe punishment. They will come to expect that submissive behavior is necessary to survive. Such perceptions, feelings, and expectations generally were characteristic of blacks in the American south during slavery and legal segregation. These types of reactions have often occurred among Mexican farm laborers as well.

Objective Situation

A paternalistic relationship develops under particular conditions of intergroup contact. Clear differences between groups in physical characteristics and in culture, together with differences in status and divergent interests, lead members of each group to see a firm distinction between the groups. Thus, for example, black slaves and later black tenant farmers in the south differed from white landowners both physically and in culture, and were much lower in economic and social status (Franklin 1994; Kolchin 1994). Moreover the economic interests of the two groups differed; lower income (or no income at all) for the blacks usually meant higher profits for the whites. Comparable differences between white landowners and Mexican farm workers in the southwest have often been present. These groups differed in physical appearance, culture (often including language), in economic and social status (education, wealth, and so on), and in economic interests (more money for one usually meant less for the other). In both cases, these differences between the groups led people in both groups to see their world in terms of a sharply separate "us" and "them."

In a paternalistic situation, the perception of the dominant group that people from the subordinate group are not an immediate threat stems primarily from the unequal and low power of members of the subordinate group. Dominant group members' apprehensions about a potential threat from subordinate group individuals, and their belief that firm control over the subordinate group must be maintained, stems from their recognition that the interests and goals of the groups differ in some important ways. The apprehensions of the dominant group may be reinforced by memories of occasional past rebellions by the subordinates, such as those by black slaves.

The fear by subordinate group members that they may be punished for expressing opposition or hostility to the dominant group is based on the greater power of the dominant group members. For example, in the segregated south the white landowner usually saw little threat from his black sharecroppers because he could evict them if he wished and had the police and courts on his side. The black sharecropper feared to be anything but subservient to the landowner; he had no sanctions with which to pressure the white man and risked losing his house, his meager livelihood, and, perhaps, his life if he showed rebelliousness.

The Mexican farm worker was often in a similar position of unequal power relative to the Anglo farm owners (Taylor 1934; Martin 1992). Often he was an illegal immigrant subject to deportation, had little facility in English, and had little information about other job opportunities. He could be fired or laid off at the whim of the employer, who usually had a replacement pool of other Mexican workers on which to draw. Thus, the Anglo farm owner usually perceived little threat from the Mexican workers, while the latter usually feared the possible sanctions that the owner could impose. (The rise of the farm workers union changed the relative power of the two groups in some places but has not generally succeeded in changing the basic owner-worker relationship.)

These immediate conditions under which particular members of the dominant and subordinate groups come into contact reflect great inequalities within the larger society. Blacks in the segregated south and Latino farm workers generally had much less education and income, less skilled occupations, and fewer positions of political influence than the whites with whom they interacted. These differences in economic and political status resulted in large part from widespread discrimination and from segregation in housing, education, and employment.

Such discrimination and segregation resulted, in turn, from the operation of existing societal institutions. The political-legal system in the south at the time of slavery, and later during the era of segregation, put all decision making authority in the hands of whites. Blacks had no representation whatever in legislatures, courts, and law enforcement agencies. The law itself gave more rights to whites than blacks. The foundation of the economic system was agricultural land owned and controlled entirely by whites. Blacks had few or no alternatives to being anything else but field hands and maids. The school system required segregation of the races and provided poorer education for black children than for white children.

These institutional structures of slavery and then of segregation had been created to serve the needs of planters who needed large numbers of cheap (or better still, free) field workers for their crops of cotton, tobacco, and other products. Africa was a potential source of such labor. The more advanced military technology and organization of Europeans and Americans gave them the physical power to force large numbers of Africans to perform the necessary labor against their wills. To maintain such a system of forced (and later, very low priced) labor, establishing the political, legal, and educational institutions to keep blacks subordinate and submissive was necessary. (Some would even argue that creating Christian churches for the blacks also functioned to keep them submissive.)

The conditions under which Latino farm workers generally have related to Anglo farm owners (conflict of interests but low relative power) and the discrimination and segregation they have often experienced have also been influenced by institutions of the larger society. The political and legal systems classify many Latino farm workers as aliens, and sometimes as being in the United States illegally. Therefore, many do not enjoy the legal rights and protections of American citizens.

The economic system of the American southwest has often had a high need for unskilled agricultural labor, but the supply of such labor usually has also been high, thus keeping wages down. Farm owners have often formed organizations to permit them to coordinate their labor policies, thus strengthening their power. The educational system has usually put little emphasis on the education of poor and often transient Latino children.

Like the institutional arrangements that determined relations between blacks and whites in the south, political and economic institutions in the southwest have been shaped by the primary conditions of contact between Anglos and Mexicans and by the conflicts of interest and relative power that resulted from these conditions. Military victory of the United States over Mexico in a war that ended in 1848 gave Anglos control over the southwest's lands ceded by Mexico to the United States. Anglo farm owners have had a long-standing need for cheap field labor. They have used their political and economic power over the years to shape the laws that have affected the flow of Mexican agricultural labor and the conditions under which they are allowed to migrate and to work (Craig 1971).

In sum, paternalistic relations occur when there is conflict of interest between two distinct ethnic groups and when one of the groups is so much more

224

powerful than the other that it can force the other group to accept a subordinate position without overt challenge or conflict.

Oppression

Subjective Reactions

The pattern of interaction that we have labeled oppression (one group dominant over another and generally unfriendly behavior) is based first, as is paternalism, on perceptions of sharply separate categories of people—"us" and "them."

In the case of early relations between whites and Native Americans (Utley 1984; Koning 1993), whites typically saw the Indians as "heathen savages," while the Indians viewed the whites as alien beings. Whereas in paternalistic situations, members of the dominant group see the subordinate group as posing little immediate threat, in the situation of oppression the dominant group perceives a substantial present threat. Thus, white settlers pushing into the western part of the American continent often saw both an economic threat from Indians, who wished to remain on the same lands that the whites wanted, and a physical threat of Indian attack. Focusing on incidents such as seemingly unprovoked massacres of whites by Indians, many whites saw the Native Americans as deceitful, treacherous, cruel, and barbaric. They heartily disliked and feared the Indians. Many came to expect that to live together peacefully with the Indians was not possible (at least as the whites wished to live) and that the only solution was to get rid of the Indians, by forcing them to move if possible, by killing them if necessary.

For their part, the Native Americans saw even greater threat from the whites. The ever-growing influx of white settlers was destroying their whole way of life. The lands on which they had lived and hunted (or, in some areas, farmed) were being seized. On the Great Plains, the giant herds of bison that had sustained Native Americans were being exterminated. Their peoples were being ravaged by diseases brought by the whites. Observing the arrogant and destructive actions of the whites, and the recurring pattern of the whites breaking their promises and treaties, Native Americans saw white behavior as reprehensible. They felt dislike, anger, and fear of the whites. Recognizing the whites' greater military power, Native Americans sometimes expected that some concessions of land and privileges (such as mining rights) were necessary. But when whites broke their agreements, Native Americans often came to expect that, no matter how desperate the odds, war was the only way by which they might preserve their way of life. Thus, a general pattern of white domination, punctuated by sporadic fighting, continued during much of the eighteenth and nineteenth centuries.

Perceptions of threat have also been important in other instances of oppression. In Croatia and in Bosnia, ethnic Serbs saw the prospect of their being a numerical minority in these newly-independent republics as threatening. They were fearful that they would be oppressed and mistreated by the majority Croats and Muslims in these areas (Cohen 1993).

Similar perceptions and fears were present in Rwanda before the ethnic massacres began (Destexhe 1995). Tutsis feared they would be mistreated by the numerically dominant Hutus, while Hutus feared that the Tutsis, who historically had subordinated them, would regain control and oppress them. In such atmospheres of fear, many Serbs in Croatia and Bosnia and many Hutus in Rwanda believed that they could best protect their own interests by using their current power superiority to get rid of the rival ethnic group. Mass expulsions and massacres—"ethnic cleansing," in the Serbian phrase—followed.

Objective Situation

Oppression tends to occur when the immediate conditions of contact between groups are characterized by strong conflicts of interest coupled with inequalities of power. In addition, the weaker group has sufficient present or potential power to be perceived as a threat by the dominant group.

The types of conflicts of interest between the dominant and the subordinate group may vary. They may center on competition for land and/or for political control. Native Americans and whites wanted to occupy the same land. Serbs and Croatians in Croatia and, later, Serbs and Muslims in Bosnia wanted political control of the same territory; the same was true of Hutus and Tutsis in Rwanda.

But other types of conflict of interest may be involved in situations of oppression. In Germany during the 1930s and 1940s, Jews did not compete for land or for political control. But the German Nazis saw Jews as having too much influence over the economy and culture, as representing an alien ideology of cosmopolitanism, as being insufficiently loyal to the nation, and as representing by their very presence an obstacle to "Aryan" racial purity (Melson 1992). From 1915 to 1916, revolutionary Turkish leaders saw similar types of conflicts of interest between themselves and the Armenians in their country. They resented the economic success of Armenians (seen as coming at the expense of Turks), saw the Armenians as disloyal in the war with Russia, and wanted a nation peopled only by ethnic Turks. They also coveted land occupied by Armenians (1992).

The conflicts of interest seen by those in the dominant group may not be objectively accurate. For example, the Nazis' perceptions of Jewish influence over the economy were grossly exaggerated and Nazi allegations of Jewish disloyalty were not consistent with widespread service by German Jews in World War I. While some objective conflicts of interest and outcomes are usually behind perceptions of competitive threats, it is the perceptions that ultimately are most important.

In situations of oppression, one side has at least a temporary power superiority. Whites in North America had greater numbers and more advanced weapons than the Indians; the Serbs in Croatia and Bosnia were supplied with heavy weapons by the former Yugoslavian army, while the Croatians and the Muslims initially had few arms; in Rwanda, the Hutus had greater numbers than the Tutsis and controlled the government at the time of the massacres. The side with the greater power acted to kill or expel the other group, which it saw as a great threat.

Conflicts of interest and inequality of power between ethnic groups in these situations were influenced by the economic, political, and other institutions of each society. The conflicts of interest between whites and Native Americans partly stemmed from their being separate peoples with different ways of life, based on different histories and cultures. The white economy was based on the cultivation of privately owned land and provided incentives for individuals to acquire their own land in "unclaimed" areas. Native Americans usually earned their livelihood through communal activities, especially hunting. The two economic systems were incompatible.

In both Rwanda and the former Yugoslavia, ethnic conflicts over control of territory were primarily a result of the instability of political institutions. In Croatia and in Bosnia, the political system of Yugoslavia had crumbled in 1991 and the component republics that constituted Yugoslavia began to declare independence. Uncertainties arose over the role and welfare of various ethnic groups in each of the republics (Croatia, Bosnia, and so on). This unstable situation led to competition among ethnic groups for political and/or military control over the areas in which they lived. Ethnocentric statements by political leaders and media in Serbia, Croatia, and elsewhere gave further impetus to the competition.

The competition among ethnic groups for political control when Yugoslavia broke up was not primarily a result of recent discrimination against, or segregation of, any group; under the communist regime great effort had been made to have all ethnic groups treated with equal respect. Each group's determination to gain political control stemmed, rather, from memories of past discrimination and mistreatment (such as atrocities committed by Croatians against Serbs in World War II) and by the fear that such discrimination might recur. The collapse of the country's political institutions revived old memories and stimulated new fears. Recent discrimination did play a role in the power inequalities that existed initially among the Serbs and other ethnic groups. The Yugoslavian Army—especially its officer corps—had a high proportion of Serbs; when the central government collapsed and ethnic conflict threatened, the Serb-dominated Army supplied the Serb militias with a variety of armaments.

In Rwanda, as in Croatia and Bosnia, a competition for power also occurred in a time of political instability. This competition was fueled by a history of discrimination by Tutsis against Hutus, extending up to recent years. Tutsis had conquered the area several centuries earlier and had imposed their military and political control to further their own interests. Tutsis owned most of the land, which was worked mainly by Hutu agricultural laborers. Wealth and influence in Rwanda was held mainly by the Tutsis. However, the numerical superiority of the Hutus made the historical Tutsi control unstable and enabled Hutus to gain control over the government in 1962. Militant Hutus used their new position of greater power to try to eliminate effective Tutsi competition and ensure the continued dominance of their own ethnic-political group.

In sum, then, oppression is likely to characterize interaction between ethnic groups when, based on societal conditions and institutions, one group sees another as a serious threat to its vital interests and uses its superior power to impose its will on the other group.

Contention

Subjective Reactions

When interaction generally is contentious (neither group is clearly dominant and there is a high level of conflict between them) members of each group have retained a sense of strong boundaries between the groups—between "us" and "them." In the examples of contention mentioned earlier—white and black firefighters contesting for jobs in Memphis (Firefighters 1987); whites and blacks in Chicago competing for the office of mayor (Holli and Green 1984); and Asian Americans and others disputing college admissions in California (Takagi 1992)—members of each group saw racial distinctions as very salient.

In each case, members of the different groups also saw harm done (or threatened) to their group. White firefighters saw blacks as taking away jobs from whites, while black firefighters saw whites as trying to limit jobs available to blacks. Black residents of Chicago saw whites as keeping political power (and patronage jobs and contracts that go with such power) for themselves, while most whites felt threatened by the prospect of a black takeover of political control in the city. Asian Americans in California perceived that many Asians had been denied admission to the top state colleges, while many whites, as well as blacks and Latinos, perceived that college admissions to members of their own ethnic group were threatened.

In situations of contention, members of each group are likely to see the other group as receiving unfair advantages. Thus, black firefighters pointed to past discrimination against their race and saw whites as trying to maintain the unfair advantages that historically they enjoyed. White firefighters, on the other hand, complained of "reverse discrimination" that might permit blacks to keep their jobs, while whites with more seniority were laid off.

In the controversy over college admissions in California, different groups used different standards of fairness in judging themselves to be victims of unfair treatment. Asian Americans pointed to the fact that, among students who were well-qualified academically, the proportion of Asians admitted was lower than that for whites; thus, they perceived unfair discrimination against Asians. But many whites, blacks, and Latinos focused on the fact that the proportion of Asians at the top California colleges was far above the proportion of Asians in the California population; thus, judging the college representation of their own groups against that of the total population, they saw their own groups being treated unfairly.

Seeing the other group as doing or threatening harm to their group, and of enjoying or seeking unfair advantages, those in each contending group tend to be fearful of harm to their interests and angry and resentful towards those in the other group. Since each group has at least some power in society (based on voting strength, the backing of law and the courts, and so on), leaders of each group are likely to expect that contentious tactics (e.g., protest marches, political lobbying, court cases, and boycotts) may help them to win concessions from the other ethnic group and/or from third parties (such as government agencies and large employers). Thus, as long as the issues are unresolved, a generally contentious pattern of interaction will continue.

Objective Situation

Contentious relationships are apt to occur in situations in which there is inequality and some conflict of interest among groups but in which neither group is powerful enough to enforce its will easily. Often there has been a reduction in power differences among groups, with a previously weak subordinate group having increased its power. These conditions were present in each of the examples of contentious interaction that we have considered.

In the case of firefighters in Memphis, there was inequality between whites and blacks in the department. Blacks were underrepresented in total numbers, did not hold many higher ranks, and had less seniority than whites. These status differences were largely caused by past and present discrimination against blacks, both directly when hiring for the fire department and indirectly by the poorer education provided to blacks. (The broader societal sources of such discrimination have been discussed earlier, in this and in previous chapters.)

There was an objective conflict of interest between blacks and whites underlying their dispute. Giving preference to blacks when hiring and job protection in order to correct for past discrimination meant fewer positions and less security for whites. However, hiring and retaining people on the basis of test scores or seniority meant more positions for whites and fewer for blacks.

While whites enjoyed more power in the fire department and in city government, the overall inequality in power had been reduced. Greater voting power of blacks and the backing of the federal courts for equality had provided blacks with levers for exerting influence.

These circumstances led, on the one hand, to dissatisfaction by the racial minority with remaining inequalities and, on the other hand, to their expectations that changes could be made. Since white firefighters also hoped to prevail in the dispute, a legal and political battle ensued. (The legal case was decided in 1984 when the Supreme Court ruled that, although the city had agreed to give blacks preference in hiring, this did not override the use of seniority in deciding layoffs.)

In Chicago, the contention between blacks and whites for the mayoralty office in 1983 took place in a city where there was considerable opportunity for contact between the races. But that contact tended to be of a superficial kind, such as riding in the same bus or shopping in the same store. While both whites and blacks shared much of American culture in common, there were some cultural and other differences that contributed to feelings of racial separation.

Largely because of the familiar effects of segregation and discrimination in Chicago, blacks generally had been lower in economic status (occupation, income) than whites and had less influence on local government decisions (on city services, taxes, schools, and so on). As the education of blacks became more like that of whites and as more blacks moved into the middle class (see chapter 5), gaps in economic status between whites and at least a portion of the black population narrowed. However, continuing inequalities between the races and continuing separation between them (in residential location, social interaction, and so on) led blacks and whites to have conflicting political interests. Each group wanted to control the policies, jobs, and contracts of city government.

In the past, blacks never had sufficient political power in Chicago to challenge white domination. A number of developments came together to raise the power potential of blacks in Chicago in the 1983 city elections. Continuing growth in the black population and an increase in voter registration among blacks had raised the number of black voters (though not to a majority). At the same time, a breakup of the once solid Democratic Party organization led to two strong white candidates (Jane Byrne and Richard M. Daley) running against each other in the Democratic primary election. Blacks, therefore, saw a realistic prospect of gaining political control of Chicago, while most whites still expected to retain their domination.

In this situation, competition for political power and for the economic benefits (such as jobs and contracts) that may accompany political power became structured along racial lines. The Democratic Party chair was quoted as saying that the campaign had become "a racial thing . . . we're fighting to keep the city the way it is" (Holli and Green 1984: 30). The racial nature of the primary election may be seen in the racial voting patterns. In heavily black wards, the black candidate, Harold Washington, received almost 90 percent of the vote. Though Washington also received support from white liberals in some wards, in many white wards he received well under 10 percent of the vote.

Washington won the Democratic primary election with a plurality of the votes; most of the white vote was split between his two opponents. In the general election, which usually was not much of a contest because the Republican Party was so weak in Chicago, large numbers of usually Democratic white voters switched their support to the Republican candidate. Black support for Washington in the general election was massive—over 99 percent in most black wards. Enough whites and Hispanics also backed Washington to give him a narrow victory in this racially based contest.

Contention among ethnic groups in California over college admissions in the 1980s and 1990s also involved issues of inequality, clear conflicts of interest, and no clear power dominance by any group.

Since 1974, the University of California had been governed by a resolution of the state legislature that recommended the ethnic mix of the student body match the ethnic mix of the state's high school graduates. At the same time, the schools were subject to federal law that prohibited basing admissions solely on race. The growth of the Asian population in the state, together with the high proportion of Asian students with top grades and test scores, created a situation in which admitting students on the basis of academic qualifications would have resulted in a great "overrepresentation" of Asians. During the mid-1980s, school administrators began to apply to Asians a higher standard for admission than they applied to other students. Thus, while Asians were "overrepresented" at top campuses (Berkeley and Los Angeles) in relation to their proportion in the California population, they were underrepresented in terms of their proportion of the best-qualified students.

The situation was reversed in the case of blacks and Latinos. Under affirmative action programs, students from these minority groups were given preference in admissions. However, their representation at California colleges still was below the proportions of these minorities in the general population.

Standards and procedures for college admissions represented a real conflict of interest among ethnic groups. Especially in a time of tight university budgets, the number of admission places was limited and therefore a greater number of places given to students from one group meant fewer places available for those from another group. This conflict of interest was not only between Asian Americans and whites. The more that Asian Americans were successful in increasing their numbers in California colleges, the fewer the places that were likely to be open to blacks, Latinos, and Native Americans under affirmative action programs.

In the disputes that erupted over college admissions, initiated primarily by Asian American groups, no one ethnic group exercised decisive power. Whites were in control of the educational and political institutions but Asian Americans and others wielded considerable influence and, moreover, had federal law and federal agencies on their side at times. Perceptions and feelings of members of all ethnic groups were influenced by accounts of the controversy in the media—including campus newspapers, local city newspapers, national newspapers and magazines (e.g., the *New York Times*, *Newsweek*), and network television. After years of lobbying by Asian American groups, and after federal agencies exerted pressures in their behalf, California universities changed their admission procedures in ways that reduced discrimination against Asian Americans. The result, as of the late 1990s, was that a larger proportion of highly qualified Asians was being admitted to the top California colleges than had been the case ten years earlier.

In summary, in situations of contention, members of each ethnic group see some threat to its interests from another group and often perceive the other group as enjoying or claiming unfair advantage. There is some inequality and conflict of interest between the groups but neither group has clearly decisive power so that there is a struggle in which each tries to improve its position.

Cooperation

The final pattern of interaction we wish to explain is that of cooperation—one in which ethnic groups have about equal control and exhibit generally friendly behavior towards one another.

Subjective Reactions

Cooperation often stems, in part, from people's perceptions that, while individuals' ethnic identities differ, they share some other important group identity. For example, Latino and Anglo students may see themselves as part of the same class project group, Asian and white garment workers may see their common membership in a labor union as important, and blacks and whites may see themselves as sharing an important identity as members of a baseball team. In Hawaii, while most people identify with a particular ethnic group (Japanese, Filipino, and so on), there is also a widespread sense of a common identity as "islanders."

Cooperative interaction may also stem from generally positive appraisals by members of each ethnic group of those in the other group. They may be seen as usually helpful—for example, as helping to get good grades for all those in a class project group, helping to get good wages and working conditions through their joint union activities, or helping a baseball team to win. Members of each group may also see the behavior of those in the other ethnic group (such as making some personal sacrifices for joint success) as praiseworthy. In Hawaii, people of various ancestries (European, Japanese, Filipino, Chinese, and so on) generally feel little threat from those with other backgrounds. They generally see those from other groups as contributing positively to the culture and prosperity of the islands (Grant and Ogawa 1993).

Given such positive intergroup perceptions, individuals from different ethnic groups are apt, in general, to like those belonging to the other groups. They will expect that behavior that is friendly and egalitarian (e.g., compromising different preferences) will result in success for the group and for themselves as group members.

Objective Situation

The generally positive reactions that people engaged in cooperative interactions have toward each other are a consequence of situations in which people are roughly equal in status and power and share similar values and interests. In Hawaii, although there are some disparities of occupation and income among ethnic groups, a number of groups (e.g., whites, Japanese, and Chinese) are roughly equal in economic status. Equally important, no one group occupies a dominant position of political power. While there are some conflicts of interest (e.g., native Hawaiians seeking restitution for past injustices), most Hawaiians see all groups as having a stake in general prosperity. Moreover, election to high office in Hawaii requires appeal to a broad spectrum of ethnic groups.

The relatively high degree of equality and common interests in Hawaii, as well as a generally high level of opportunity for contact across group lines, reflects a society with relatively little ethnic segregation or discrimination. However, this was not always the case. During the late nineteenth and early twentieth centuries, the economy and the government of the islands were controlled by white (American) sugar planters, who not only kept themselves apart from other ethnic groups (brought in as agricultural laborers), but who also discriminated among these groups (Japanese, Filipinos, Chinese, and so on) with respect to wages and kept them segregated from each other. After World War II, however, people from a variety of ethnic groups formed coalitions to challenge and to overthrow the ruling oligarchy. A multiethnic labor force was drawn together, in part through the rapid growth of the International Longshoremen's and Warehousemen's Union. A coalition of mixed ethnicity, including Japanese Americans, whites, and others, gained control of the Democratic Party and of both houses of the legislature. These economic and political changes, together with other changes in the economy (especially the rise of the tourism industry, which overshadowed the sugar industry) contributed to a set of economic and political institutions that greatly reduced the ethnic segregation and discrimination that had existed earlier. The existence of a good public education system that served a variety of ethnic groups also contributed to these positive changes.

The cooperative relations that often exist in the other settings we have discussed—class projects, unions, and sports teams—also arise out of opportunities for contact in situations of equal status and of shared values and interest. In their positions as members of a school group project, or as workers and union members, or as players on a team, people from various ethnic groups are equal in status and power in those roles. In each case they also share common values and common interests—whether these be good marks, high wages, or winning a sports contest. Their equality and common interests in these situations reflect situations in which there is little or no segregation or discrimination.

The absence of segregation and discrimination in these situations is a consequence of particular societal institutions. That Anglo and Latino students, or black and white students, are together in classes is a consequence of the way in which the educational system has been run—for example, assigning students to multiethnic schools, placing students in multiethnic classes within schools, and organizing students in multiethnic work groups within classes. The fact that Asians and whites, or blacks and whites, or those from any other ethnic groups, are members of the same union may have resulted, in part, from laws that make unions easier to organize, from the organization of jobs in an industry that results in Asian and white workers doing the same types of work, and from the policies of a union that, unlike many unions in the past, encourages a multiracial membership.

Similarly, absence of segregation and discrimination with respect to black and white players on a sports team may be a consequence of the policies of their own team and of the league to which it belongs. For example, until the late 1940s the major professional baseball leagues prohibited blacks from playing with whites, and blacks, therefore, played separately in what were then called the Negro leagues (Ward 1994). Changes in the institution of professional baseball were a prerequisite to the occurrence of cooperative interaction between whites and blacks on these teams.

The changes in particular institutions within society are, in turn, a result of broader social changes. In the example of professional baseball just mentioned, changes in the racial policies of this institution were a consequence of broader historical events and societal changes. These events included a war by the United States against the racist Nazi regime in Germany that made racism at home seem less defensible and increases in the relative power of blacks in the United States, as they moved out of the south, increased their voting strength and economic resources, and organized themselves more effectively.

In sum, interactions of cooperation are based on individuals' perceptions of some degree of shared identity with those in other ethnic groups and on generally positive appraisals of members of the other groups. Such views of the other group arise out of situations in which members of different ethnic groups are roughly equal in status and power and share similar interests.

Changes in Relations Between Groups

The relationship between ethnic groups may change over time. For example, the relationship between whites and Native Americans changed from primarily ones of oppression during the early nineteenth century to primarily ones of paternalism during the early twentieth century. The relations between blacks and whites in the United States changed from primarily one of paternalism in the late nineteenth century to a more mixed pattern of relations, often including ones of contention, by the late twentieth century.

Figure 8.2 indicates that the most fundamental sources of relations between groups are the primary conditions of contact. These include historical events, the environment, the skills and organization of the groups, their cultures, and the size and location of groups. Changes in such primary conditions of contact will result in changes in relations between the groups. For example, the change in relations between Native Americans and whites, from oppression to paternalism, resulted most fundamentally from the military defeat of the Indians; from the change in the basis of Indian subsistence from primarily hunting to one of farming, ranching, and reliance on government payments; and from the relocation of most Native Americans on reservations. These changes in the primary conditions of contact meant that the most basic conflict of interest between native Americans and whites—that over control of large areas of the American continent—was no longer relevant. Also, the Indians, now largely dependent on the federal government, had lost what little power they once had.

Changes in these basic conditions resulted in changes in the institutions of many Native American tribes. Whereas formerly they had been autonomous, now the federal government exercised control over their political, economic, and educational affairs. The changed primary conditions of contact and institutions led to more cultural similarity in some ways (e.g., use of the English language) but also to a breakdown of some positive aspects of tribal culture, to great economic and other inequalities between Native Americans and whites, and to some new conflicts of interest between Indians and whites (e.g., over federal efforts to eradicate Indian culture in education).

The new circumstances led to a change in the perceptions that many whites had of Indians, from an earlier image of a people who were dangerous but proud and fiercely independent to that of a lazy, drunken, and dependent people. Many Indians continued to see whites as arrogant and untrustworthy, but now saw them also as a potential source of some benefits. Interaction, once often violent, now was most often one of "benevolent" dominance by white government officials, while Indians usually were passive. (Recently, however, many Native Americans have become more active in their own governance and in asserting their rights and interests.)

To help explain the changed relations of blacks and whites in the United States between the late nineteenth and late twentieth centuries, we also need to look at changes in the primary conditions of contact between these groups. This period saw a major change in the way Americans earned their livings. New machinery greatly reduced the need for labor on farms, while the invention of new products (automobiles, airplanes, and so on) and the growth of world trade contributed to a great expansion of industry. These developments led to a vast migration of blacks out of the rural south to the growing industrial cities of the north and midwest. Blacks began to acquire more education and to organize more effectively for the betterment of their race. The advent of the World War II created the need to employ blacks in relatively skilled positions in war industries and to use blacks alongside whites in combat.

Such changes in the primary conditions of contact between blacks and whites in the United States reduced the importance of some conflicts of interest (e.g., those between white landowners and black farm laborers in the south) and created some new correspondence of interests (e.g., between black and white workers in the same industries). These changes also increased the relative power of blacks somewhat—as shown, for example, in the ability of black leaders during World War II to win a fair employment decree from President Roosevelt by threatening a mass protest march on Washington.

The result of such changes in the basic conditions of contact was change in the nation's institutions. Restrictions on voting began to ease. Many blacks were included in unions, especially the new industrial unions. More jobs were opened to blacks. Legal segregation of schools was ended, along with segregation in the armed services.

Thus, segregation and discrimination by race was reduced, though not eliminated. These changes, in turn, created more opportunity for interpersonal contact between blacks and whites in some settings (e.g., school and work); created greater cultural similarities (e.g., in life style) between some middle-class blacks and whites; reduced racial inequalities of power (e.g., election of more black officials); reduced economic inequalities between whites and those blacks who were able to take advantage of new opportunities; and greatly reduced transmission of blatantly racist information by the media.

These new circumstances contributed to changed attitudes and behaviors among both blacks and whites. Most whites no longer thought of blacks as biologically inferior to whites but many now saw blacks as a threat to their neighborhoods and/or their jobs. Blacks now expected that challenging injustices would often result in beneficial changes. Because blacks had increased power and conflicts of interests still existed, contentious interaction occurred more often than in previous times (e.g., disputes about job equality and contests for political office). However, other types of interaction—especially paternalism (e.g., where white business owners employed black workers) and cooperation (e.g., where blacks and whites participated together in sports teams)—were also frequent.

The main direction of causation and change is from the primary conditions of group contact to interaction between groups (i.e., from left to right in figure 8.2). However, the amount and types of interaction that occur can "feed back" to change attitudes and even conditions of contact (i.e., from right to left in figure 8.2). For example, if blacks become more assertive and make some demands in their interactions with whites, many whites will see them as more of a threat. Demands by blacks—say, for a greater share of jobs—may lead to more conflict of interest with whites than before but negotiations may lead to less discrimination and perhaps to some institutional changes (e.g., in seniority systems) that affect inequality.

To take another example, cooperative interaction between blacks and whites is likely to result in more positive perceptions and feelings on both sides (such as seeing members of the other group as helpful). Cooperative interaction may also lead people in both groups to arrange further opportunities for contact and may, over time, tend to increase cultural similarity, as the members of each group adopt some of the values and customs of the other group. Cooperative interaction may also lead members of both groups to engage in further joint activities in which they will share the benefits of success (i.e., raise the correspondence of their interests) and may even promote greater equality of status, by encouraging more equal participation in various tasks and decisions. Cooperation between racial groups, if widespread enough, also could result in political coalitions and other coalitions (e.g., labor unions) that may change social institutions in ways that make them less segregated and less discriminatory.

Summary

This chapter has presented an overview of the determinants of relations between ethnic groups. Our aim has been to account for a variety of types of relations (e.g., conflictive and cooperative) and to consider the key determinants of these relations, including variables at both the individual and the societal levels.

Interaction between members of different ethnic groups may be described in terms of the amount and the nature of the interaction. The nature of the interaction may be described by two dimensions: a) the extent to which one group dominates the other (at one extreme there is complete dominance, at the other complete equality of influence), and b) the degree of friendliness (total friendliness at one extreme and total unfriendliness at the other). If we categorize any given interaction between ethnic groups as usually closer to one end or the other end of each of these dimensions, we have four general types of interaction: 1) one group is dominant and relations generally are friendly, or at least not overtly unfriendly (paternalism); 2) one group is dominant and relations generally are unfriendly (oppression); 3) the groups are roughly equal in influence and relations are generally unfriendly (contention); and 4) the groups are roughly equal in influence and relations are generally friendly (cooperation).

To help explain the amount and type of interaction that occurs between ethnic groups, we presented a theoretical framework or model. The actions of members of each group depend first on their perceptions of and feelings toward the other group and their expectancies about the outcomes of their own actions. More positive perceptions and feelings toward the other group, and expectancies that positive behaviors will have good outcomes, are increased by favorable conditions of interpersonal contact, especially by cultural similarity, equality, and compatible goals or interests. However, when conflict of interest is high, greater equality of power leads to more contentious behavior.

Opportunity for contact, as well as the conditions of interpersonal contact, are strongly influenced by two key processes: segregation and discrimination. More segregation and more discrimination result in less opportunity for intergroup contact, more cultural dissimilarity, more inequality, and more conflict of interest.

The amount of segregation and discrimination that exists in a society is determined primarily by political, economic, educational, and other types of institutions. By their organization, rules, and policies, social institutions determine where members of different groups will be admitted or welcomed, where they will not, and how they will be treated when they are present. The relative size of different groups, and their cultures, may also have effects on segregation and discrimination. As a minority group becomes larger and is more culturally different, segregation and discrimination tend to increase.

To explain the structures and functionings of societal institutions, as they are relevant to intergroup relations, we need to examine the basic conflicts of interest and the relative power of ethnic groups, when they first come into contact and later. Conflicts of interest and initial relative power are shaped by the environment and technology, by the cultures of each group, by each group's skills and organization, by the groups' relative sizes, and by historical events.

The general model of interethnic relations was used to explain each of the four types of interaction described earlier. Paternalism (dominance by one side but little overtly unfriendly behavior) was illustrated by relations in the American south and southwest between white landowners, on the one hand, and black slaves (later tenants) and Mexican agricultural workers, on the other hand. This type of relationship is based, in large part, on fear of punishment by members of the submissive group if they challenge the wishes of the dominant group. The reactions of both groups derive from a situation of considerable conflict of interest, of economic inequality, and of greatly unequal power. These conditions are mainly shaped by segregation and discrimination fostered by societal institutions. These institutions have been initially created to serve the interests of the more powerful group.

Oppressive relations (dominance by one group, unfriendly behavior) were illustrated by the relations between Native Americans and whites during white settlement of the North American continent; between Serbs, Croatians, and Muslims in Bosnia; and between Hutus and Tutsis in Rwanda. Oppressive relations usually occur when a dominant group is fearful of the other group and believes it must treat the other group harshly in order to maintain control. The fears of the dominant group arise from a situation in which there is substantial conflict of interest (and perhaps cultural dissimilarity) between ethnic groups and in which the dominant group enjoys greater power. However, the weaker group has sufficient present or potential power so that it contends or is a potential contender for dominance. The situation in which oppression occurs is strongly influenced by discrimination and/or separation between the groups. However, the institutions that affect the groups' relationships—especially the political institutions—are likely to be unstable. This instability gives rise to a belief among members of the dominant group that they must suppress or eliminate the other ethnic group in order to preserve their own dominance and interests.

Contention between groups (roughly equal power and generally unfriendly behavior) was illustrated by relations between blacks and whites contending for jobs as firefighters in Memphis, Tennessee; blacks and whites competing in an election for mayor in Chicago; and Asian Americans and other groups disputing admission standards at Californian colleges. In contentious relationships, members of different ethnic groups believe that harm is being done to or threatened to their group by actions of another group. They also are likely to see the other group as enjoying or asking for unfair advantages. These perceptions lead them to be angry at and resentful of the other group. Also, members of each group expect that their own contentious efforts (e.g., protests, lobbying, boycotts) have a good chance of winning at least some of their objectives.

Contentious interaction is most likely to occur in situations where there is inequality in benefits (such as jobs, political offices, or college admissions) between groups, usually largely due to past and/or present discrimination practiced by the institutions of society. There is an objective conflict of interest between ethnic groups, often stemming from efforts by a disadvantaged group to get a bigger share of social benefits. No group is powerful enough to get its way easily. Therefore, members of each group contend against the other to advance or to defend their own interests.

Finally, cooperation between ethnic groups (roughly equal influence and generally friendly behavior) was illustrated by ethnic relations in Hawaii; Latino and Anglo students in a class-project group; Asian and white workers in a labor union; and black and white players on a baseball team. Cooperative relationships are facilitated by people from different ethnic groups sharing an important identity (e.g., members of a union or a team). Cooperative interaction is also likely to occur when members of each ethnic group see those in another group as helpful in reaching their own goals. As a result, they will tend to like members of the other group and expect that friendly, egalitarian behavior will aid their own and the larger group's success.

The positive perceptions and feelings of people in cooperative interactions result from situations in which people are roughly equal in status and power and also share similar values and interests. These conditions derive from broader social institutions that do not support ethnic segregation or discrimination. Such egalitarian social institutions are likely to be initially formed when conflicts of interest between ethnic groups are low and/or when no one group is powerful enough to impose its will on others.

The type of interaction that occurs between members of different ethnic groups may differ for different individuals—for example, some blacks and some whites have a primarily contentious relationship, while others have a primarily cooperative relationship. Moreover, the amount and predominant type of interaction between ethnic groups may change over time—for example, from oppression to paternalism or from cooperation to oppression. Such changes are the result of changes in the subjective and objective variables shown in the model. The key initial determinants in the chain of causation are the extent of conflict (versus correspondence) of interests and the relative power of the groups.

9) Visions of Society I: Assimilation

In previous chapters we have tried to explain the types of relationships that occur between different ethnic groups. Now we ask: how may our understanding be used to change relations between groups in desirable ways—for example, to reduce conflict and promote harmony.

People differ with respect to what kinds of relations between ethnic groups they think are most desirable. We may distinguish two major visions of the kind of ethnic group relations we should try to attain in America (and in other nations as well). The first is one of assimilation, in which various ethnic groups are gradually integrated into a single society. The second vision is one of pluralism, in which various ethnic groups retain separate identities and separate communal lives indefinitely.

In terms of the general types of relationships shown in figure 8.1 (chapter 8), people from different ethnic groups would exercise equal power both in situations of ideal assimilation and of ideal pluralism. But under assimilation they would do so as individuals; under pluralism, power would be exercised by ethnic groups acting collectively. Interactions among those from different ethnic groups would be very friendly and often intimate under assimilation, while they would be more distant and impersonal under pluralism.

We will not attempt to decide whether the assimilationist vision or the pluralist vision of society is a more desirable one. That choice depends to some extent on personal values. What we will do in this chapter and the next is to see what kinds of policies and changes would be necessary for each pattern of ethnic relations to be realized, at least to some extent. We will also discuss the problems that must be faced when trying to achieve either assimilation or pluralism and the probable consequences (some of them unintended) of each of these patterns of ethnic relations. Our discussion will be guided by the analyses of interethnic relations that have been presented throughout the book, and especially by the model of ethnic group relations presented in chapter 8. We will try to use what we have learned about how ethnic group relations are determined to devise ways to change and improve them.

In this chapter, we will discuss policies intended to promote assimilation. In the next chapter we will discuss policies that promote pluralism. The next chapter also considers some related approaches to ethnic relations—multiculturalism and cosmopolitanism.

Assimilation as a National Goal

The assimilationist vision (McLemore 1994) is a nation in which the ethnic identity of each person would not be of major importance in affecting his or her relationships with others. People of all ethnic backgrounds would share a common culture, which has been formed by the merging of many diverse cultures. In such a society, people from different ethnic backgrounds would all stand equal under the law and enjoy the same rights. They would live in the same neighborhoods, work together, worship together, and play together. They would belong to the same organizations and clubs, would be friends, would socialize together, and would even marry each other. (Their pattern of interaction would conform to the pattern we have called cooperation.) People of different ethnic backgrounds would, over time, gradually merge into a single inclusive community—for example, Americans.

For a long time, from the eighteenth to the early twentieth centuries, most Americans accepted assimilation as a national goal. They saw America as a "melting pot" in which people of diverse backgrounds from all over the world were being forged into a new common national identity—Americans (Glazer 1993).

Up until the end of the Second World War, discussions of assimilation focused on the incorporation of European immigrants into America. Little attention was paid to possible assimilation of racial minorities, which had remained generally separate from other Americans. However, the civil rights movement, which grew in strength after World War II and reached the peak of its influence and success in the 1960s, pressed for the full integration of blacks (and other racial minorities) into American life. Together with their white allies, African Americans, led by Martin Luther King Jr., fought to be integrated into all areas of American life—in schools, work places, neighborhoods, the armed forces, public accommodations, and so on. Those of all races who continue to favor assimilation argue that only by gradually eliminating divisions based on ethnic background can we promote equality and harmony within the nation.

Criticisms of Assimilation

Assimilation as a process and an ideal has been the target of much criticism, both in the past and the present. In the early part of the 1900s, pressures on immigrants and their children to be rapidly "Americanized" met with resistance among many people. These individuals wished to preserve their ethnic heritage and identity at the same time that they became Americans (Newman 1973). They saw the "melting pot" as leading to the loss of their own national heritage and culture and to the homogenization of all people into an Anglo-Saxon mold.

In more recent times, the idea of assimilation in the United States has been attacked by some as representing an attempt to impose a dominant European culture on people whose ancestors came from other parts of the world. Thus, any effort to assimilate people into a single American culture is viewed as a kind of oppressive cultural imperialism (Hilliard 1988).

Critics also view assimilation as an idea and a goal that has not worked well in practice. In particular, they point out that racial minorities in the United States—especially African Americans, but also Mexican Americans, Chinese Americans, and others—historically were not assimilated into the mainstream of the economic, political, and social life of the country. Moreover, they point out that despite the legal equality now enjoyed by racial minorities, segregation of schools, neighborhoods, and the social life of America, especially for African Americans, continues at a high level. These critics believe that the vision of a society in which ethnic groups are assimilated into a single people is naive and unrealistic (West 1993).

Nathan Glazer has remarked, "Assimilation is not today a popular term. . . . Indeed in recent years it has been taken for granted that assimilation . . . is to be rejected. Our ethnic and racial reality, we are told, does not exhibit the effects of assimilation; our social science should not expect it; and as an ideal, it is somewhat disreputable, opposed to the reality of both individual and group difference and to the claims that such differences should be recognized and celebrated" (1993: 123). Yet, Glazer goes on to assert that ". . . assimilation is still the most powerful force affecting the racial and ethnic elements of the United States" (1993: 123). Similarly, in an article titled "Assimilation's Quiet Tide," Richard Alba writes about the continuing importance of assimilation, despite its recent unpopularity among many intellectuals, stating that "assimilation has become America's dirty little secret" (1995: 3).

Evidence Concerning Assimilation

Assimilation Around the World

Is the assimilation of distinct ethnic groups into a single society an outcome that, realistically, is possible and expectable? Historically, the evidence from countries around the world is mixed. In some places, such as England and France, separate ethnic groups—for example, Celts, Angles, Saxons, Normans, and Norsemen in England—blended over time into one people. This blending has also been true in other places—such as in Mexico, and to some extent in Brazil, Cuba, and Puerto Rico—where the original groups were of different races (Shibutani and Kwan 1965).

However, in other parts of the world, different ethnic groups have lived separately (and usually unequally) in the same society for many centuries without appreciable assimilation occurring. Examples include northern Ireland (Scotch descent Protestants and native Irish descent Catholics), Bosnia (Serbs, Croats, and Muslims), and Cyprus (Greeks and Turks).

Writing about ethnic division and ethnic assimilation around the world, Shibutani and Kwan (1965) see a general tendency for greater contact among different ethnic groups to lead to the acculturation of minority groups, which, in turn, leads to fuller economic and social assimilation. Sometimes, as in the case of Jews throughout much of European history, the minority remains separate from the rest of society for many centuries. But when conditions change, assimilation may occur rapidly, as it did for Jews in Europe and in the United States, who—once they were granted the rights of full citizens—rapidly assimilated culturally and socially.

Several different aspects or types of assimilation may be distinguished (Gordon 1964). "Cultural assimilation" occurs when members of one ethnic group adopt the culture (e.g., the language, style of dress, music, values, and family structure) of the larger society. "Structural assimilation" occurs when members of a group participate in the formal and informal organizations and social life of the larger society. For example, group members may engage in the same occupational roles in the same work places as others, take part in the same political activities, schools, hospitals, theaters, and so on. Such activities may, in turn, lead to close personal relationships (such as friendships, home visits, and shared recreational activities) outside one's own ethnic group. Such intimate interactions usually contribute to other aspects of assimilation, including marriage with members of other ethnic groups (marital assimilation).

Often cultural assimilation occurs first, is followed by structural assimilation, and finally by marital assimilation. But this progression does not always occur, especially in the short run. An ethnic group may adopt most or all of the culture of the larger society but legal and social barriers may prevent members of this group from fully entering into the life of the wider community.

Assimilation in the United States

To what extent has assimilation occurred in the United States? We may consider the evidence briefly, first for Americans of European descent and then for Americans of non-European descent.

Americans of European Descent

Ethnic differences and identities among the majority of Americans whose ancestors came from various European countries—Italy, Poland, Ireland, Germany, and so on—appear to be fading (Alba 1990). Some cultural differences—for example, in food tastes—continue and many Americans have some nostalgia for their country or countries of origin. But use of foreign languages has declined precipitously among white Americans and the great majority share a common culture with regard to a wide range of activities such as music, entertainment, sports, and even foods. Richard Alba comments: ". . . it is difficult to see the cultural expressions of [ethnic] identity as more than a fragile and thin layer alloyed to a larger body of American culture . . ." (1990: 121). Differences in the socioeconomic position of European descent groups have also been greatly reduced. Educational and occupational attainments for those of eastern and southern European heritage are now similar to those of British ancestry (1990). Residential separation into ethnic enclaves has greatly declined as well.

Most significantly, marriage across ethnic lines for those of European descent now is common. Among younger Americans of the major European ancestry groups, a substantial majority is married to someone not of their own ancestry group. One study showed, for example, that among those born after 1950, 60.0 percent of Irish, 75.0 percent of Italian, and 82.3 percent of Polish descent were married to someone not from their own ethnic heritage (1990: 13). Most white Americans now have ancestors from two or more European countries; "old country" ethnic identity is weakening and identity as just "American" is becoming more common (Alba 1990; Waters 1990).

Intermarriage across religious lines has also become common. A national survey of American Catholics in 1995 found that 29 percent were married to a non-Catholic spouse (Davidson 1995). These survey data also indicate that the percentage who have out-married is higher among younger Catholics; while only 16 percent of the pre-Vatican II generation (prior to the early 1960s) have nonCatholic spouses, 40 percent of the youngest age group are married to non-Catholics (Davidson et al. 1997). Among Jews in the United States, inter-marriage also has risen dramatically. Among Jews married before 1965, only 9 percent married someone not of their own faith; among those Jews married after 1985, over half (52 percent) married someone of another religion (Steinfels 1992).

Overall, then, while some ethnic and especially religious differences remain among Americans of European descent, a substantial amount of assimilation—cultural, structural, and marital—has occurred and continues to occur. Alba comments: "Ethnic distinctions based on European ancestry, once quite prominent in the social landscape, are fading into the background" (1990: 3).

Americans of Non-European Descent

Questions about the realism of assimilation have been raised primarily with respect to those whose ancestry is non-European—that is, Latinos, Asian Americans, African Americans, and Native Americans. Differences among these groups and whites with respect to socioeconomic position have decreased considerably in recent decades (Farley 1993). However, substantial proportions of Latinos, African Americans, and Native Americans remain below whites in education and especially in their occupational level and income. Asian Americans are a notable exception; on average, they are about on a par with white Americans in their socioeconomic position (see chapters 5 and 7 for comparisons of these groups with respect to occupation, income, and education). Considerable residential segregation also exists, especially for African Americans (Farley and Frey 1994).

Culturally, there are differences between members of each of these groups and white Americans. Latinos and Asian immigrants are likely to be bilingual (speaking both English and a "native" language). Many members of these groups, as well as African Americans and Native Americans, also retain some cultural differences from white Americans, such as use of a different English dialect (primarily among some blacks), different musical tastes, and having extended families. However, what members of these groups and white Americans share in culture tends to be greater than their differences (Gordon 1964; Williams 1990; Kitano and Daniels 1988). Thus, after reviewing the evidence on cultural assimilation of African Americans, McLemore concludes that "although African Americans are not simply and totally identical to middle-class Anglo-Americans in culture, their level of cultural assimilation is high" (1994: 330). For those non-European groups whose numbers have been swelled by recent immigration (i.e., Latinos and Asians), cultural assimilation, including use of English, appears to increase greatly with successive generations in the United States (Kitano and Daniels 1988; Williams 1990; de la Garza 1992).

When judging the extent of assimilation of particular groups into American society, the amount of marital assimilation is especially significant.

Among Americans of Latino origin, in 1992 there were 4.5 million couples in which the husband or wife were of Latino origin. In 35 percent of these couples, one spouse was of Latino origin and the other was not (U.S. Bureau of the Census 1992). Marger comments: "This indicates significant marital assimilation" (1994: 326). Intermarriage within the largest Latino group, Mexican Americans, has been found to increase as families have been in the United States for more generations (Murguia 1982). A number of writers have seen marital assimilation among Mexican Americans as comparable to that of European immigrants. Thus, Schoen and Cohen assert that the assimilation of Mexican Americans "appears to be very much in the tradition of earlier American immigration" (1980: 365).

Marital assimilation also is substantial among Asian Americans. Among Chinese Americans, one study indicates that over 30 percent are married to nonChinese, most often whites. There has been, moreover, a great increase in out-marriage with successive generations since immigration. One study found that, among third-generation Chinese women, over 70 percent married nonChinese men (Kitano and Young 1982).

The picture regarding out-marriage among Japanese Americans is similar. Marriage to non-Japanese, usually to whites, has increased rapidly with each generation since immigration. Japanese Americans are now as likely to marry outside their group as within it. For example, in Los Angeles between 1975 and 1984, over 50 percent of Japanese Americans married outside their ethnic group (Kitano and Daniels 1988). The high rate of intermarriage between Japanese and white Americans is rather remarkable, considering that Japanese were confined in camps during World War II and that there were laws forbidding marriage between Japanese and whites in many states, including California, until 1948. Tinker comments:

> The United States has, with reason, been called a racist society and the discrimination against Japanese in this country was systematic and severe through World War II. The data on intermarriage indicate a very high rate of marital assimilation now, however, suggesting that the racial boundary that was once thought so important that it was used to justify major national policies, such as immigration laws and the whole process of relocation, has almost faded away. (1982: 64)

When considering intermarriage between Native Americans and others, it should be noted that only a minority of those who consider themselves Native Americans are "full-blooded" members of this group. (Most of those who consider themselves black and many who consider themselves white are also actually of mixed race biologically.) Substantial and increasing proportions of Native Americans are marrying non-Indian spouses. One study found that in states that have relatively large numbers of people who define themselves as Native Americans, 37 percent of Native American men marry women of different races (usually white). In states with smaller numbers of Native Americans, about 62 percent of Native American men marry women of other races (Sandefur and McKinnell 1986).

The one minority racial group in America for which intermarriage is low is African Americans. About 97 percent of all married blacks in the United States have a black spouse (Tucker and Mitchell-Kernan 1990) and about 98 percent of all married whites have a white spouse (Sandefur and McKinnell 1986). However, while black-white marriages are still uncommon, there has been a noticeable increase in such marriages in recent decades. In 1963, there were 1.4 black-white marriages per one thousand marriages; this rose to 2.6 per thousand marriages in 1970 and to nearly 4 per thousand by 1990 (Monahan 1976; Wilkerson 1991). For young African Americans, the frequency of intermarriage is much greater. In recent years, about 5 percent of marriages by blacks were to a white partner; and ten percent of African American men under thirty-four are married to white women (Alba 1995).

What overall conclusions about the assimilation of non-European groups may be drawn? Culturally, while differences among groups remain, all groups appear to have adopted much of the national culture. Structurally, the picture is mixed; differences in socioeconomic positions among groups generally have narrowed but African Americans, Latinos, and Native Americans (though not Asian Americans) remain considerably below whites in occupation and income. Residential segregation is high for African Americans, though much lower for Latinos and Asians.

Most striking is the evidence of substantial marital assimilation among most of the non-European groups. Blacks are the exception. However, despite the small numbers of black-white marriages, a recent large percentage increase in such marriages raises the intriguing possibility that a more substantial amount of intermarriage between blacks and others may occur in the not-too-distant future. Marriage between such groups as Jews and Christians, and Japanese and whites also was fairly rare less than a century ago and then rapidly accelerated. A similar increase in black-white marriage (and between blacks and those in other groups) may possibly occur, especially if the socioeconomic gap between blacks and other Americans continues to be reduced.

Conditions Necessary for Assimilation

The evidence just reviewed indicates that assimilation of ethnic groups into a broader society can occur in various countries around the world, including the

United States. But assimilation occurs at different speeds for different groups and sometimes hardly occurs at all. What are the conditions under which assimilation of a minority group into the broader society is most likely to take place?

A number of circumstances relevant to the occurrence of assimilation have been described by Yinger (1985). He states that, in general, more assimilation occurs for a group that is small relative to others; is geographically dispersed; has only a small percentage of immigrants in its ranks; cannot travel easily back and forth to its homeland; and has few members who wish to return to their homeland eventually. In addition, greater similarity between members of the group and others in the community—with respect to such characteristics as race, religion, and physical appearance—facilitates assimilation. Also, the more diverse the group is in occupation and social class and the more that social mobility in the society is possible (thus, the group is not trapped in a limited economic niche), the more rapid assimilation is likely to be.

Shibutani and Kwan emphasize the importance of equal treatment of ethnic minorities:

> If persons of different ethnic minorities are treated in the same manner, they eventually recognize their common fate, and social distance is reduced among them. When subject peoples are not treated alike, as in indirect rule, they retain their sense of indirect rule and remain ethnocentric. As members of ethnic minorities become acculturated, they are approached with more respect, and the more successful come to conceive of themselves as part of the larger community. As they develop a stake in the existing system, they become assimilated to it; if they are accepted, they become integrated into it. (1965: 578)

In earlier chapters we have discussed the kinds of social conditions that affect individuals' behavior towards those in other ethnic groups. These analyses may be applied to understanding the conditions necessary for intergroup interactions that reflect assimilation.

Our analyses of ethnic relations (see chapter 8) suggest that people will form intimate relationships with those from other ethnic backgrounds when: 1) they see the others as part of some group to which they also belong; and 2) they have positive appraisals of the others. Such perceptions are likely to occur under circumstances that include (among others) the following key conditions: 1) equality among those from different ethnic groups; 2) the absence of segregation of members of different ethnic groups; and 3) nondiscrimination among those from different ethnic groups.

We next discuss possible ways in which these types of perceptions and these key social conditions necessary for assimilation might be promoted.

Changing Perceived Group Boundaries

People sometimes see those in another ethnic group as complete outsiders, almost as though they were alien beings. Where this is the case, they are likely to be suspicious and distrustful of the others. Feelings of empathy and norms of fair treatment and mutual aid that apply to someone defined as "one of us" are not extended to the "outsider." Thus, in order for people from different ethnic backgrounds to develop close relationships, they must see each other as sharing at least some common social identity.

If people could be categorized only by their ethnicity, then those of different ethnicity would always be outsiders to each person. But there are many other bases on which people may be categorized—by gender, age, occupation, political party or ideology, religion, recreational interests, geographical location, and so on. While ethnic background will, no doubt, usually be one basis for grouping people, other categorizations may be seen as equally or more important. For example, an individual may see those of different ethnic backgrounds as part of his school, his team, his occupation, his work organization, his military unit, his union, his church congregation, or his political party. The more that members of another ethnic group are seen as belonging to one or more groups that are important to a person, the more likely that person is to feel and act positively toward the other group. In fact, encouraging people to see those with different ethnic backgrounds as part of their own group(s) can be a key to reducing intergroup conflict (Hogg and Abrams 1988).

We have seen (chapter 2) that individuals place people in the same category the more they: a) are seen as similar; b) are seen as sharing a common fate; c) communicate frequently with each other; and d) are physically close, especially with a clear spatial boundary around them.

To widen perceptions of group boundaries, there is a necessity to promote greater actual and perceived similarities among people of diverse backgrounds. This is especially true with respect to characteristics that affect how they relate to each other, such as their languages and norms of behavior.

People become more similar the more they interact with each other and the more they have common experiences. The more that people from different ethnic groups grow up in the same neighborhoods, attend the same schools, get the same amount of education, work in similar occupations, and are exposed to the same mass media, the more likely they are to be similar in culture and in social status, and to be seen as such.

We have noted that once perceptual boundaries between groups are drawn, there is a tendency for people to exaggerate the similarities within each group and the differences between groups. Thus, in order to help widen perceived group boundaries, emphasizing the things that people share in common—such as their concern for family, their efforts for success, their desire for acceptance and respect—would be helpful. Schools and the media can be especially helpful in showing such similarities, often in dramatic fashion.

To broaden peoples' perceptions of the boundaries of their in-group(s)—those who are included in "us"—increasing both proximity and communication among people of diverse ethnic backgrounds would be desirable. More racial and ethnic mixing of neighborhoods, schools, church congregations, sports teams, and a variety of other types of social settings would result in greater physical proximity. But physical proximity does not always result in a high level of communication; thus, not only to have people from different backgrounds close to each other but also to create conditions that encourage communication is necessary. Arranging for people from various ethnic groups to participate in cooperative activities (in schools, work organizations, community organizations, and so on) is a particularly good way to increase communication across ethnic lines.

Encouraging perceptions that people share a common fate can also contribute to their sense that they constitute a single group. One important way to do this is to promote similar treatment of people with different backgrounds. This approach requires reducing as much as possible discrimination in housing, employment, and other areas of life based on ethnicity or other nonrelevant characteristics. The government may also have to reduce or eliminate differential treatment of various ethnic or other groups.

Sharing a common fate is a result not only of similar treatment by others but of having common goals and of being dependent on each other for reaching these goals. When people of varied ethnic background are involved in cooperative activities, they not only communicate more, as already noted; they also share common goals, the attainment of which depends on joint effort. Such activities may include classroom projects in school, athletic teams, work projects, and activities in the community, unions, and politics. In addition, the common goals that all Americans share (e.g., economic prosperity, preservation of the environment, reducing crime and drug addiction) may be emphasized as foci for cooperative effort.

The combination of the possible changes just outlined—greater proximity and communication among people of different backgrounds, promotion and highlighting of their similarities, nondiscrimination in their treatment by others, and promotion of activities with shared goals requiring joint effort—is likely to blur the sense of separateness felt by people with different ethnic backgrounds and increase the perception of a broader, more inclusive "us."

Changing Appraisals of Other Groups

Even if people from different ethnic backgrounds share some common social identities, they are likely to see each other at least partly in terms of ethnic group identity. Thus, if close friendly relationships among people of different ethnic backgrounds are to develop, members of each group must see the other group in positive ways.

We have seen (chapter 3) that feelings and behaviors toward those in another ethnic group are strongly affected by certain key appraisals of that group. A man who perceives that the other group is doing, or may do, harm to himself or his own group is likely to be angry and fearful and may react by avoiding, or being aggressive towards members of that group. A woman who sees members of another group as having undeserved advantages will tend to be angry at members of that group and to protest against what she sees as an unfair situation. A person who perceives the behavior of members of another group as reprehensible (i.e., as violating her standards of proper behavior) will probably feel dislike for that group and will tend to avoid its members.

How can appraisals of other ethnic groups be modified so that they are less negative and more positive? One possible approach is to provide information and present persuasive arguments. For example, in schools, in churches, on television, and elsewhere, teachers, clergymen, public officials, and public service groups can provide information that shows minority groups as helpful rather than as a threat to the dominant group, as not enjoying unfair advantages, and as often acting in praiseworthy ways that are consistent with community norms. For example, perceptions of certain threats from minority groups may be modified by information showing that Latino immigrants contribute to general economic prosperity and that home prices do not necessarily fall when blacks move into a neighborhood. Perceptions of minority people as criminals, content to live on welfare, unconcerned with family, and so on, can be countered by showing the lives of some of the great majority of hard-working, law-abiding people in these groups, people who are struggling to improve their lives and those of their children.

However, while providing favorable information about various ethnic groups may be of some value, such efforts are not likely to have a major impact on most people's perceptions. Public information programs in the media and school programs that provide students with information about various groups in our society have had only limited success in changing negative stereotypes (Atkins 1981; Stephan and Stephan 1984). Images of other ethnic groups, once acquired, are difficult to change. Moreover, given the present conditions in our society, there are real differences between ethnic groups that lead people to hear negative information about a particular group (e.g., that it has a high rate of crime or out-of-wedlock births).

Another way that has often been suggested for changing people's perceptions of those in other groups is promoting greater interpersonal contact. When reviewing evidence on the effectiveness of this approach (chapter 4), we found that more contact generally leads to more positive attitudes and behaviors only under favorable conditions, including equal status, compatible interests, and similarities in values and behavior. Thus, to change perceptions of those in other ethnic groups, the conditions under which members of different groups live and come into contact in society must be changed. Next we consider how some negative perceptions of other groups might change as certain key social conditions change.

Perceived Threats

Many types of perceived threats from other ethnic groups would be greatly reduced as a result of greater economic equality among groups. For example, if fewer blacks and Latinos lived in poverty, they would be much less likely to be involved in street crimes and, therefore, whites would be less likely to see them as posing physical threats. If there was greater economic equality, whites also would be less likely to see themselves suffering from higher taxes in order to provide benefits to black and Latino poor. They would be less likely to see blacks and other minorities as causing deterioration of their neighborhoods, since the link between ethnicity and "lower class" status would no longer be strong. As for African American and Latinos, they would be less likely than at present to see themselves being "kept down" by the dominant white group.

We have seen, however, that greater equality among ethnic groups may lead to greater perceptions of threat if there is a high level of competition for scarce resources, such as jobs and political positions, among groups (see chapter 5). Therefore, trying to reduce conflicts of interest and promoting shared interests across ethnic lines is important. For example, including people from different ethnic groups in the same labor union or political coalition will increase shared interests.

Perceptions of Unfair Advantage

Perceptions that members of another ethnic group are enjoying unfair advantages, relative to oneself or one's own group, may be modified by reducing or eliminating any actual patterns of preference to one group over another. There is now wide consensus among Americans on the principle that employers, the government, schools, and other institutions should not give preference to whites over people from other racial groups. The more that verbal agreement to this principle of nondiscrimination against minorities is translated into actual nondiscriminatory behavior, the less likely blacks, Latinos, Asians, and other minorities will perceive whites as receiving unfair advantages.

There is much less agreement among Americans concerning whether minorities—especially African Americans, Latinos, and Native Americans—should be given preferential treatment with regard to jobs, contracts, college admissions, and other benefits. We will discuss this difficult issue at greater length later in the chapter.

Nonpreferential measures (e.g., improved education or more overall job opportunities) that reduce inequalities between ethnic groups may also be expected to reduce perceptions of unfair advantage. For example, if differences between racial groups in occupation and income were greatly reduced, previously low-status groups would be less likely to see themselves as being treated unfairly. Greater equality would also remove much of the rationale for giving compensatory preferences to minority groups; if these types of preferences were not needed, members of the white majority would be much less likely to perceive themselves as being unfairly disadvantaged.

Perceptions of Reprehensible Behavior

Reducing economic inequalities would also be likely to decrease appraisals by members of more affluent ethnic groups that the behavior of other groups is reprehensible. The types of behaviors that are often condemned—for example, committing robberies, having children outside marriage, living on welfare—are the result, to a large extent, of poverty and the lack of jobs that pay enough to support a family. If low-income blacks, Mexicans, Puerto Ricans, Native Americans, and so on were able to have jobs and incomes that were roughly comparable to those of whites and Asians, the types of "lower-class" behaviors that run counter to the norms of most middle-class people would undoubtedly decrease sharply. Behavior among the great majority of those in minority groups would become similar to that now common among middle-class whites, blacks, Latinos, and others—that is, generally consistent with conventional norms.

Behaviors by members of particular ethnic groups that seem repugnant to those in another ethnic group may also be affected by cultural differences. For example, economic and other conditions in black inner city areas have led to a "culture of poverty" that accepts such behaviors as violence and pregnancy outside marriage as normal, if not acceptable. In addition to changing the economic conditions that spawn such a culture of poverty, providing greater exposure to the broader culture to poor members of minority groups is possible. Reducing racial and economic segregation of neighborhoods and schools, thus providing more contact with a wider range of people, would be an important step. Community programs, such as Boys Clubs and Girls Clubs, that teach young people to work towards conventional goals, and which reward them for doing so, may be helpful. Greater exposure of minority youth to adults from their own ethnic group and background who have succeeded in business and the professions may also contribute to an acceptance of goals and behavior that are valued in the broader society.

Reducing Inequality

When discussing ways in which perceptions of other ethnic groups can be made more positive, we have emphasized the importance of reducing inequalities—especially economic inequalities—between ethnic groups. The achievement of substantial reductions in economic inequality poses a difficult challenge. It is a complex topic that can be addressed only briefly here (for more extensive discussions, see Atkinson 1983 and Braun 1997). However, we may consider some of the main directions of policy that would be necessary to move toward greater equality.

Better Schooling

The primary requirement for greater economic equality is better education for minority groups—especially African Americans, Latinos, and Native Americans. As we have seen (chapter 5), changes in technology and the development of a global economy have greatly reduced the number of good-paying jobs available to, and lowered the pay for, people with little education and few skills. As we have also seen (chapter 7), African Americans, Latinos, and Native Americans are getting fewer years of formal education, and the quality of their education is much poorer than that of whites and Asian Americans. Until these disparities in education are greatly reduced, large economic inequalities will continue to exist among ethnic groups.

How can these gaps in education be narrowed? Our review of research on education (see chapter 7) indicates that minority students often do not take many of the academic courses (science, math, and so on) that they need to go to college or to perform highly-skilled jobs. Public schools may need to modify their structures—for example, by making academic "tracks" more flexible—in order to provide solid academic courses to more minority students, just as many Catholic schools have done.

Expectations for minority students by administrators and teachers have often been low, accompanied by lax academic standards and discipline. Teachers and administrators, confronted with students who come to school with poorly-developed academic skills, whose subcultures and perhaps languages are different from those of middle-class school personnel, and who sometimes are difficult to control, often give up on serious attempts to educate poor minority students. Yet, despite the special problems that they often present, there is evidence that poor minority children can succeed in school.

What appears to be necessary is high academic standards, firm discipline, high expectations, and support and encouragement from parents and school personnel to do well. Racial mixing, though sometimes helpful in maintaining high standards, does not appear to be essential. Predominantly minority schools can be good schools.

In addition to the key elements already mentioned, a number of other policies may be helpful. Preschool programs, (e.g., Head Start) that prepare children for school by giving them relevant cognitive and social skills help children, espe-

255

cially poor minority children, to do better in school. Such programs can be expanded and improved. Students who do not speak English well need to be helped with bilingual programs that permit them to learn basic skills, while making as rapid a transition to all-English as possible.

For some minority students, especially those in high school, low motivation impedes their learning. Sometimes low motivation arises from the perception that chances for success for people of their own ethnic and class background are poor. In the long run, such perceptions are best changed by increasing opportunities for people of all ethnic groups and reducing inequality. However, acquainting poor minority students with some of the many adults of their own ethnic groups and backgrounds who have "made it"—doctors, lawyers, engineers, businessmen, professors, military officers, and so on—would be helpful. Such exposure would help counter the perception among many minority youth that the only way for them to succeed is to be a professional athlete or entertainer (neither of which requires academic success).

Many poor minority students need more counseling about the specific type of academic work that is needed to qualify for the kinds of occupations to which they aspire. In addition, prestigious minority figures—such as ministers, politicians, and heads of ethnic organizations—can help to convince minority students that doing well in school is not "acting white" (as some think) but, rather, showing that their ethnic group is just as capable as whites.

Last, but not least, young people need help to pay for the schooling they may want. A college education, and graduate and professional education beyond college, have become more expensive, even at public institutions. Since minority people generally are poorer than others, cost tends to be a greater obstacle to higher education for them. Cost is also often a serious obstacle for those who want technical or vocational education beyond high school. If our society is serious about reducing inequality among ethnic groups, helping young people from all groups to pay for schooling beyond high school seems necessary. Besides the benefits of such a policy for social peace, it may pay dividends in greater national prosperity and higher tax collections—just as the GI Bill, which paid for the education of war veterans, did after World War II.

Other Measures

In addition to the crucial necessity of improving education, a number of other policies have been proposed to help reduce poverty and economic inequality. Many of these proposals—like many ideas for improving education—are not aimed exclusively at minorities but, since more minority people are poor, would help them disproportionately.

Some proposals aim at raising the incomes of people who work at relatively low-wage jobs. These include raising the minimum wage, providing some government income supplements through "earned income credits," and providing low-cost health insurance to all workers (Ellwood 1988; Reich 1991). Some have suggested that wages for low-skill workers may be raised also by limiting the number of low-skill immigrants, on the grounds that an oversupply of workers lowers general wage levels (Reich 1991), although the validity of this argument has been disputed (Simon 1990).

Many ideas have also been advanced to try to get more people who are on welfare into the work force (Heineman 1987; Stoesz and Karger 1992; Bane and Ellwood 1994). In 1996, changes in federal law have restricted the availability of welfare—for example, requiring states to limit benefits to a maximum of five years or less. To enable former welfare recipients to get and keep jobs, basic education and job training, as well as care for young children, needs to be available.

One of the main factors associated with low-income families, particularly among African Americans, is the high incidence of one-parent families (see chapter 5). Reducing the proportion of one-parent families would be a formidable task. There is widespread agreement that the shortage of well-paying jobs for young black men, which makes it difficult for them to support a family, is a major contributor to the problem. Thus, better education for minority youth to give them the skills to get good-paying jobs (as discussed earlier) is essential not only to provide higher incomes for individuals but also to encourage the formation of two-parent (usually two-income) families. Changes in welfare laws that would discourage young women from having children outside of marriage (e.g., not permitting young women under the age of twenty-one to have separate households) and that would encourage family formation (e.g., permitting two-parent families to get welfare payments under some circumstances) have also been proposed (Stoesz and Karger 1992).

Finally, economic inequalities may be reduced by decreasing discrimination against minorities and/or by giving some preferences to minorities (in education, jobs, promotions) in order to help narrow existing disparities. We turn next to the subject of discrimination.

Reducing Discrimination

Discrimination is, as we have discussed, one of the basic processes that affect relations among ethnic groups. The greater the discrimination against a given group, the more the inequality, the more the conflicts of interest between groups, and the more those who suffer discrimination are angry at their unfair treatment. Close, friendly egalitarian relations among those in different ethnic groups is not possible so long as significant discrimination exists.

While discrimination occurs in many settings, discrimination with respect to jobs (hiring and promotions) and discrimination in the criminal justice system have especially important consequences and deserve particular attention here (discrimination in housing, also very important, will be discussed in the next section on segregation).

Jobs

Discrimination in hiring, pay, and promotion on the basis of race, ethnicity, religion, or gender is against the law (see chapter 6). However, as we have noted, the relevant laws are hard to enforce.

Often, to determine and to prove that discrimination against minorities is taking place is hard. Among other problems, determining whether particular educational credentials, tests, or other requirements are necessary for performance of a job or are intended, instead, to screen out most minority group members, is difficult. Moreover, legal action on behalf of those alleging discrimination has often been slow; large backlogs of individual cases have piled up and the government has filed relatively few "class action" suits on behalf of groups of people who may have suffered discrimination by employers (Rose 1994).

Affirmative Action

Programs that use "affirmative action" to increase the numbers of minority persons being hired and promoted—including the setting of numerical goals and timetables for reaching the goals—have been promoted by federal agencies, and by some other public and private organizations, for several major reasons. One reason is that setting numerical goals (e.g., based on the proportion of minority persons in a given labor market who are qualified to do a specific type of work) is a way of overcoming the difficulties of enforcing laws that bar discrimination against minorities. Rather than having to go through the difficult process of finding and proving discrimination, enforcement agencies need only see whether the proportion of minority people hired or promoted match the proportion in a specified "pool" of potential workers.

A second major reason behind efforts to set numerical goals for minority workers is that such procedures often help to achieve more rapid results, in terms of increasing the proportion of minority workers in a given type of job, than would merely achieving a nondiscriminatory "color-blind" situation. Because members of some minorities—African Americans, Latinos, and Native Americans—generally have lesser amounts of education, and because they do not score as high on application tests, they may, in general, rank lower on criteria generally used for some types of jobs, as well as for university admissions. They may also have other disadvantages, such as limited ability to use standard English. Thus, simply eliminating discrimination based on race or ethnicity—even if this were achieved completely—would leave a substantial disparity in the representation of various ethnic groups in jobs and in higher education, at least for a considerable period of time. Affirmative action programs—especially in their emphasis on numerical goals and timetables—aim at speeding up the process by which the ethnic gap is closed (Taylor 1991; Edwards 1995).

What are the effects of affirmative action programs likely to be on relations between whites and minorities, especially African Americans and Latinos? Are such programs likely to contribute to, or be an obstacle to, closer, more friendly relations—that is, to assimilation?

In some ways, programs that give minority people special aid or preferences in getting good educations and good jobs are likely to improve intergroup relations—at least in the long run. As more blacks, Latinos, and other minority group members become more like whites (and Asians) in their education, occupations, and incomes, they are more likely to be seen by whites as "like us," rather than as outsiders; they are apt to be more similar culturally to the majority; and they are less likely to be seen as threats to engage in "street crime" or to degrade neighborhoods. These greater objective similarities and more positive perceptions should encourage greater willingness among both whites and members of ethnic minorities to have close and friendly contacts (for discussions of these and other possible advantages of affirmative action programs, see Howard-Pitney 1990; Ezorsky 1993).

On the other hand, programs that give preferences to minority groups may have some negative effects on intergroup relations. First, blacks, Latinos, and other minority group members who attain good educations and jobs may have their attainments devalued by others as being due to special preference. Thus, they may be seen as not really equal to their white or Asian peers. Secondly, conflicts of interest—real and perceived—are created between minority groups and whites. To the extent that the number of available jobs or promotions or school admissions are limited (and they almost always are limited to some extent), the more that one group gets, the less the other group gets. Minority groups are likely to be seen by many whites as threatening to take what might otherwise be theirs. Preferences given to minorities over whites are seen by many whites as being unfair advantages. These perceptions result in strong feelings of resentment, anger, and dislike directed at the minorities (as well as at authorities). The experience in many nations—such as India, Nigeria, and Malaysia—that have given preferential treatment to previously disadvantaged groups shows a consistent pattern of widespread hostility among those who see themselves as newly disadvantaged by "affirmative action" programs. Such hostility has sometimes resulted in violence and even civil war (see Horowitz 1985; Sowell 1990).

Overall, the advantages of affirmative action programs in accelerating the achievement of greater economic equality for minorities must be weighed against the hostilities among ethnic groups that such programs—in so far as they involve preferential treatment—inevitably produce.

Other Approaches

Affirmative action programs, and their pros and cons, need to be considered in the context of possible alternatives. What other options for reducing discrimination *against* minorities and/or accelerating economic equality are available? While there are no quick, magical, or completely satisfactory solutions, a number of other possibilities have been discussed (Burstein 1994; D'Souza 1995). More vigorous enforcement of laws against discrimination in hiring and promotions—including more class action suits in behalf of groups of people who have suffered discrimination—may be helpful to further reduce discrimination against minorities. In addition, requiring decisions on hiring, pay, and promotions to be based on objective criteria (years of education, relevant tests, credentials, amount of experience, and so on) may reduce the scope for ethnic bias to affect decisions. Objective criteria would have to be carefully chosen so that only ones that are genuinely important for job or school performance would be used.

Another approach (not inconsistent with those just mentioned) is to give special assistance to some people not on the basis of their race or ethnicity but on the basis of economic need (W. Wilson 1987). For example, special job training or financial aid for higher education can be offered to low-income people regardless of their ethnic identity. Advocates of such a class-based approach argue that it would not only avoid the conflicts resulting from race-based programs, but would be much more likely to get the political support needed to be sustained. Programs based on economic need might also unite power people from all ethnic groups in common interests. Thus, they would contribute to more positive feelings and relationships among people from various groups.

Criminal Justice System

The extent to which the criminal justice system discriminates against blacks and other minorities has been the subject of extensive debate and of somewhat conflicting findings (see chapter 6). Regardless of how widespread discrimination is by the police and by other parts of the criminal justice system (prosecutors, juries, and so on), such discrimination clearly does occur at times. Moreover, the widespread perception by minorities of discrimination against them by the police and by others in the legal system has generated widespread anger that has sometimes erupted in riots and has alienated minorities from the rest of society.

There have been some efforts to regulate the behavior of police, of prosecutors, and of judges to make the operation of the justice system more fair in general and less discriminatory against minority groups in particular.

Police Behavior

One important reform effort has been to try to control the use of deadly force by police. In response to protests by community groups (often minority groups) over incidents where police allegedly shot a civilian without adequate justification, and in response also to lawsuits, most police departments (especially in big cities) have adopted rules that restrict the use of guns to situations where an officer believes that he or another person is "in danger of death or grievous bodily harm." Also, police officers are now usually required to file a written report after firing a weapon and such reports are reviewed by their superiors (Geller and Scott 1992; Walker 1993). The effort to control use of deadly force by police has been, asserts Samuel Walker, "the great success story in the long effort to control police discretion" (1993: 25). The number of people shot and killed by police was reduced by about 30 percent between the early 1970s and the late 1980s. Moreover, the ratio between blacks and whites shot and killed by police was cut in half, from about six blacks for every white to about three to one (1993: 26).

While some police behaviors (use of deadly force, obtaining evidence, advising suspects of their rights) have increasingly been subject to standard rules imposed by the courts or by police departments themselves, other types of police actions remain almost entirely subject to the discretion of the individual officer. These include, for example, decisions about whether to stop or to frisk an individual, whether to detain or arrest someone, and, if an arrest is made, what charge to bring. Walker advocates a greater development of rules governing such actions, especially arrests: ". . . the critical arrest decision represents a major gap in the entire area of discretion control . . . there is no compelling reason why this should be the case. The domestic violence policies (concerning arrests) provide a model for possible reform. Departments could draft similar policies designed to guide officers' discretion" (1993: 41).

In addition to greater development and enforcement of rules to govern police actions, other measures to reduce arbitrary or discriminatory actions by police have been proposed, and in some cases, implemented. One important measure is to increase the representation of minorities on police forces. Significant increases in the proportion of blacks and Latinos on the police forces of many cities (e.g., Atlanta, Detroit, Chicago, Los Angeles) have occurred (U.S. Department of Justice 1990). However, the proportions of police forces composed of minorities in most cases remain below the minority proportion in the general population.

Another measure that can help to reduce discrimination against minorities (as well as restraining abuses by police more generally) is the creation of civilian review panels (Walker 1993). Such community groups, which usually include members of minority groups, provide an independent outside monitoring of police actions and exert pressure for changes when abuses occur.

Other Aspects of Criminal Justice System
In addition to efforts to impose more uniformity on the actions of police, there have been attempts to provide more uniformity (thus, less discrimination) in other procedures of the criminal justice system.

Supreme Court rulings in the 1960s and 1970s mandated that every criminal defendant has a right to an attorney; that a person being interrogated by the police has a right to have an attorney present; and that even defendants in misdemeanor cases are entitled to a lawyer, if there is a possibility of imprisonment. These rulings provided more equal legal protection for poor people, many of whom are members of minority groups. More recent Supreme Court decisions that prohibit the use of race to arbitrarily exclude jurors should also increase the chances for members of racial minorities to get a fair trial.

Fairness in providing bail for persons awaiting trial has long been a concern to many because those who are poor (disproportionately minorities) are less able than the wealthy to afford bail. The 1966 Federal Bail Reform Act and laws in many states prior to 1970 were intended to make bail more widely available. The impact of this first bail reform movement was mixed. The percentage of defendants detained in prison prior to trial decreased and there appeared to be some reduction in racial discrimination. But for those accused of a serious crime, having a prior record, weak family ties, and no employment record, bail continued to be hard to obtain (Thomas 1976; Walker 1993). In more recent years, great public concern about crime and a more conservative ethos has led Congress to change federal bail rules in a more restrictive direction. Specifically, the new laws make it easier for judges to deny bail, especially in the case of drug arrests.

Another aspect of the criminal justice system that some people believe discriminates against the poor and minorities is the widespread system of "plea bargaining" by which a person accused of a crime may plead guilty to a lesser charge or plead guilty in exchange for the promise of a lighter sentence. Attempts to limit or regulate plea bargaining have resulted in higher rates of case dismissals but in *more* disparity in sentencing among those persons going to trial. Thus, efforts to reform plea bargaining have not resulted in more uniform treatment of defendants (Eisenstein and Flemming 1988; Walker 1993).

The greatest efforts by those who want to improve the criminal justice system have been directed to changing sentencing practices. The use of indeterminate sentences (and discretionary parole) have been criticized by some people concerned that criminals are often treated too leniently. Others have charged that poor people, and especially members of minorities, are often incarcerated, or given longer sentences, than others who commit the same types of crimes but are given probation or lighter sentences. In response to such concerns, a number of states (e.g., Minnesota, California, and Washington) have adopted sentencing guidelines for judges, indicating appropriate sentences (or a range of sentences) for those committing particular types of crimes and having particular types of past criminal records. Similar guidelines now operate for federal courts. Studies of the effects of sentencing guidelines in Minnesota, Washington, and Oregon indicate that racial disparities in sentencing declined under the guidelines. However, blacks still tended to get harsher sentences than whites, due, at least in part, to their being more likely to have a prior criminal record (Tonry 1993).

262

While sentencing guidelines appear to have reduced racial differences in sentencing in those states where they have been used, most convicted offenders in the United States still face indeterminate sentences (Walker 1993). Use of sentencing guidelines by more states, especially as developed by politically independent sentencing commissions (Tonry 1993), appears to be an effective way to reduce discrimination in the criminal justice system.

Overall, we may conclude, echoing Walker (1993), that positive changes in the criminal justice system, while difficult to effect, are possible and that some progress toward reducing racial discrimination in these systems has been made. At the same time, more needs to be done to reduce both the reality and perception of discrimination.

Reducing Segregation

Segregation of ethnic groups reduces opportunities for intergroup contact, leads to greater cultural dissimilarities, and contributes to inequality. Thus, to promote close friendly relations across ethnic boundaries, reducing the physical separation of groups is important.

In the United States, racial segregation has been reduced or eliminated in a few institutional settings—most notably in the military services and in professional sports. Racial integration was accomplished in these settings soon after World War II. However, separation of those of different races—especially of blacks from others—remains widespread in many other major areas of American life, including schools and neighborhoods.

Schools

Desegregation of schools is not a panacea for producing either high academic achievement among minority students or more positive racial attitudes. But under the right conditions, attending racially mixed schools does tend to raise the achievement of minority students (chapter 7) and to result in more positive racial attitudes and behaviors by students from all groups (chapter 4). Proponents of school integration have argued that it is essential for children to learn to get along with those from other racial groups if a harmonious adult society is ever to be achieved.

In the southern United States, segregation of the public schools had been legally required until the Supreme Court ruled this practice unconstitutional in 1954. After delays in implementing the court decision, segregation of black children from white children declined sharply in the south during the 1960s and 1970s; though some segregation remains—including placement in different tracks within schools—blacks and whites have become more mixed in

263

southern schools than elsewhere in the United States (Jaynes and Williams 1989).

Outside the south, where school segregation was also common (and often the result of design by school boards), efforts to reduce such segregation have been widespread. Many school districts adopted (by court order or voluntarily) plans to bus students in order to achieve better racial balance.

Despite the efforts to promote desegregation of schools throughout the country, segregation of African American children from European American children remains widespread. While segregation of blacks decreased substantially from 1968 to 1972 as a result of the end of legal segregation in the south, it has not changed much since then; in 1988, about one-third of black students in the United States attended schools that were over 90 percent minority. Segregation of Latino students from their European American peers also is substantial and actually increased from 1968 to 1992 (Farley 1995).

Programs to bus students have sometimes succeeded in producing greater racial mixing in schools, especially in smaller cities that do not have large minority populations (U.S. Commission on Civil Rights 1976). But busing programs have often run into a number of difficulties. First, they have been strongly opposed by many white parents (and by some black parents) for a variety of reasons, some based on prejudice and some on practical problems such as long bus rides. In some cities, such as Boston, busing led to heightened hostility and conflict between the races. Secondly—and a more basic obstacle to success—busing programs have often not succeeded in producing student bodies that are (or that remain) racially mixed. In some cities, busing has contributed to "white flight" from the central city to the suburbs, where children can attend neighborhood schools (F. Wilson 1985). Where such moves by whites occur on a large scale, the children who remain in central city schools are predominately from minority groups.

The problem of "white flight" is only part of a much larger problem of the racial composition of the areas from which schools draw their students. In many large cities, 70 to 95 percent of the students are from minority groups (Pisko and Stern 1985). Thus, even if the small minority of white students is dispersed by busing, little meaningful desegregation is possible.

Some school districts have used plans that rely on parental choice, rather than on mandatory busing, to promote school desegregation. Many of these districts have created "magnet schools," which emphasize particular subject areas (such as music, arts, or science) in order to attract a diverse student body. While some of these programs have had some success (Rossell 1990), this approach does not reduce the high level of segregation in most central city schools.

A more general potential solution to the problem of school segregation is to mix students from predominately minority central cities with mostly white students from suburbs. In a few metropolitan areas, such as Louisville and St. Louis, courts have ordered desegregation plans covering an entire metropolitan area. But the Supreme Court has ruled that cross-district busing is not required unless governments have deliberately created school segregation between city and suburbs. In most cases, therefore, it is not possible now to transfer students across district lines.

How can greater desegregation of schools be accomplished? There are two major options. The first is to continue to try to bus students across district lines— usually from central city to suburbs or vice versa—to achieve better racial balance. While theoretically possible, such an approach would involve a number of serious difficulties. It would require, first, amassing enough political clout to overcome present legal obstacles to cross-district plans. It would, undoubtedly, arouse widespread and often passionate opposition from many parents and probably lead to greater racial tensions—at least in the short run. The financial costs of busing large numbers of students would be substantial. Given large differences in the social class background of inner city and suburban children, and probable widespread resistance to such a cross-district program, the academic and social outcomes of desegregation in these circumstances would be questionable.

The other major option for attaining racially desegregated schools is to first desegregate the neighborhoods in which people live. Such an option would avoid the difficulties of cross-district busing just mentioned. However, desegregation of neighborhoods presents formidable difficulties as well. It is a long-range approach that offers possible long-range, but not quick, results (see the next section of this chapter for further discussion of neighborhood desegregation).

Colleges

While the racial composition of schools below the college level is heavily influenced by residential patterns, students generally attend colleges that are distant from their home neighborhoods. Yet there is some racial segregation of blacks at the college level, most of it voluntary. About 80 percent of African American college students attend integrated schools, where they are, on average, over 7 percent of the total student body. The other 20 percent of African American college students attend one of ninety-nine "historically black" schools, forty-three of which are branches of public systems, and almost all of which are in southern or border states. These schools now have small minorities of white students (Hacker 1992).

Predominately black institutions provide an atmosphere in which many black students appear to feel comfortable and their retention rates appear to be generally higher than those for black students at predominately white schools. The continued existence of these predominately black colleges is staunchly defended by their administrators, faculty, students, and alumni. Yet, some observers see a system of separate black colleges as a relic of the white supremacy era of legal segregation. They maintain that these racially separate institutions should be phased out in order to move towards a more racially integrated and equal society (Roebuck and Murty 1993).

265

In assessing this issue, recognizing the difficulties that many black students experience when they attend predominately white colleges is important. Some black students who have academic problems receive inadequate help or guidance. Often, black students feel isolated and unwelcome by many, if not most, whites. In reaction to such feelings, many black students associate primarily with other blacks and sometimes actively seek to separate themselves from whites—in separate dorms, a separate student union, and so on (Hacker 1992). If the potential benefits of racial integration are to be realized, some observers have noted, predominately white colleges need to become places in which blacks and other minorities feel more at home and have friendly contacts more often with students outside their own racial group. How to accomplish this is a complex subject. Here we may merely point out that the research on intergroup contact (see chapter 4) offers some guidelines. These include arranging situations that bring people from different groups into face-to-face contact to work cooperatively on common tasks (class projects, community projects, teams, and so on), which bring rewards to all.

Neighborhoods

As we have noted, racially segregated neighborhoods lead to segregated schools. Also, concentration in segregated areas of central cities often restricts minorities to areas where employment opportunities are low. Both African Americans and Latinos have higher unemployment rates where they are more concentrated in central cities (Farley 1987). Ethnic segregation of neighborhoods may also contribute to the development and continuation of a "culture of poverty" among a poor "underclass" who are isolated from the mainstream of society (W. Wilson 1987).

Segregation of one ethnic group from another in neighborhoods is usually measured by the "index of dissimilarity," sometimes referred to as the "segregation index." This index, based on information on the residents of each city block or each census tract, can range from 0 (no segregation) to 100 (total segregation). The index number indicates what percentage of a given group (e.g., whites or blacks) would have to move to another block or census tract for there to be zero segregation. The segregation indices showing the separation of blacks from non-blacks in American cities generally have been high. For all metropolitan areas with substantial black populations, the average segregation index fell slightly from 1980 to 1990, but was 65 (towards the high end of the segregation scale) in 1990. Neighborhood segregation of blacks from non-blacks has been highest in old industrial areas of the east and midwest (e.g., Detroit, Chicago, Cleveland, Newark, and Philadelphia) and in retirement communities. It is lowest in the west, in smaller cities, in cities with small black populations, and in university towns and cities around military posts (Farley and Frey 1994).

Residential segregation of Latinos from non-Latinos and of Asians from non-Asians is considerably less than that of blacks from others. While there are a few areas in which segregation index scores are high for these groups (especially for Latinos), the average index of segregation for these groups in 1990 (42 for Latinos, 41 for Asians) was over 20 points lower than the average score for blacks (Farley and Frey 1994). Moreover, Latinos and Asians are more separated residentially from blacks than they are from whites. Thus, the problem of residential segregation is primarily a problem of the separation of African Americans from the rest of American society, including other minorities (Massey and Denton 1993).

How can segregation of neighborhoods—especially of African Americans from other Americans—be reduced? Raising the income of African Americans, thus making a wider range of housing choices affordable for them, can help some (Galster and Keeney 1988). But there is ample evidence that racial segregation is not primarily due to income differences between blacks and others (Darden 1987; Clark 1988). Put another way, if where people lived depended on their incomes, there would be a great deal more racial integration of neighborhoods.

We have noted earlier (see chapter 6) that historically the policies of banks, realtors, and governments (both local and federal) were intended to keep blacks and whites in separate neighborhoods. While racial discrimination in housing now is illegal, advocates of increased neighborhood integration urge that these laws be more strictly enforced so that banks make loans without regard to race, and realtors do not "steer" clients to the "appropriate" area for their race, or attempt to frighten white homeowners into panic selling of their homes on rumors of a neighborhood racial "turnover." Amendments to the federal Fair Housing Act of 1968 aid in enforcement of the law by making it easier to sue people who allegedly discriminate, increasing monetary damages that can be collected, and setting up a system of administrative law judges to review complaints.

Governments and other public bodies, such as school boards, can also fashion their own actions in ways that reduce, rather than increase, neighborhood segregation. These include such actions as locating public housing projects in racially mixed, rather than all-black, areas; placing highways in locations that do not separate black areas from other areas; and locating schools in racially mixed rather than single-race areas, thus encouraging people of several races to stay in these areas.

Policies to promote school desegregation may also be shaped in ways that aid, rather than harm, neighborhood integration. In some cases, efforts to desegregate a city's schools have led to busing white students out of racially integrated neighborhood schools in order to distribute them more evenly around the city, thus leading inadvertently to whites moving away from racially integrated neighborhoods. Some advocates of urban integration have urged that the policies to desegregate schools and neighborhoods need to be more closely coordinated (Orfield 1981).

While racial segregation in neighborhoods has been greatly facilitated by the past actions of banks, realtors, and governments, its fundamental causes lie in the preferences and choices of whites. As we have seen (see chapter 5), a large majority of white Americans say they would not mind if a black family with the same income and education as themselves moved into their block. However, when asked about living in a neighborhood where blacks are more than just a small proportion, whites express increasing reluctance as the stated proportion of blacks increases. When asked about situations in which blacks would constitute more than 30 percent of the neighborhood, more whites state they would move out of such an area than would be willing to move in (Farley et al. 1978; Farley et al. 1994). Such expressed preferences parallel events in neighborhoods in many cities around the United States; once blacks became more than a small proportion of residents in a neighborhood, the neighborhood often has "turned over" in just a few years and become almost all black.

Why are most whites (and many Latinos and Asians) reluctant to live in a neighborhood that has a high percentage of blacks? While there are a variety of reasons, key elements in this reluctance appear to be fear of crime and concern about possible deterioration of the neighborhood, lowered property values, and (for those with children) poor quality of neighborhood schools (Saltman 1990; Farley et al. 1994).

In their concern about such possible problems, whites often do not distinguish much between lower-class and middle-class blacks as potential neighbors. Commenting about whites' reactions to both African Americans and Latinos, Martin Jankowski states: "Most people believe that the lower class of these groups is primarily responsible for crime, but since they also believe that the majority of the group is lower class and they have no quick way to determine whether a member of this group is middle class or not, they prejudge anyone who looks like a member of the group as a potential threat" (1995: 92). In addition, whites may fear that middle-class blacks in their neighborhood may be followed by lower-class blacks.

How can greater racial integration of neighborhoods be achieved? One general strategy is to facilitate the movement of more black families out of heavily-black inner cities to primarily white sections of metropolitan areas. In several cities, including Chicago, Illinois, and Louisville, Kentucky, thousands of low-income black families have been assisted by government housing programs to move to scattered locations in the suburbs. In the Washington, D.C., metropolitan area, under a joint program of local governments and realtors, minority brokers were employed to help blacks obtain housing in primarily white areas. In Ohio, a government housing agency gave financial incentives in mortgages for buyers who made moves that increased racial integration. As a result of this program, several formerly all-white neighborhoods became racially integrated (Saltman 1990).

The greatest challenge is to keep neighborhoods that have reached some degree of racial integration (occasionally due to some organized programs, more often by uncoordinated actions of individuals) from "turning over" and becoming all-black. In metropolitan areas throughout the United States, organizations have been formed in racially mixed neighborhoods to try to maintain integration. Such efforts at neighborhood racial stabilization have, in fact, assumed the dimensions of a social movement, with national organizations such as "National Neighbors" providing some coordination for local efforts (1990). In some cases, such as neighborhoods in Indianapolis, Rochester, and Milwaukee, community organizations have been fairly successful in maintaining racially integrated neighborhoods. In other neighborhoods, such as ones in Hartford, Chicago, Los Angeles, and Washington, D.C., community organizations that formed to try to preserve racial integration have not been successful, despite gallant efforts (Molotch 1972; Goodwin 1979; Saltman 1990).

What is necessary for success in preserving integrated neighborhoods? In her book *A Fragile Movement: The Struggle for Neighborhood Stabilization* (1990), Juliet Saltman reviews many such efforts. Her review indicates that success in maintaining racial integration in a neighborhood requires that a number of key conditions be met.

First, present white residents must be discouraged from panic selling and the area must be "marketed" to potential new white residents. Banning solicitation of present homeowners by realtors and "for sale" signs on lawns, pressuring real estate agencies to show houses to white buyers, pressuring banks to make loans to whites in the area, even direct advertising for new white residents by the community organization, are examples of steps that have been taken to reassure old residents and to attract new white residents.

Secondly, the neighborhood must be made as attractive as possible in every way so that whites, as well as blacks, will want to stay or to move in. Programs to maintain or improve homes, lawns, streets, parks and playgrounds, and commercial areas, and to prevent crime can help make a neighborhood desirable.

Two conditions appear to be especially crucial in making integrated neighborhoods attractive. One is that neighborhood schools also be racially integrated, rather than predominately black. The second is that there should not be a high concentration of public housing, because "of the common *perception* of public housing as associated with poverty, welfare, and blackness" (Saltman 1990: 396). Saltman refers to primarily black schools and a high concentration of public housing as "killer variables"—that is, as killing the chances for maintaining racial integration of neighborhoods.

A third condition that may be necessary to prevent neighborhood racial "turnover" is to provide white residents with assurance that they will not suffer substantial loss of their property value by remaining in, or moving into, the area. A program to provide such assurance was set up by Oak Park, Illinois, as part of that community's effort to preserve neighborhood integration. The program reimburses residents for any losses incurred in the sale of their homes after they have lived in them for five years (Saltman 1990).

While the efforts of a group in a single neighborhood to preserve racial integration can be important, such efforts cannot succeed in isolation from events in the total city or metropolitan area. If nearby residential areas are all-white and nearby schools are all-white, white buyers may "play it safe" by choosing an all-white area with an all-white school, rather than an integrated area with an uncertain future. But if governments and school boards have succeeded in "opening up" the whole city or whole metropolitan area to at least some black residents, and have racially integrated the entire school system, white buyers have no reliable "havens" to which to withdraw. In such circumstances maintaining diversity in an area that already is substantially integrated becomes much more possible.

Other actions by governmental units beyond the neighborhood also may either help or impede stable integration. Actions that can impede stable integration of a particular neighborhood include public projects (such as highways), zoning changes, or concentration of public housing projects that make a neighborhood less desirable. Actions by government that can aid stable integration include public projects (e.g., parks), zoning changes, or crime fighting programs that make a neighborhood more attractive and programs (such as the one in Ohio, previously mentioned) that provide financial incentives for moves by homeowners that increase racial integration.

In sum, the most fundamental barrier to reducing racial segregation of neighborhoods is the association in the minds of many people between a neighborhood having a high proportion of blacks and it being an undesirable neighborhood in other ways—high crime, deteriorated housing, bad schools, and so on. Programs that have attempted to keep neighborhoods from "turning over" as blacks begin to move in have tried to show residents that racially mixed neighborhoods can be good neighborhoods. To be successful, such programs also reassure people that racial integration is not necessarily just a phase for a neighborhood on its way to becoming totally black.

Maintaining racially integrated neighborhoods as desirable neighborhoods is a goal that seems possible to achieve (though usually with difficulty) only where both whites and blacks in the neighborhood are from stable middle-class or working-class families. Where residents are from a population in which crime, drug usage, single-parent families, and other social problems are at high levels, then a neighborhood will not be a desirable place to live.

The social problems that make a neighborhood seem undesirable are primarily a result of inequality and poverty. Segregation helps to create inequality but, once created, inequality feeds back to maintain segregation. Thus, the ultimate solutions to racial segregation of neighborhoods also require solutions to the problems of inequality and poverty.

Other Settings

Outside the crucial areas of schools and neighborhoods, where decreases in segregation have been modest, progress in reducing segregation has been mixed. As noted earlier, dramatic reductions in racial segregation have occurred in the military services and in professional sports. Also, many other occupations that once were largely segregated—such as bank clerks, sales clerks, and police officers—also have been substantially integrated. However, separation of racial/ethnic groups, especially of blacks from others, continues in some institutional settings. Most notable are the churches. Sunday morning has been described as the most segregated time in American life (see chapter 6), since most congregations are composed almost exclusively of those from a single racial/ethnic group. Yet, those from different racial groups often share similar religious beliefs and may even belong to the same overall church or denomination (Catholic, Baptist, Methodist, and so on). Although a few efforts have been made to promote racial integration in churches, little progress has been made. However, given the similarity of religious beliefs across racial groups, there appears to be a great unrealized potential for racial integration in the churches. Since religion is important in the life of most Americans, such integration could contribute much to better race relations in the nation.

Summary

This chapter has considered one possible vision of society, one in which various ethnic/racial groups are steadily assimilated into a single people. Historically, assimilation has been promoted as necessary for harmony and unity within society. However, critics have attacked policies that promote assimilation as stifling diverse cultures and as being unrealistic in the light of continued separation and inequality among racial/ethnic groups.

In some countries, different ethnic groups have merged over time into a single nation; in other countries, different ethnic groups have remained separate over long time spans. In the United States, assimilation has progressed rather far among those whose families came originally from various parts of Europe. Among Americans of non-European origin (African Americans, Latinos, Asians, and Native Americans), cultural assimilation also is substantial and, though important socioeconomic differences among groups remain, some movement of non-Europeans into mainstream occupations, business firms, and other institutions has occurred. Also, a substantial and generally increasing amount of intermarriage has occurred between Latinos, Asians, and Native Americans, on the one hand, and whites, on the other hand. Marital assimilation among blacks is much lower, though some increase is apparent. Residential segregation from other ethnic groups is also greatest among blacks. Overall, considerable assimilation appears to be occurring among Americans of non-European descent, but is slowest for African Americans.

The extent to which people from any given ethnic group will form intimate, friendly relationships with those from other ethnic backgrounds (i.e., the extent to which assimilation occurs) depends on a number of conditions that have been discussed in earlier chapters. Subjectively, intimate positive relationships are most likely when people see those with other ethnic backgrounds as sharing membership with them in important groups (such as occupational or religious groups). Positive appraisals of the other ethnic group—as helpful rather than harmful, as not getting unfair advantages, and as behaving in praiseworthy ways—also contribute to friendly, equal relationships.

Favorable views of members of another ethnic group are likely to result from particular societal conditions. The basic conditions needed to produce positive intergroup perceptions and relationships include: equality between groups; lack of discrimination between groups; and absence of segregation of members of one group from another.

To reduce inequality among racial/ethnic groups, improving the education of minority youth is especially crucial. Possible ways to improve minority education were discussed. Other ways to improve the socioeconomic status of minorities include reform of the welfare system, strengthening the family, and reducing job discrimination.

Affirmative action programs that give preference to minorities may accelerate the achievement of economic equality for minorities but often evoke hostility among whites. Alternatives to programs that give preference to minorities include more vigorous enforcement of laws that prohibit discrimination against minorities; use of relevant objective criteria (e.g., years of education) in job decisions; and providing special help based on income rather than on race or ethnicity.

Many members of minority groups have been especially angered by what they perceive as discrimination against them by police and others in the criminal justice system. The widespread adoption by police departments of stricter rules governing the use of deadly force has led to a reduction in the disproportionate number of blacks shot and killed by police. However, police officers still have very wide discretion with respect to many other decisions—such as when to stop or to arrest persons. Some have advocated that police departments adopt more rules to curb arbitrary and possible discriminatory actions by police. Other policies that have been suggested to reduce possible police discrimination against minorities include further increases in the proportion of minority police officers and the creation of more civilian boards to review police actions.

Court decisions have provided lawyers for poor defendants and have prohibited use of race when excluding potential jurors. Among other changes in the criminal justice system, sentencing reform has been most significant. In states that have adopted sentencing guidelines, racial disparities in sentences have declined. However, most convicted offenders in the United States still face indeterminate sentences. Overall, while some progress has been made, more needs to be done to reduce real and perceived discrimination in the criminal justice system.

Racial segregation in the United States has been greatly reduced in some institutions, such as the military and professional sports, but remains substantial in other important settings, most notably schools and neighborhoods. Busing to achieve racial balance in public schools has often been ineffective, mainly because there are few whites living in many central cities where larger numbers of minority children live. One strategy for increasing school racial integration is to bus students across district lines (merging central city and suburban student bodies) but this solution faces serious political and legal obstacles. Another strategy is to try to desegregate neighborhoods, thus producing racially mixed neighborhood schools.

Residential segregation is more pronounced for African Americans than for other minority groups. Advocates of desegregation urge that laws barring discriminatory practices by realtors and banks be enforced more vigorously. Governments and school boards also can take a variety of actions that tend to reduce neighborhood segregation—such as placing schools and housing projects in racially mixed areas.

The most fundamental cause of neighborhood segregation is the fear of many whites that substantially integrated neighborhoods may bring a variety of problems, including high crime, neighborhood deterioration, and poor schools. Neighborhood organizations in some cities have attempted to promote racial integration by showing that a racially mixed neighborhood can be a good place to live. This can be true only when poverty and the social problems that derive from poverty are not widespread in a neighborhood. Thus, solving the problem of neighborhood segregation ultimately requires solving the problems of inequality and poverty as well.

Overall, we may conclude that the United States has continued to move in the direction of greater assimilation of diverse ethnic groups into a common society. However, movement in this direction has been limited, especially for African Americans. Further progress in reducing inequality, discrimination, and segregation would be likely to result in a further reduction of barriers separating ethnic groups.

10) Visions of Society II: Pluralism, Multiculturalism, and Cosmopolitanism

Some people believe that for various ethnic groups to merge into a single society is not desirable or not feasible. As we noted in chapter 9, critics of assimilation assert that the cultures and the social solidarities of various ethnic groups should be preserved. Some also say that to expect genuine assimilation to occur is unrealistic, especially for minority racial groups, in the foreseeable future. Many of these critics of assimilation advocate ethnic pluralism instead.

In this chapter, we will discuss the alternative of ethnic pluralism. We will also consider a popular set of ideas, called multiculturalism, that overlaps but is not identical to ethnic pluralism. Finally, we will briefly discuss a policy approach labeled cosmopolitanism that attempts to combine ideas from both the assimilationist and the pluralist perspectives.

Pluralism

The word "plural" signifies many, rather than one. In an ethnically pluralist society, each ethnic group retains its own identity and its own community life.

As with the concept of assimilation, the concept of pluralism may be applied to culture and/or to the patterns of interaction across ethnic group lines, that is, to social structure (Farley 1995). Cultural pluralism refers to a situation in which each ethnic group preserves its own traditions, language, customs, and life style.

Structural pluralism refers to a situation in which each ethnic group also maintains its own separate communal life.

Under structural pluralism, a group maintains, as far as possible, its own institutions—such as local government, schools, medical facilities, banks, stores, recreational places, and social clubs. Members of the group have intimate social contacts—friendships and especially marriage—within their own ethnic group. Their contacts with those from other ethnic groups are primarily instrumental ones—those necessary for making a living and for meeting other practical needs (e.g., some shopping, obtaining some professional services).

Since a distinctive culture will continue to be maintained only by a cohesive social group, those who advocate cultural pluralism also tend to promote some degree of structural pluralism. By living together in separate neighborhoods, and having their own institutions and organizations (religious, professional, youth, and so on), those in given ethnic groups communicate primarily with each other. The model of relations between ethnic groups presented in figure 8.2 (chapter 8) indicates that segregation of an ethnic group from other groups affects the opportunities for and conditions of interpersonal contact between those belonging to different groups. Communication focused within an ethnic group promotes cultural similarity and interdependence among group members, leading to a strong sense of group belonging. Physical proximity, especially in informal settings, also provides frequent opportunities for close friendly interactions within the group.

In many countries around the world—such as Northern Ireland, Malaysia, Lebanon, Israel, Switzerland, Bosnia, Canada, and Iraq—there has been a high degree of both cultural and structural pluralism (Lijphart 1977; Farnen 1994; Gurr and Harff 1994). In these countries, different ethnic groups have retained their own languages, customs, and traditions—often over many hundreds of years. They also live fairly separate lives from the other groups. Often there has been some geographic separation, with one ethnic group concentrated in a certain region or in a certain area of a city and other groups located in other places. Beyond any geographical separation that may be present, each group usually has maintained control of some institutions (local government, schools, religious places, and so on). Most significant, while economic contacts with those from other groups frequently occur, intimate social relationships in pluralistic societies have been kept almost exclusively within each ethnic group.

In the United States, while assimilationist ideas were dominant throughout most of American history, surges of support for some degree of pluralism have occurred (Newman 1973). Pressures for "Americanization" in the early part of the 1900s met with resistance from many Americans who wished to preserve their ethnic identity. These "cultural pluralists" wanted to retain their ethnic heritage at the same time that they became Americans. For example, in answer to an attack on "hyphenated Americanism," led by former President Theodore Roosevelt, John Dewey said in 1916: "The fact is, the genuine American, the typical

American, is himself a hyphenated character. It does not mean he is part American and some foreign ingredient is added. It means that . . . he is international and interracial in his make-up" (Kallen 1924: 131–132).

Cultural pluralists argued then, and continue to argue, that the preservation of ethnic culture and ethnic identity has positive value, both for those in minority groups and for the total society. Members of particular ethnic groups can derive pride and dignity from the history, culture, and heroes of their own people. The society as a whole benefits also, from the stimulation and dynamism of cultural diversity. As opposed to the "melting pot" metaphor of assimilation, pluralists offer images of the salad bowl, and of the orchestra—each having separate and diverse components but producing a pleasing whole (Gutmann 1994; Choi et al. 1995).

Support for pluralism in the United States usually has focused on its cultural aspect, rather than on structural pluralism. For example, in the "ethnic revival" among whites in the 1970s (Novak 1971), many people rediscovered or reemphasized their cultural roots in such countries as Italy, Ireland, or Poland, but there was little interest in promoting social separation of the group from other Americans.

However, there have been instances in which some members of a particular group have promoted institutional and social separation of their group from other Americans. Some religious groups—such as the Mormons and the Amish—have preferred at various times to live separately from others. Some individuals from particular European countries—for example, Norway, Sweden, Germany, and Switzerland—established small separate and ethnically homogenous communities in the United States during the eighteenth and nineteenth centuries (e.g., see Schelbert 1970; Wheeler 1986). However, such efforts to create ethnically separate communities were relatively few and the communities usually did not continue to stay ethnically separate.

A more significant movement to promote separation arose among African Americans in the 1970s. Although the civil rights movement succeeded in winning legal equality for blacks, many African Americans felt frustrated by slow progress in achieving economic and social equality with whites. Some black activists rejected racial integration as a goal and argued for separate black institutions instead (Carmichael and Hamilton 1967). They believed that African Americans were more likely to achieve economic and social progress by relying on their own efforts and their own institutions than by waiting for the uncertain prospect of being integrated into mainstream American society. Though most African Americans still hold to the vision of a racially integrated society (Kilson and Cottingham 1991), separatist views continue to be represented among some African American intellectuals, as well as in the religious group the Nation of Islam.

Those who argue for promoting solidarity within and some separation of a particular ethnic group point out that a separate cohesive group is often able to

277

exercise power in pursuit of its collective self-interest. For example, geographical concentration in a particular neighborhood or inner city permits an ethnic group to elect its own representatives, run its own schools, control local government jobs and contracts, support ethnic businesses and banks, and so on.

Support for ethnic pluralism, especially cultural pluralism, has received further impetus from recent demographic changes in the American population. Following changes in immigration laws in 1965, there was a dramatic increase in immigrants from Asia and from Latin America (U.S. Bureau of the Census 1993). Most of these new immigrants were racially different from the white majority and many wished to retain cultures that differed from that of white Americans.

Problems in Plural Societies

In a few countries, ethnic pluralism has coexisted, at least in some periods, with considerable socioeconomic equality and with social peace. For example, in Belgium (Flemish and French speakers) and in Switzerland (German, French, and Italian speakers), members of different ethnic groups have been roughly equal in social class (Lijphart 1977). In Switzerland, where ethnic groups tend to be geographically separate and enjoy considerable autonomy, relations between these groups have been generally (though not always) harmonious (Schmid 1981). In Belgium, though ethnic frictions have been greater, Flemish and French speakers have lived in relative peace (Covell 1993).

However, inequality and intergroup conflict appear to be common in ethnically plural societies.

Inequality

The socioeconomic positions of different ethnic groups within the same society rarely are equal (Horowitz 1985). Usually the disparities are sizable. Economic inequality between groups is illustrated by historical differences between ethnic Malays and ethnic Chinese in Malaysia. Though they were favored by the Malay-dominated government for the civil service, most Malays were farmers or unskilled workers, while Chinese dominated all other parts of the economy; the average Chinese earned about three times the income of the average Malay (Esman 1972). In Lebanon, a similar difference has long existed between more affluent Christians and mostly poor Moslems (Meo 1965). In Quebec, Canada, French speakers generally have been poorer than people of English origin, while in Northern Ireland, Catholics have lagged behind Protestants in occupational level and income.

Inequality may be due, at least in part, to discrimination by a dominant ethnic group against another group. For example, industry in Quebec was long controlled by people of English heritage who reserved the most skilled and desirable jobs for themselves and gave the more menial jobs to French speakers (Royal Commission 1969). A similar situation existed in Northern Ireland, where British-origin Protestants controlled industry and discriminated against indigenous Catholic workers (Osborne and Cormack 1991).

Inequality also may be the result, at least in part, of cultural differences between groups. Members of a particular ethnic group may do well economically because they have business skills; have close-knit families who cooperate in small businesses; place high value on education and work; or are proficient in the language used most in business or government. For example, the economic success of Chinese in Malaysia appears to have been aided by their business skills and willingness to devote long hours to work. Many indigenous Malays perceive ethnic Chinese as working harder than themselves, but do not want to emulate the Chinese because they see a single-minded emphasis on economic activity as ignoring religious values and personal cultivation (Wilson 1967).

In Nigeria, the Ibo and Efik peoples have had different attitudes toward education and work; the generally wealthier Ibo believe work to be a more important goal (Horowitz 1985). An interesting example of the possible impact of culture comes from a region of India where Bengali people grew vegetables for sale but neighboring Assamese people did not take advantage of this profitable business; the Assamese felt that carrying vegetables to market on their heads (as the Bengali farmers did) was incompatible with their dignity (Nair 1962).

Sometimes the cultural advantages of nondominant minorities—for example, Chinese in the Philippines, Tamils in Sri Lanka, Armenians in Turkey, Jews in Europe—have led to them having greater economic success than other ethnic groups, often despite discrimination against them. However, when this has happened, the dominant group has sometimes acted to remove the economic advantages of the minority group (e.g., by barring them from certain occupations or confiscating their wealth). In general, if people organize and compete economically on the basis of ethnicity, those groups with the most political and economic resources will enjoy the most influence and will be the most favored by society (Patterson 1977).

Some scholars have seen the existence of economic inequality in ethnically plural societies as inevitable. They suggest that separation of ethnic groups and cultural diversity tend to produce and perpetuate such inequality (Bullivant 1981).

Ethnic Conflict

Another common problem in plural societies is that of interethnic conflict. Often such conflict has led to violence.

Conflicts between ethnic groups have arisen over a variety of issues. These include distribution of material benefits, political influence, use of languages, symbols of inclusion and prestige, and autonomy.

Material Benefits

Some ethnic conflicts have stemmed, at least in part, from inequalities and the efforts of disadvantaged groups to change the distribution of wealth and jobs (Gurr 1993). In Malaysia, indigenous Malays have resented the greater wealth and key economic positions (e.g., in banking and trade) of ethnic Chinese (Esman 1994). In Fiji, native Fijians have complained about the greater economic success of ethnic Indians (Premdas 1993). In Northern Ireland, the greater wealth of British-origin Protestants and their control of the best jobs helped to fuel insurrection among indigenous Catholics (Probert 1978).

In addition to overall inequalities, direct economic competition between members of different ethnic groups can lead to conflict. In some cases workers compete for jobs. For example, black laborers in Guyana (South America) have rioted against further immigration by Indian workers whom they saw as under-cutting their wages and Burmese longshoremen have battled Indian workers who were brought in to replace them (Horowitz 1985).

Competition between businessmen from different groups also may contribute to ethnic conflicts (1985). In Uganda, African traders were outspokenly hostile to ethnic Indians (many of whom were business rivals) and African entrepreneurs in Kenya sponsored measures to restrict activities by ethnic Indians in trade and commerce. In the Philippines, Filipino businessmen pressed for government actions against ethnic Chinese merchants.

In many countries ethnic groups have disputed over the allocation of government jobs, government contracts, and admissions to public universities. In some countries, such as Sri Lanka and Malaysia, members of minority groups have complained against preferences given to members of the dominant majority. In other places, such as India, disgruntled members of the dominant majority have protested, and sometimes rioted against, policies that give preferences to previously disadvantaged minorities.

Control of Government

An issue that sometimes leads to major conflicts between ethnic groups is that of control of government. For example, in Lebanon a struggle by Moslems to overthrow Christian control of the country's government led to a long bloody civil war (Hiro 1993). In Rwanda, conflict over which ethnic group would govern led to mass murder by Hutus of Tutsis (Destexhe 1995). In Indonesia, an attempted but unsuccessful coup against the government was followed by widespread violence against Chinese because the Chinese were accused of supporting those who attempted to overthrow the government (Enloe 1973).

In some countries, political parties have been formed along ethnic lines. For example, Hindu, Muslim, and Sikh parties in India, Tamil and Sinhalese parties in Sri Lanka, and an Ibo party in Nigeria have represented the interests of their ethnic groups. When politics is conducted on an ethnic basis, conflict between ethnic groups tends to become more intense (Horowitz 1985).

Language Usage

Disputes between ethnic groups also may arise over the use of languages. The issue is usually whether only one language, or more than one language, may be used in government agencies, business firms, and schools. Sometimes, rules about language give one ethnic group an advantage over another for economic success. For example, when Sinhalese was made the official language of Sri Lanka, those who spoke Tamil were at a disadvantage in qualifying for government positions or for university admissions. This language policy helped to produce an armed Tamil rebellion (Hellman-Rajanayagam 1994). In Quebec, Canada, the exclusive use of English by most big business firms effectively shut off many well-paying jobs to French speakers for decades and fueled anti-English and separatist sentiments.

Disputes about the use of language are not based merely on considerations of practical advantage. Being able to use the language of one's own ethnic group in the larger society also may have great symbolic significance. To many members of minority groups, it signifies the worth and prestige of their ethnic group. Many members of dominant groups, on the other hand, feel that the status of their group and their place in society is threatened by the widespread use of other languages.

Prestige and Dominance

Beyond the specific issue of language, there is often a more general competition between ethnic groups for prestige and dominance within a society (Horowitz 1985). Conflicts sometimes arise that have little to do with material advantages but relate to symbols of prestige, such as the name of a town and whether that signifies "ownership" by one ethnic group or another. Such issues arise especially between an ethnic group that is native to an area and a group that has migrated to that area. In cities around the world such as Bombay, India; Phnom Penh, Cambodia; Karachi, Pakistan; Singapore; and Miami, Florida, groups with longer residence have resisted "domination" by immigrants and sometimes have claimed the right to expel them. Immigrant groups, on the other hand, have struggled for equal rights and sometimes for exemption from local customs or requirements (1985).

Autonomy

In many countries, ethnic groups have fought for greater autonomy in running their affairs or even for political separation from the larger political unit (Gurr 1993; Gurr and Harff 1994). Kurds in Iraq, Serbs in Croatia, black Christians in Sudan, Basques in Spain, and Tigreans in Ethiopia are examples of ethnic groups that have waged armed rebellion in their attempts to win greater autonomy or separate states. Usually, serious movements for autonomy arise among ethnic groups that are geographically separate from other groups. Ethnic groups whose primary goal is greater autonomy may be contrasted with those ethnic groups, such as Koreans in Japan and North Africans in France, whose primary aim is for more equality within the larger society (Gurr 1993).

Critics of pluralism—especially of those versions that promote a high level of separation between ethnic groups—point to the many examples and varied bases of ethnic conflict within a society. They are alarmed too, by recent events—such as those in Bosnia and Rwanda—in which conflicts between ethnic groups have erupted into violence and have torn nations apart. They question whether ethnic pluralism in the United States as well is compatible with harmony and unity (Schlesinger 1992; Hughes 1993).

Does Pluralism Require Conflict?

In the long run, the very existence of ethnic pluralism may depend to some extent, on inequality, discrimination, and conflict (Bullivant 1981). Solidarity within an ethnic group partly results from the positive satisfactions its members derive from the culture and social life of the group. But in-group solidarity is often a consequence of inequality and conflict of interest between ethnic groups and discrimination against certain groups (see chapter 2). Faced with hostility and mistreatment by outsiders, members of an ethnic group see themselves in terms of their ethnic identity and close ranks against their antagonists. For example, among Jews throughout earlier centuries in Europe, a strong sense of Jewish identity and strong cohesion in Jewish communities was partly a reaction to the frequent segregation and persecution of Jews.

If members of an ethnic group are not targets of discrimination, are equal to those in other groups, and are not in conflict with them, then the solidarity of their own group may decline. Thus, for example, after Jews gained greater equality in Europe and in the United States, the religious and ethnic affiliations of many Jews weakened and their assimilation into the larger society greatly increased. Similarly, once discrimination against Japanese-Americans declined sharply after World War II, members of this ethnic community increasingly assimilated into the broader society. Since ethnic pluralism depends on the continued existence of separate and cohesive ethnic groups, the long-term continuation of a pluralistic society—except perhaps one based on territorially separate groups—seems most likely when ethnic inequality and conflict are high.

Trying to Overcome Problems of Pluralism

Are the problems of inequality and of conflict that tend to be found in pluralistic societies inevitable? Are there ways in which these tendencies can be countered? We will discuss first some strategies for reducing inequality and then some strategies for reducing conflict in ethnically pluralistic societies.

Reducing Inequality

Two major approaches that have been proposed for reducing economic inequality in a pluralistic society are 1) to mandate preferences for disadvantaged groups; and 2) to change social institutions to better accommodate cultural diversity and thus enable all ethnic groups to succeed economically.

Group Preferences

One way to reduce inequality between ethnic groups is to distribute economic benefits and opportunities in ways that help disadvantaged groups. In many countries, including the United States, governments have given preferences in hiring, promoting, college admissions, giving contracts, making loans, and so on, to members of particular ethnic groups.

We discussed such "affirmative action" programs when we considered the topic of assimilation (chapter 9), since they may be intended as a nonpermanent measure to encourage assimilation. However, programs that allocate positions or rewards proportionally on the basis of ethnicity can also be used as a permanent strategy for dealing with ethnic pluralism. Such programs can help to reduce inequalities among ethnic groups. However, they have some drawbacks, particularly in producing resentments among those ethnic groups that are disadvantaged under the ethnic preference system (see section on Reducing Disparity in Rewards later in this chapter).

Accommodating Diversity

Another approach to fostering greater economic equality is to encourage key social institutions, especially schools and businesses, to be more accepting of cultural diversity. Such openness to diversity may make it more possible for ethnic minorities to achieve economic success.

Efforts to have schools, businesses, and other organizations adapt to and make use of cultural diversity have, in fact, become widespread in the United States in recent years. We may look first at efforts to be more accepting of cultural diversity in schools and then at similar efforts in work organizations.

Schools

Differences in the amount and quality of education obtained by different racial and ethnic groups (see chapter 7) is a major obstacle to achieving economic equality among these groups. Some educators and others have argued that most American schools pay insufficient attention to cultural differences among ethnic groups that may affect student learning. They urge that schools should pay greater attention to diverse ethnic experiences and should tailor their curricula and their teaching methods to the experiences, the interests, the learning styles, and the languages of minority students (Bennett 1990; Gollnick and Chinn 1990; Diaz 1992).

Curricula in a variety of subjects may be broadened to include more material on the experiences, the history, and the contributions of various ethnic groups to America and to the world. Schools also may be flexible in permitting the use of languages other than English, both in bilingual programs for students whose first language is not English, and in some nonacademic activities such as clubs and entertainments. If students from minority groups feel that the school accepts their culture, rather than treating them as outsiders, they may feel more involved in the school environment.

Minority students may also have learning styles that differ from those of middle-class white children. For example, Latino and African American children tend more than white children to prefer working in groups rather than as individuals. Teachers have been urged to use methods in the classroom that take into account the different learning styles and special needs of minority students. Among the teaching techniques that have been suggested as especially appropriate for minority students are the following: motivating students by involving them in activities that relate to their own personal and cultural experiences; structuring activities so that students can progress at their own rates without being compared to others; and having students work in cooperative groups rather than only as individuals (Bennett 1990).

The overall goal, states Christine Bennett, "is . . . a supportive noncompetitive communal learning environment where individual students are encouraged to put forth their best efforts and achieve their highest potential. Student diversity is accommodated without sacrificing quality in what students are learning" (1990: 266).

Work Organizations

Many business and government executives have seen their workforce become more ethnically and culturally diverse in recent years. This trend has occurred because of affirmative action programs and because of demographic changes in the United States. The minority share of the total workforce, which was about 17 percent in the late 1980s, is estimated to grow to about 25 percent by the year 2000. By the end of the century, minorities and women are expected to compose about 75 percent of new entrants to the work force (Loden and Rosener 1991; D'Souza 1995).

Cultural diversity may be valuable for organizations in a number of ways. People from different cultural groups have distinctive ways of thinking and acting and, therefore, can make distinctive contributions (Jackson and Ruderman 1995). Taylor Cox writes: "Diverse groups have a broader and richer base of experience from which to approach a problem. If persons from different sociocultural identity groups tend to hold different attitudes and perspectives on issues, then cultural diversity should increase team creativity and innovation" (1993: 33–34).

However, organization executives traditionally have attempted to shape their employees into a single mold. People have been encouraged to look, talk, think, and behave according to a single set of norms—usually those of white males.

Recent literature concerned with managing diversity in the work force has urged organization leaders to recognize, respect, and adapt to cultural differences among their employees (Loden and Rosener 1991; Thiederman 1991; Jackson 1992). For example, they may permit the use of languages other than English on the job, if this policy is compatible with work efficiency. Of course, where customers or clients often speak a non-English language, knowledge and use of such languages by employees may be very useful in accomplishing the organization's goals.

Even when employees are all speaking the same language, they may differ in their pronunciations and in their styles of communication. For example, people from different cultural backgrounds may differ with respect to how much they initiate discussion versus listening and responding; the extent to which they rely on factual data versus intuitive judgments; and the amount of confrontational versus compliant behavior shown in conflict situations (Loden and Rosener 1991: 87–90). Managers have been urged to familiarize employees with ethnic pronunciations and to recognize and respect differences of communication style. Loden and Rosener state: "If we think of them as contrasting behaviors that are equally effective, then we will have less difficulty dealing with diverse styles" (1991: 90).

Organizations also may respect and adapt to other types of cultural differences. They may permit employees to wear clothing or hair styles that are culturally distinctive. Work schedules can be arranged to permit worship on nonstandard days (such as for Muslims and Jews). If tests are used for hiring or promotion, they may be screened to omit any items that culturally biased. Alternatively, some promoters of cultural diversity in the workplace have advocated giving up standard tests and even standard performance evaluations. One idea is to individualize performance evaluations, in order to take account of each individual's particular background.

Many large corporations and government agencies have instituted programs to encourage and adapt to cultural diversity in their organizations. A few well-known corporate examples are Apple Computer, AT&T, Coca-Cola, General Motors, Goodyear, IBM, Motorola, and Proctor and Gamble.

Some features of cultural diversity programs have met criticism. Critics are dubious about whether a variety of ethnic cultures necessarily contributes to better productivity or higher morale. They question especially the removal of uniform standards of performance, suggesting that such action may cause efficiency to suffer (MacDonald 1993; Sowell 1993; D'Souza 1995). However, the critics acknowledge that there is a need for organizations, both in their own interests and those of society, to review possibly arbitrary standards, rules, and procedures that may disadvantage members of ethnic minorities.

Accommodating cultural diversity in societal institutions is not likely to eliminate completely the economic inequality that may be due to cultural differences. But such accommodation to, and even welcoming of, diversity—especially in schools, business firms, and government—may help to reduce such economic inequality.

Promoting Harmony

When trying to reduce conflict between groups in an ethnically pluralistic society, several types of strategies are possible. The general approach that has received most attention focuses on arranging the political system in ways that will give minority groups more say in societal decisions. A second strategy would try to reduce disparities in rewards received by different ethnic groups and thereby reduce dissatisfaction that may lead to conflict.

Political Changes

A key problem in an ethnically divided society is that political decision making may be monopolized by the ethnic group that constitutes a numerical majority. Members of a group that is a numerical minority may feel condemned to having no say in decisions, including those policy decisions that may directly affect them. Such perceived, and actual, lack of meaningful political participation tends to fuel frustration, dissatisfaction, and ethnic conflict.

Partition

One possible "solution" is a redrawing of national boundaries—for example, by partition of a country—so as to create a new nation in which the ethnic group that had been a minority is now in the majority. Since partition rarely, if ever, results in ethnically homogeneous states, it rarely solves the problem of interethnic conflict—though it may improve the position of a particular minority (McGarry and O'Leary 1993). For example, the partition of Ireland left an Irish-Catholic minority in Northern Ireland and resulted in continuing strife between Irish-Catholics and British-Protestants there.

Given that one or more ethnic groups is a numerical minority in a given society, two major types of political arrangements that may give minorities greater influence are: a) providing some autonomy for ethnic groups, and b) electoral systems that maximize minority representation.

Local Autonomy

Since the ratio of different ethnic groups may vary across different geographical areas, a group that is a numerical minority in a country as a whole may be a majority in certain areas. If many government functions in a country are assigned to separate geographic regions, people in each ethnic group can make decisions about their own affairs in those areas where they predominate. They also are able to gain a greater share of benefits provided by local government, including government jobs. In addition, conflict between ethnic groups is dispersed away from a single point (the central government) and may become *intra*ethnic (for control of local decisions) as well as *inter*ethnic (Horowitz 1985; McGarry and O'Leary 1993).

Decision-making functions of government have been divided among different areas in many countries, often in an attempt to reduce ethnic conflict. Sometimes the process has been called federalization, sometimes cantonization ("canton" is a term referring to a local region in Switzerland). Examples of countries in which considerable regional autonomy exists include Switzerland, Spain, Nigeria, Canada, and Belgium.

In the United States, considerable decentralization of government power also exists (with fifty separate state governments), although the political system was not designed with a focus on ethnic minorities. No single ethnic group that is a minority in the United States as a whole is a numerical majority in any state (though a cluster of minorities constitutes a numerical majority in Hawaii). However, minority groups—especially African Americans and Latinos—are a majority in some cities and towns and control the local governments and schools in those places.

In some cases, arrangements that provide some local autonomy appear to have contributed to a reduction in ethnic conflict. This appears to have been true, for example, in Switzerland, Belgium, and for a period in Nigeria (Lijphart 1977;

Horowitz 1985). However, federalist arrangements have not often succeeded in avoiding ethnic conflict. McGarry and O'Leary state: "Unfortunately, federalism has a poor track record as a conflict-regulating device in multiethnic states, even where it allows a degree of minority self-government. Democratic federations have broken down throughout Asia and Africa" (1993: 34). They go on to point out the emergence of secessionist movements—for example in India and Canada, and the actual secession of Slovakia from the former federalist Czechoslovakia.

Conflicts between ethnic groups may continue in federalist systems because some important decisions continue to be made at the level of the central government, which the majority group controls. The frustrations of a minority group with central government actions may lead it to try to withdraw the area it controls from the larger nation. Sometimes a secession proceeds peacefully, as when Slovakia broke off from Czechoslovakia. In other cases, however, attempted secession is met with harsh reactions from the central government. This has occurred, for example, in India, where secessionist movements in Kashmir and Punjab have resulted in violent struggles. If federalism is to be not merely a step to secession, different ethnic groups must have shared interests in the undivided state (Horowitz 1985). If, for example, they are dependent on each other for economic prosperity or for adequate defense, their incentives to remain unified may be greater than the incentives to divide.

Electoral Systems

Within a given political entity, whether it is a central government or a more local government, electoral systems may differ in ways that affect the influence of minorities and the degree of cooperation versus conflict across ethnic lines. Among the ways in which electoral systems may differ are: who qualifies to fill particular positions; what the sizes of election districts are; how many choices voters make; and how the winners are selected.

A. Who Qualifies for Office

In order to ensure that each ethnic group has a share of political power, particular offices may be reserved for persons from this group. Thus, in Lebanon, the president had to be Christian, the prime minister had to be Moslem, and a fixed number of seats in the legislature were reserved for members of each group (Hudson 1968). Another device is to rotate occupancy of an office. This procedure was used, for example, in what was formerly Yugoslavia where for a time representatives of different ethnic regions took turns in the presidency (Cohen 1993).

B. Size and Shape of Election Districts

The size and shape of election districts may affect the representation of minorities. For example, if a city has a population composed of 70 percent of group A and 30 percent of group B, it may create ten large election districts, in nine of which (90 percent) group A constitutes a majority; or it may create twenty smaller districts, in fourteen of which (70 percent) group A has a majority. Assuming that people tend to vote along ethnic lines, in the second case the two ethnic groups would be represented in proportion to their share in the population, while in the first case the smaller group would be underrepresented. In the United States, use of large election districts sometimes has diluted the voting strength and representation of minority groups—for example, of Latinos in Los Angeles.

C. Voter Choices

Usually, voters are asked to select one person among a number of candidates. A variant of this method is to have each voter indicate not only her first choice but her second and third choices as well. When no candidate receives a majority of first choices, second and third choices are assigned to the two top candidates in order to select a winner.

This voting system tends to encourage accommodation between ethnic groups. The method was adopted for presidential elections in Sri Lanka, under their constitution of 1978 (de Silva 1979). Tamils in Sri Lanka are a numerical minority and—given considerable ethnic hostility in the country—a Tamil candidate is unlikely to win a presidential election. But the second- or even third-preference choices of Tamil voters are valuable to Sinhalese candidates. Therefore, those candidates have a strong incentive to reach out to Tamil voters by moderating their positions on ethnically controversial issues.

D. Selection of Winners

Even when voters indicate only their first choices among candidates, the rules by which winners are selected may differ. The person who wins the most votes in each district may be declared the winner, as is true in American elections. Alternatively, under a system of proportional representation, each party may be allotted a number of seats proportional to its share of the vote across all districts. For example, in Guyana (South America), Afro-Guyanese and other minorities obtained a greater number of seats in the legislature after proportional representation was introduced (Newman 1964).

Another type of selection formula that may increase the influence of minority voters was used in Nigeria. To be elected president there, a candidate had to win a plurality of votes in the whole country, plus at least 25 percent of the vote in at least two-thirds of the nineteen states. Since minority groups were numerous in many states, a successful candidate had to appeal to a number of ethnic groups. As Horowitz comments, "It is not surprising that the president emerged as a conciliatory pan-ethnic figure" (Horowitz 1985: 638).

Power-Sharing

Federalism and certain electoral arrangements may be part of a broader system of power-sharing or "consociationalism" (Lijphart 1977; Kellas 1991). A power-sharing system usually will have most or all of the following institutional arrangements:

1. A "grand coalition" of leaders of all ethnic groups who make up the central government.
2. A proportional representation electoral system and a proportional system for sharing public spending and public jobs, based on the size of each group.
3. A "mutual veto" system, whereby each ethnic group can veto government decisions on issues of vital concern to it.
4. Some autonomy for each ethnic group, either through local government or through control of certain institutions (e.g., its own schools).

Although power-sharing is intended to dampen ethnic conflict, in many cases it has not accomplished this goal. Reviewing the experiences of various countries that have tried power-sharing, or consociationalism, Kellas states:

> Despite some successes, the history of consociationalism has been marked by tensions and breakdown. Belgium has continuing strife (mostly nonviolent) between Flemings and Walloons; Switzerland had a civil war in the nineteenth century and a separatist movement in the Jura in the 1970s; in Northern Ireland power-sharing collapsed in 1974 after a brief existence; in Lebanon, Nigeria, Sri Lanka and Malaysia there are only echoes of consociationalism while civil wars rage or have just recently ceased. Of the consociational countries still operating as such, perhaps only Switzerland is entirely politically stable today. (1991: 138)

Some writers have suggested that a successful power-sharing system requires certain conditions. Among the conditions proposed are: 1) that the ethnic groups do not change in their relative size or importance; 2) that the leaders of each ethnic group are able to control extremists within their own group; 3) that no single ethnic group be a majority, which might tempt it to try to dominate others; 4) that there be some overarching loyalty to the state among all groups. (Lijphart 1977; Kellas 1991).

Overall, it appears that power-sharing is difficult to maintain successfully over a long period in many societies but that it may be a useful strategy under favorable conditions.

Power-sharing in the United States

The applicability of ethnic power-sharing to the United States seems limited at present. The United States has a federal system composed of fifty states but one ethnic group (non-Hispanic whites) is a dominant majority in almost every state. An exception is Hawaii, where no single ethnic group is dominant.

Some political power is exercised by ethnic minorities at the level of city government. African Americans constitute a numerical majority in many large American cities—for example, Atlanta, Detroit, Newark, and Gary—and in many smaller cities and towns as well. By controlling many municipal governments, school boards, and other public bodies, African Americans often are able to exercise autonomy over many of their own affairs. The same is true for Hispanics in a number of towns and cities. Many Native Americans control their own tribal governments on reservations or at least share power with agencies of the federal government.

Of the electoral and other techniques used to disperse power among ethnic groups in other countries, some are presently or potentially applicable to the United States. In some American cities the number of electoral districts has been increased, creating more areas that minority people can control. Other techniques to increase minority power (e.g., asking voters to give second and third preferences) also could be used at various levels of government.

However, some important power-sharing mechanisms used in other countries do not seem applicable to the American scene, at least at the present time. Most notable of these is proportional representation. When this idea has been proposed (e.g., Guinier 1992), it has met almost uniform opposition among leaders of the dominant political establishment. Moreover, the Supreme Court has ruled that creating congressional districts for the primary purpose of providing minority representation is unconstitutional.

Demographic trends may result in a greater degree of ethnic power-sharing in the United States than has been true to date. Demographic trends indicate that, by the latter part of the twenty-first century, whites may no longer constitute a clear numerical majority in the United States (Murdock 1995). Whites are likely to lose their majority status even sooner in some states, such as California and New Mexico. As the relative size of the African American, Hispanic, and Asian American populations grow, they will inevitably—separately and in potential coalitions with each other—share power more with white Americans. Greater power-sharing among ethnic groups could lead to fewer frustrations among minorities and to less conflict among ethnic groups. On the other hand, a changed distribution of power, and possible new coalitions (e.g., between whites and Asians on some economic issues) might produce some new conflicts.

Reducing Disparity in Rewards

Conflict between ethnic groups may result, in part, from economic inequalities (see chapter 5). To reduce such inequalities, government programs that give preferences to particular ethnic groups have been adopted in many countries (Horowitz 1985). Often, such programs, discussed earlier in this

291

chapter, have been intended primarily to reduce ethnic conflicts. Discussing the thinking behind the adoption of such programs, Donald Horowitz states: "Beneath preferential policies lies a mind-set that sees ethnic conflict as the product of economic differences and ethnic harmony as the result of proportional distribution of all groups at all levels and in all functions of a society (1985: 659).

Programs that give preference in hiring, university admissions, contracts, and so on, to disadvantaged minorities have had some success in reducing inequality. For example, preferential programs succeeded in greatly increasing the proportion of previously underrepresented ethnic groups among university students in Tanzania, Malaysia, and Sri Lanka. In Malaysia and in India, employment preferences have been used to open many jobs to members of disadvantaged groups. In Indonesia and Malaysia, members of disadvantaged groups have obtained more business licenses, contracts, and ownership under preferential programs. In the United States, preferential programs in a variety of areas—including employment, college admissions, and contracts—have helped many members of minority groups (Belz 1991).

While preferential programs have been intended, by reducing inequality, to also reduce ethnic conflict, their short-run effect has been, instead, to increase such conflict. For example, in Sri Lanka the adoption of school admission quotas that favored Sinhalese led to a wave of violence by Tamils. In Malaysia, preferences for ethnic Malays in school admissions led many ethnic Chinese to join a guerrilla movement that was trying to overthrow the government. In India, members of higher-caste groups have rioted to protest preferences given to the lower castes for schooling and jobs. In the United States, affirmative action programs that give preferences to minorities for jobs, school admissions, or business contracts have led to resentment among many whites (Sniderman and Piazza 1993).

If preferential programs were able to achieve economic equality, they might be expected, in the long run, to have their intended effect of ethnic harmony. This achievement would be especially likely if, once equality had been achieved, preferences could then be dropped. However, if society continued to be characterized by cultural pluralism, such cultural differences might tend to lead to inequalities again (see earlier discussion of the relation of cultural differences to inequality). Thus, to continue preferences in order to ensure equality of result might be necessary. In that case, the resentments and conflicts stemming from preferences would be likely to continue as well.

Although preferences for disadvantaged groups have been widely used as a way to effect change rapidly, there are, of course, other ways to promote economic equality. These include the strategy discussed earlier in this chapter of making schools and business organizations more accommodating of cultural diversity, thus making it easier for ethnic minorities to succeed (for further discussion of ways to reduce inequality, see chapter 9).

Multiculturalism

Many of the arguments in favor of cultural pluralism have been advanced in recent years under the name of multiculturalism.

Multiculturalism is a set of ideas that affirms the value of cultural differences within a society. Policies that affirm and support cultural differences have been adopted in many countries including Canada, Australia, and Great Britain and multicultural ideas have been influential in the United States as well (Bullivant 1981; Shafir 1995).

The approach has been applied most prominently in education. In the United States, multiculturalists argue that American schools traditionally have presented a distorted "Eurocentric" picture of American history and society that ignores or downplays the role of non-white segments of the population. They advocate a school curriculum that includes the values, experiences, and contributions of the varied ethnic groups within American society. Such a curriculum, they say, will lead students from historically disadvantaged groups to have more pride in their own groups and will lead all students to have a better understanding and appreciation of the different ethnic groups with whom they must live in our society (Sleeter and Grant 1988; Bennett 1990; Levine 1996). The curriculum in many American schools, from kindergarten to graduate school, has, in fact, changed in recent years to give students a more multicultural perspective.

Greater respect for and affirmation of cultural differences has also been urged for the workplace. Influenced by these ideas (and by pressures from minority groups and from government) many corporations have instituted programs to encourage and respect diversity in their organization. The multicultural perspective has also been applied to other areas of society, such as the arts and entertainment, where greater attention to distinctive cultures has been urged and sometimes realized.

Beyond the call for greater attention to and appreciation of diverse cultures, a number of other themes are frequently expressed by multiculturalists. One is an emphasis on the past and present oppression of ethnic minorities—African Americans, Latinos, Asian Americans, Native Americans—by white Americans. A related emphasis is on the inequality that exists among ethnic groups and the need for greater equality to be achieved. Thus, multiculturalism for most of its adherents is not only about culture. Pressures for assimilation are seen as part of an oppressive white-dominated society. Multiculturalism usually means not only respect and support for minority group cultures but also support for equality—in wealth, power, prestige—among these groups (Vega and Greene 1993).

Multiculturalists usually are suspicious of talk about national unity or patriotism. They tend to see the nation as being dominated by a white power structure that is fundamentally not responsive to the interests of minority groups. Multiculturalists wish to see harmony among various ethnic groups. But they are

293

acutely aware of the conflicts of interest among ethnic groups (especially between whites and minorities) and see a need to struggle to obtain the rights and benefits for minority groups that they see as appropriate.

In its emphasis on the value of cultural diversity, multiculturalism is similar to pluralism. But while pluralists advocate a permanent cultural and physical separation of ethnic groups within a society, multiculturalists generally have not been clear about this point. They usually do not specify whether separate cultures are (or should be) a temporary or a permanent feature of a society. Is it expectable and/or desirable for example, that the cultures of various ethnic groups eventually will blend into a common American culture? Or should we expect and/or try to arrange a preservation of separate cultures indefinitely? Writing about what he describes as an ambiguity in multicultural thought on this point, Shafir states: "This ambiguity allows multiculturalism to cohabit with either assimilation or segmentation, though permanent compromise between these views is unlikely" (1995: 11).

Criticism of Multiculturalism

The call of multiculturalists to foster understanding of, and respect for, different cultures meets with almost universal agreement. However, critics have raised a number of serious objections to multiculturalism, as it has been advocated and/or put into practice (e.g., see Schlesinger 1992; Hughes 1993; Hollinger 1995; Lind 1995).

Objections have been raised, first, to what some see as an excessive and arbitrary focus on particular types of ethnic identities and affiliations, to the exclusion of other identities and affiliations. Critics object to the widespread classification of Americans into five ethnic groups (white, African American, Hispanic, Asian American, and Native American) as artificial. They point out that this classification was introduced rather arbitrarily by government officials in the 1970s and does not correspond either to cultural differences (e.g., Asian Americans come from widely different cultures) or to racial differences (e.g., Hispanics may be white, black, or Indian). Furthermore, forcing people into these arbitrary categories ignores the large and growing number of people who are of mixed ethnic backgrounds.

Under the guise of multiculturalism, allege critics, there is sometimes a particularistic emphasis that focuses on and glorifies the culture of a particular group. "Afrocentric" curricula have been especially criticized from this perspective. More generally, multiculturalism has been criticized as elevating ethnic identities and affiliations as primary while downgrading or ignoring other social ties—for example, those of religion or social class—that may be of equal or greater significance to many people. Critics have argued, too, that there is a common American culture that is ignored by the multiculturalists.

The effects of multiculturalism on society also have been questioned. An emphasis on retaining ethnic culture and community may, some suggest, tend to retard the economic mobility of young people (Bullivant 1981). Most of the criticism has focused on the effects of multiculturalism on unity in the broader society. Multiculturalists have been pictured by critics as treating the United States as a kind of federation of ethnic groups rather than as a unified nation. It is said that, while deeply suspicious of the nation-state (which they see as racist and oppressive), multiculturalists wish to use the state, nevertheless, to benefit specific groups, without concern for the wider national interest.

These critics see a common national identity and a sense of common national interest as important to the solution of important social problems, such as inequality. They are concerned that a multiculturalism that emphasizes differences among Americans can lead to division, separatism, and conflict. They point to the conflicts between ethnic groups in many countries—such as Canada, Sri Lanka, Northern Ireland, Sudan, Nigeria, and Belgium—as warnings of the dangers of promoting ethnic differences. Thus, these critics argue that, while cultural differences should be respected, we should not promote their continuation, especially not with taxpayers' money.

Advocates of multiculturalism respond to such criticisms by noting first that the existence of separate and culturally different ethnic groups in society is a reality. They maintain that multiculturalism encourages not ethnic conflict but rather a harmony based on respect for these cultural differences.

Cosmopolitanism: A Middle Way?

A model of pluralism and a model of assimilation are both consistent in part with the present reality in America (see chapter 9). A considerable amount of assimilation has occurred and is continuing, not only among those of European descent but also among non-whites. At the same time, there continues to be considerable inequality, physical and social separation, and some cultural differences among ethnic groups, especially between non-white groups and whites. A continuing flow of new immigrants contributes to continuation of such differences.

We have seen that the assimilation model and the pluralist model each has some advantages and some disadvantages as a guide for policy. The integrated society envisioned by the assimilation model would be one in which segregation, discrimination, inequality, and ethnic strife would fade away. But achieving such a society would be very difficult, at least in the short run, and efforts to reduce the importance of ethnic affiliations might reduce the group pride and cohesion that are important to many people belonging to minority groups.

A pluralist society (and some of the multiculturalist versions of such a society) would increase the pride, solidarity, and strength of ethnic groups as they try to gain respect and fair treatment. However, by promoting separation and cultural dissimilarity among ethnic groups, pluralism may make it more difficult to

295

achieve economic equality. Pluralism also may lead to ethnic conflicts and to lack of unity within the larger society.

In view of the limitations of both models, we may consider a policy approach that attempts to move beyond either pure assimilation or pure pluralism. When describing this approach, which is an amalgam of the ideas of a number of recent writers (Walzer 1992; Fuchs 1993; Hollinger 1995; Kymlicka 1995; Lind 1995), we will use the label "cosmopolitanism," a term introduced by David Hollinger (1995).

The "cosmopolitan" approach, first, sees the United States as home to a diversity of ethnic cultures. People would be encouraged to learn about and respect different cultures. There would be no effort by government or any other official agency to assimilate those of any ethnic group into a single culture, nor, conversely, any official efforts to promote the continuation of given cultures. It would be anticipated that, especially with a continuing flow of immigration, cultural diversity would continue indefinitely.

However, cultural diversity would not be equated with distinctions between the five ethnic categories (white, Hispanic, African American, Asian American, Native American) used frequently in recent years. These are seen as too arbitrary and restrictive, not allowing, for example, for cultural differences within each broad category and not taking into account of various ethnic mixtures. Moreover, cultural background, and the racial or national ethnic identity often associated with culture, would not be seen as necessarily the "master identity" for any individual. Rather, each individual would be seen, and would be treated, in terms of a variety of social identities and affiliations (many "we"s), of which ethnic identity is one—sometimes the one most important to the individual, often not.

In the cosmopolitan society, while ethnic and cultural differences would be respected, the focus would be on individuals, rather than on ethnic groups. Positions and rewards would be given on an individual basis, rather than allotted on the basis of group membership. Because social rewards would not be given on the basis of ethnic group identity, members of each ethnic group would have relatively little reason to feel threat from other ethnic groups or to feel that they were being unfairly disadvantaged.

There still would be conflicts between various groups within society, including different ethnic groups, since there will always be differences of interest and viewpoint among those in different groups. Struggles to reduce discrimination against minority groups and to promote greater equality among ethnic groups would continue to be waged. But conflicts within society would not be waged solely, or even primarily, along ethnic lines. Because individuals have multiple identities that are important to them—including those of income, occupation, industry, religion, region, and gender—the lines of conflict and of cooperation would fluctuate, depending on the issues at hand. The fact that such "cross-cutting cleavages" exist and that social conflict need not always be along the lines of ethnicity means that overall social conflicts will be less bitter; one's adversary today is one's ally tomorrow (see also Blau 1994). Moreover, the fact that conflict and alliances may occur in terms of social ties other than ethnicity—especially on the basis of class similarities and differences—will often increase the power and the chances for success of those fighting to achieve greater equality.

While those promoting a cosmopolitan approach foresee the continuation of ethnic affiliations and of conflicts among various groups in society, they also emphasize the importance of forces that tie people from different groups together. These integrative forces, which are seen as creating a needed unity, include shared political principles, shared interests, and common culture.

In many countries—for example, Germany and Japan—being a full fledged citizen of society depends on ethnicity or "blood." But in the United States, from its beginning, what was important to be an American was not ethnicity but acceptance of American ideals. Philip Gleason states: "To be or to become an American a person did not have to be any particular national, linguistic, religious, or ethnic background. All he had to do was to commit himself to the political ideology centered on the abstract ideas of liberty, equality, and republicanism" (1980: 32) To be sure, during much of American history, some racial minorities—including blacks and some Asians—were not admitted to full citizenship. But the principle that belief in a common set of ideals, rather than ethnicity, defines Americans is an important one that now has been extended in law to those from all ethnic groups. Moreover, the great majority of Americans from every ethnic background do share important ideals—such as freedom, democracy, and equality of opportunity.

Americans from varied ethnic groups also share many common interests. These include a robust economy, a clean environment (including clean air and drinking water), low crime, good public education, low drug usage, strong families, freedom from foreign and domestic terrorism, and civil peace. An example of the way in which shared interests can promote unity across racial lines is provided by a recent study of a community in California (Horton 1995). An influx of Chinese residents into that community initially led to hostility by whites ("Anglos") and by Latinos towards the new arrivals. Soon, however, Anglos, Latinos, and Chinese residents found that they shared common interests in opposing local developers' plans for changes in land use. Members of all three ethnic groups cooperated in controlling what they saw as the undesirable land uses. These shared interests, and resulting cooperation, helped to produce much greater harmony among the groups. Advocates of the cosmopolitan approach to ethnic relations argue that, while we cannot ignore the genuine conflicts of interest that exist among groups (including ethnic groups) in our society, this should not blind us to important common interests that we share.

Finally, cosmopolitan thinkers assert that, while particular ethnic cultures exist and may continue to exist within the United States, there is also a common American culture that unites Americans. Michael Lind states: "A real nation is a concrete historical community, defined primarily by a common language, common folkways, and a common vernacular culture. Such an extrapolitical American nation exists today, and has existed in one form or another for hundreds of years. Most Americans, of all races, are born and acculturated into the American nation; most immigrants and their descendants will be assimilated into it" (1995: 5). Lind is referring only to *cultural*, and not to other types of assimilation. However, a common culture is seen by him and others as an important element in promoting national unity.

National unity is seen by the "cosmopolitans" as important for achieving national purposes that are important to those of all ethnic groups. They point out, for example, that the success of the civil rights movement in the 1960s, and broad initial support for the War on Poverty in the 1960s and 1970s, were based, in part, on a widespread sense of shared problems and shared national interest—that what affects some Americans affects all Americans (Hollinger 1995).

They argue also that strong allegiance to the nation is not incompatible with strong ethnic identities (an idea that is consistent with research summarized in chapter 2). For hyphenated Americans, states Michael Walzer, the hyphen works "like a plus sign" (1992: 45); that is, the individual can be both French (or Korean or Mexican) *and* American. The cosmopolitan would be likely to use this perspective to interpret the national motto of the United States, "E Pluribus Unum"—From Many, One. They would say that the separate components should not be forgotten; but neither should the oneness.

Summary

Cultural pluralism refers to a situation in which various ethnic groups in a society preserve their own cultures. Cultural pluralism usually is supported by some degree of structural pluralism, whereby each group also maintains a separate communal life. Both types of pluralism are present in many countries, including the United States to some extent. Advocates of ethnic pluralism maintain that cultural diversity and cohesive ethnic groups benefit both the members of each group and society as a whole.

Two problems that often occur in ethnically pluralistic societies are inequality and conflict. Inequality between ethnic groups may result, in part, from discrimination. It also may result from cultural differences—for example, in skills or values that give some groups an economic advantage over others. Also, if people organize and compete on the basis of ethnicity, groups with the greatest power will become most advantaged economically.

Efforts to reduce inequality between ethnic groups have included a) giving preferences for jobs and other economic opportunities to members of disadvantaged groups; and b) making institutions such as schools and businesses more accommodating of cultural diversity.

Conflicts, often violent, are common in ethnically plural societies. Conflicts may occur over a wide variety of issues. These include the distribution of wealth; competition for jobs and business; preferences given to some groups; use of languages; control of government; and group autonomy.

Some countries have tried to reduce interethnic conflict by devising a political system that shares power among major groups. While power-sharing has had some success in reducing ethnic conflict for a time in some countries, it has often failed to bring sustained peace between groups.

Since interethnic conflict may result, in part, from economic inequalities, some countries have adopted programs to give economic preferences to disadvantaged groups. While such programs should, in principle, reduce conflict eventually, in the short run they have resulted in greater ethnic antagonisms.

Some of the basic ideas of ethnic pluralism have been advocated recently under the name of multiculturalism. Multiculturalists believe that greater understanding and appreciation of ethnic and cultural differences will bring greater pride to members of disadvantaged groups and will help members of all groups to function more effectively in a multicultural society. Critics of multiculturalists have accused them of focusing too much on ethnic identities while downgrading or ignoring other social ties, such as religion and social class, and of fostering ethnic divisions, rather than unity. An emerging set of ideas called cosmopolitanism attempts to find a middle ground between pluralist and assimilationist models of society. This approach would respect cultural diversity but such diversity would not be officially promoted. People would be seen in terms of a variety of social identities and affiliations of which ethnicity is only one. Social rewards would be given on the basis of individual characteristics rather than group membership. A common American culture is recognized and common interests that cross ethnic lines would be emphasized.

References

Abelson, R. P., and Levi, A. (1985). Decision-making and Decision Theory. In G. Lindzey and E. Aronson (eds.), *Handbook of Social Psychology*, 3d ed., vol. 1. New York: Random House.

Abrams, D. (1992). Processes of Social Identification. In G.M. Breakwell (ed.), *Social Psychology of Identity and the Self-Concept*. London, England: Surrey University Press.

Abramson, H. (1980). Assimilation and Pluralism. In S. Thernstorm (ed.), *Harvard Encyclopedia of American Ethnic Groups*. Cambridge, MA: Harvard University Press.

Adorno, T. W., Frenkel-Brunswik, E., Levinson, D. J., and Sanford, R. N. (1950). *The Authoritarian Personality*. New York: Harper and Row.

Adorno, T. W., and Golden, R. M. (1986). Patterns of Ethnic Marriage in the United States. *Social Forces, 65*, 202–223.

Ajzen, I., and Fishbein, M. (1980). *Understanding Attitudes and Predicting Social Behavior*. Englewood Cliffs, NJ: Prentice-Hall.

Alba, R. D. (1990). *Ethnic Identity: The Transformation of White America*. New Haven, CT: Yale University Press.

———. (1995). Assimilation's Quiet Tide. *The Public Interest*, Spring, 3–17.

Allen, T. W. (1994). *The Invention of the White Race*. London: Verso.

Allport, G. W. (1954). *The Nature of Prejudice*. Garden City, NY: Addison-Wesley.

Amir, Y. (1969). Contact Hypothesis in Ethnic Relations. *Psychological Bulletin, 71*, 319–342.

———. (1976). The Role of Intergroup Contact in Change of Prejudice and Ethnic Relations. In P. Katz (ed.), *Toward the Elimination of Racism*. New York: Pergamon.

Amir, Y., and Ben-Ari, R. (1985). International Tourism, Ethnic Contact, and Attitude Change. *Journal of Social Issues, 41*, 105–116.

Amir, Y., Ben-Ari, R., Bizman, A., and Rivner, M. (1982). Objective versus Subjective Aspects of Interpersonal Relations Between Jews and Arabs. *Journal of Conflict Resolution, 26*, 485–506.

Anant, S. S. (1972). *The Changing Concept of Caste in India*. Delhi: Vikas.

Apostle, R. A., Glock, C. Y., Piazza, T., and Suelzle, M. (1983). *Anatomy of Racial Attitudes*. Berkeley: University of California Press.

Arasaratnam, S. (1987). Sinhala-Tamil Relations in Modern Sri Lanka (Ceylon). In J. Boucher, D. Landis, and K.A. Clark (eds.), *Ethnic Conflict: International Perspectives*. Newbury Park, CA: Sage.

Armor, D. J. (1992). Why Is Black Educational Achievement Rising? *Public Interest, 108*, 65–80.

Aronson, E. (1972). *The Social Animal*. San Francisco: Freeman.

———. (1984). Modifying the Environment of the Desegregated Classroom. In A. J. Stewart (ed.), *Motivation and Society*. San Francisco: Jossey-Bass.

Aronson, E., and Gonzalez, A. (1988). Desegregation, Jigsaw, and the Mexican-American Experience. In P. A. Katz and D. A. Taylor (eds.), *Eliminating Racism*. New York: Plenum Press.

Atkins, C. K. (1981). Mass Media Campaign Effectiveness. In R. E. Rice and W. J. Paisley (eds.), *Public Communication Campaigns*. Beverly Hills: Sage.

Atkinson, A. B. (1983). *Social Justice and Public Policy*. Cambridge, MA: Massachusetts Institute of Technology Press.

Baker, K. A., and de Kanter, A. A., (eds.) (1983). *Bilingual Education: A Reappraisal of Federal Policy*. Lexington, MA: Lexington Books.

Bandura, A. (1977). *Social Learning Theory*. Englewood Cliffs, NJ: Prentice-Hall.

Bane, M. J., and Ellwood, D. T. (1994). *Welfare Realities: From Rhetoric To Reform*. Cambridge, MA: Harvard University Press.

Banton, M. (1967). *Race Relations*. New York: Basic Books.

——. (1995). Rational Choice Theories. *American Behavioral Scientist, 38*, 478–497.

Barany, G. (1974). Magyar Jew or Jewish Magyar? (To the Question of Jewish Assimilation in Hungary). *Canadian-American Slavic Studies, 8*.

Bargal, D. (1990). Contact Is Not Enough: The Contribution of Lewinian Theory to Intergroup Workshops Involving Arab Palestinians and Jewish Youth in Israel. *International Journal of Group Tensions, 20*, 179–192.

Bargal, D., and Bar, H. (1992). A Lewinian Approach to Intergroup Workshops for Arab-Palestinian and Jewish Youth. *Journal of Social Issues, 48*, 139–154.

Barlow, H.D. (1993). *Introduction to Criminology*, 6th ed. New York: HarperCollins.

Bar-Tal, D., Graumann, C. F., Kruglanski, A. W., and Stroebe, W. (1989). *Stereotyping and Prejudice: Changing Conceptions*. New York: Springer-Verlag.

Bealer, R. C., Willits, F. K., and Bender, G. (1963). Religious Exogamy. *Sociology and Social Research, 48*, 69–79.

Bean, F. (1989). *Opening and Closing the Doors*. Washington, DC: University Press of America.

Beck, E. M. (1980). Discrimination and White Economic Loss. *Social Forces 59*, 148–168.

Becker, G. S. (1971). *The Economics of Discrimination*. Chicago: University of Chicago Press.

Beers, D., and Hembree, D. (1987). The New Atlanta: A Tale of Two Cities. *The Nation,* March 21, 357–360.

Belanger, S., and Pinard, M. (1991). Ethnic Movements and the Competition Model: Some Missing Links. *American Sociological Review, 56*, 446–457.

Belz, H. (1991). *Equality Transformed: A Quarter Century of Affirmative Action*. New Brunswick, NJ: Transactions Publishers.

Bem, D. J. (1970). *Beliefs, Attitudes, and Human Affairs*. Belmont, CA: Brooks/Cole.

Ben-Ari, R., and Amir, Y. (1986). Contact between Arab and Jewish Youth in Israel: Reality and Potential. In M. Hewstone and R. Brown (eds.), *Contact and Conflict in Intergroup Encounters*. Oxford, England: Blackwell.

Bendick, M., Jackson, C. W., Reinoso, V. A., and Hodges, L. E. (1993). Discrimination Against Latino Job Applicants: A Controlled Experiment. In J. A. Kromkowski (ed.), *Race and Ethnic Relations 93/94*. Guiford, CT: Dushkin.

Bennett, C. I. (1990). *Comprehensive Multicultural Education: Theory and Practice*, 3d. ed. Boston: Allyn and Bacon.

Berg, I. (1971). *Education and Jobs: The Great Training Robbery*. Boston: Beacon.

Berry, J., Trimble, J., and Olmedo, E. (1986). Assessment of Acculturation. In W. Lonner and J. Berry, (eds.), *Field Methods in Cross-Cultural Research*. Newbury Park, CA: Sage.

Billig, M. (1976). *Social Psychology and Intergroup Relations*. London: Academic Press.

Bjerstedt, A. (1962). Informational and Non-Informational Determinants of Nationality Stereotypes. *Journal of Social Issues, 18*, 24–29.

Bjork, J., and Goodman, A. E. (1993). *Yugoslavia, 1991–92: Could Diplomacy Have Prevented a Tragedy?* Washington, DC: School of Foreign Service, Georgetown University.

Blalock, H. M., Jr. (1967). *Toward a Theory of Minority-Group Relations*. New York: Wiley.

——. (1982) *Race and Ethnic Relations*. Englewood Cliffs, NJ: Prentice-Hall.

Blalock, H. M., Jr., and Wilken, P. H. (1979). *Intergroup Processes: A Micro-Macro Perspective*. New York: Free Press.

Blaney, N. T., Rosenfield, S. C., Aronson, E., and Sikes, J. (1977). Interdependence in the Classroom: A Field Study. *Journal of Educational Psychology, 69*, 139–146.

Blau, P. M. (1964). *Exchange and Power in Social Life*. New York: John Wiley and Sons.

——. (1994). *Structural Contexts of Opportunities*. Chicago: University of Chicago Press.

Blau, P. M., and Schwartz, J. E. (1984). *Crosscutting Social Circles*. Orlando, FL: Academic Press.

Blauner, R. (1969). Internal Colonialism and Ghetto Revolt. *Social Problems, 16*, 393–408.

Bloom, B. S. (1976). *Human Characteristics and School Learning*. New York: McGraw-Hill.

Blumstein, A. (1993). Making Rationality Relevant. *Criminology, 31*, 1–16.

Bobo, L. (1983). Whites' Opposition to Busing: Symbolic Racism or Realistic Group Conflict? *Journal of Personality and Social Psychology, 45*, 1196–1210.

Bonacich, E. (1972). A Theory of Ethnic Antagonism: The Split-Labor Market. *American Sociological Review, 37*, 547–559.

———. (1973). A Theory of Middleman Minorities. *American Sociological Review, 38*, 583–594.

———. (1995). Garment Contractors and Liquor Store Owners: Asian Relations with Latinos and African Americans. Paper given at Annual Meeting of American Sociological Association, August, Washington, DC.

Bornman, E., and Mynhardt, J. C. (1991). Social Identification and Intergroup Contact in South Africa with Specific Reference to the Work Situation. *Genetic, Social, and General Psychology Monographs, 117*, 437–462.

Boucher, J. D., Landis, D., and Clarke, K. A. (1987). *Ethnic Conflict: International Perspectives*. Newbury Park, CA: Sage.

Bradburn, N. M., Sudman, S., and Gockel, G. L. (1971). *Side By Side: Integrated Neighborhoods in America*. Chicago: Quadrangle Books.

Bradley, L., and Bradley, G. (1977). The Academic Achievement of Blacks in Desegregated Schools. *Review of Educational Research, 47*, 399–499.

Brass, P. R. (1991). *Ethnicity and Nationalism: Theory and Comparison*. New Delhi, India: Sage.

Braun, D. (1997). *The Rich Get Richer: The Rise of Income Inequality in the United States and the World*, 2d ed., Chicago: Nelson-Hall.

Brehm, J. W., and Cohen, A. R. (1962). *Explorations in Cognitive Dissonance*. New York: Wiley.

Brewer, M. B. (1979). In-Group Bias in the Minimal Intergroup Situation: A Cognitive-Motivational Analysis. *Psychological Bulletin, 86*, 307–334.

Brewer, M. B., and Miller, N. (1984). Beyond the Contact Hypothesis: Theoretical Perspectives on Desegregation. In N. Miller and M. B. Brewer (eds.), *Groups in Contact: The Psychology of Desegregation*. Orlando, FL: Academic Press.

———. (1988). Contact and Cooperation: When Do They Work? In P. A. Katz and D. A. Taylor (eds.), *Eliminating Racism: Profiles in Controversy*. New York: Plenum.

Brislin, R. W. (1968). Contact as a Variable in Intergroup Interaction. *Journal of Social Psychology, 76*, 149–154.

Brislin, R. W., and Pederson, P. (1976). *Cross-Cultural Orientation Programs*. New York: Gardner Press.

Brislin, R. W., Cushner, K., Cherrie, C., and Young, M. (1986). *Intercultural Interactions: A Practical Guide*. Newbury Park, CA: Sage.

Brookover, W. B. (1979). *School Social Systems and Student Achievement: Schools Can Make a Difference*. New York: Praeger.

Brooks, D. (1975). *Race and Labor in London Transport*. London: Oxford University Press.

Brophy, J. E. (1983). Research on the Self-Fulfilling Prophecy and Teacher Expectations. *Journal of Educational Psychology, 75*, 327–346.

Brown, D. L., and Fuguitt, G. V. (1972). Percent Non-White and Racial Disparity in Non-Metropolitan Cities in the South. *Social Science Quarterly, 53*, 573–582.

Browning, R. P., Marshall, D. R., and Tabb, D. H. (eds.) (1990). *Racial Politics in American Cities*. New York: Longman.

Bullivant, B. M. (1981). *The Pluralist Dilemma in Education: Six Case Studies*. Sydney: George Allen and Unwin.

Bullock, C. S., III. (1976). *Social Desegregation: Interracial Contact and Prejudice*. Houston, TX: University of Houston.

Burkey, R. M. (1978). *Ethnic and Racial Groups: The Dynamics of Dominance*. Menlo Park, CA: Cummings.

Burkholz, H. (1980). The Latinization of Miami. *New York Times Magazine,* September 21, 44–47, 84–88, 98–100.

Burma, J. H. (1963). Interethnic Marriage in Los Angeles, 1948–1959. *Social Forces, 42,* 159–165.

Burstein, P. (ed.) (1994). *Equal Employment Opportunity.* New York: Aldine DeGruyter.

Byrne, D. (1971). *The Attraction Paradigm.* New York: Academic Press.

Cain, B. E. (1992). Voting Rights and Democratic Theory: Toward a Color-Blind Society? In B. Grofman and C. Davidson (eds.), *Controversies in Minority Voting.* Washington, DC: Brookings Institution.

Campbell, A. (1971). *White Attitudes Toward Black People.* Ann Arbor, MI: Institute for Social Research.

Cancio, A. S, Evans, T. D., and Maume, D. J., Jr. (1996). Reconsidering the Declining Significance of Race: Racial Differences in Early Career Wages. *American Sociological Review, 61,* 541–556.

Carmichael, S., and Hamilton, C. V. (1967). *Black Power: The Politics of Liberation in America.* New York: Vintage Books.

Carnoy, M. (1994). *Faded Dreams: The Politics and Economics of Race in America.* Cambridge, England: Cambridge University Press.

Case, C. E., Greeley, A. M., and Fuchs, S. (1989). Social Determinants of Racial Prejudice. *Sociological Perspectives, 32,* 469–483.

Cashman, G. (1993). *What Causes War?: An Introduction to Theories of International Conflict.* New York: Lexington Books.

Chadwick, B. A., Bahr, H. M., and Day, R. C. (1971). Correlates of Attitudes Favorable to Racial Discrimination Among High School Students. *Social Science Quarterly, 51,* 873–888.

Chavez, L. (1991). *Out of the Barrio: Toward a New Politics of Hispanic Assimilation.* New York: Basic Books.

Choi, J. M., Callaghan, K. A., and Murphy, J. W. (1995). *The Politics of Culture: Race, Violence, and Democracy.* Westport, CT: Praeger.

Chu, D., and Griffey, D. (1985). The Contact Theory of Racial Integration: The Case of Sport. *Sociology of Sport Journal, 2,* 323–333.

Clark, W. A. V. (1988). Understanding Racial Segregation in American Cities. *Population Research and Policy Review, 8,* 193–197.

Cohen, E. G. (1984). The Desegregated School: Problems in Status, Power, and Inter-Ethnic Climate. In N. Miller and M. B. Brewer (eds.), *Groups in Contact.* New York: Academic Press.

Cohen, L. J. (1993). *Broken Bonds: The Disintegration of Yugoslavia.* Boulder, CO: Westview Press.

Coleman, J. S., Campbell, E. Q., Hobson, C. J., McPartland, J., Mood, A. M., Weinfeld, F. D., and York, R. L. (1966). *Equality of Educational Opportunity.* Washington, DC: U.S. Government Printing Office.

Coleman, J. S., Hoffer, T., and Kilgore, S. (1982). *High School Achievement: Public, Catholic and Private Schools Compared.* New York: Basic Books.

Conklin, J. E. (1995). *Criminology,* 5th ed. Boston: Allyn and Bacon.

Cook, S. W. (1962). The Systematic Analysis of Socially Significant Events: A Strategy for Social Research. *Journal of Social Issues, 18,* 66–84.

——. (1978). Interpersonal and Attitudinal Outcomes in Cooperating Interracial Groups. *Journal of Research and Development in Education, 12,* 97–113.

——. (1984). Cooperative Interaction in Multiethnic Contexts. In N. Miller and M. B. Brewer (eds.), *Groups in Contact.* New York: Academic Press.

Cornell, S. (1988). *The Return of the Native: American Indian Political Resurgence.* New York: Oxford University Press.

Council of Chief State of School Officers (1990). *School Success for Limited English Proficient Students: The Challenge and State Response.* Washington, DC: Council of Chief State School Officers.

304

Covell, M. (1993). Belgium: The Variability of Ethnic Relations. In J. McGarry and B. O'Leary (eds.), *The Politics of Ethnic Conflict Regulation*. London: Routledge.

Cox, O. C. (1948). *Caste, Class, and Race*. Garden City, NY: Doubleday.

Cox, T. (1993). *Cultural Diversity in Organizations: Theory, Research, and Practice*. San Francisco: Berrett-Koehler.

Craig, R. B. (1971). *The Bracero Program: Interest Groups and Foreign Policy*. Austin: University of Texas Press.

Crain, R. L., and Mahard, R. E. (1982). *Desegregation Plans That Raise Black Achievement: A Review of the Research*. Rand Note. Santa Monica, CA: Rand Corporation.

———. (1983). The Effect of Research Methodology on Desegregation-Achievement Studies: A Meta-Analysis. *American Journal of Sociology, 88*, 839–854.

Cross, W. E. (1971). The Negro-to-Black Conversion Experience. *Black World*, 13–27.

Culture Wars: Opposing Views. (1994). San Diego, CA: Greenhaven Press.

Cunningham, W. H., and Cunningham, I. (1981). *Marketing: A Managerial Approach*. Cincinatti, OH: South-Western Publishing.

D'Alessio, S. J., and Stolzenberg, L. (1991). Anti-Semitism in America: The Dynamics of Prejudice. *Social Inquiry, 61*, 359–366.

Darby, J. (1976). *Conflict in Northern Ireland: The Development of a Polarized Community*. Dublin: Gill and Macmillan.

Darden, J. T. (1987). Choosing Neighbors and Neighborhoods: The Role of Race in Housing Preference. In G. E. Tobin (ed.), *Divided Neighborhoods*. Newbury Park, CA: Sage.

Dashevsky, A., and Shapiro, H. M. (1974). *Ethnic Identification Among American Jews: Socialization and Social Structure*. Lanham, MD: University Press of America.

Davidson, J. D. (1995). *National Poll Results: Catholic Pluralism Project*. W. Lafayette, IN: Department of Sociology and Anthropology, Purdue University.

Davidson, J. D., Williams, A., Lammana, R. A., Stenftenagel, J., Weigert, K., Whalen, W. J., and Wittberg, S. P. (1997). *Catholic Faith and Morals*. Huntington, IN: Our Sunday Visitor Books.

Davine, V. R., and Bills, D. B. (1992). Changing Attitudes Toward Race-Related Issues: Is a Sociological Perspective Effective? Paper presented at annual meeting of American Sociological Association, Pittsburgh, August.

de la Garza, R. (1992). *Latino Voices: Mexican, Puerto Rican and Cuban Perspectives in American Politics*. Boulder, CO: Westview Press.

De Ridder, R., and Tripathi, R. C. (1992). *Norm Violation and Intergroup Relations*. Oxford, England: Clarendon Press.

Der-Karabetian, A. (1980). Relation of Two Cultural Identities of Armenian-Americans. *Psychological Reports, 47*, 123–128.

de Silva, C. R. (1979). The Constitution of the Second Republic of Sri Lanka (1978) and Its Significance. *Journal of Commonwealth and Comparative Politics, 17*, 192–209.

de Sola Pool, I. (1965). Effects of Cross-National Contact on National and International Images. In H. C. Kelman (ed.), *International Behavior*. New York: Holt, Rinehart, and Winston.

Destexhe, A. (1995). *Rwanda and Genocide in the Twentieth Century*. New York: New York University Press.

Deutsch, M. (1968). The Effects of Cooperation Upon Group Process. In D. Cartwright and A. Zander (eds.), *Group Dynamics: Research and Theory*, 3d ed. New York: Harper and Row.

Deutsch, M., and Collins, M. E. (1951). *Interracial Housing*. Minneapolis: University of Minnesota Press.

Deyhle, D. (1992). Constructing Failure and Maintaining Cultural Identity: Navajo and Ute School Leavers. *Journal of American Indian Education, 31*, 24–27, 45.

Diaz, C. (ed.) (1992). *Multicultural Education for the 21st Century*. Washington, DC: National Education Association.

Dijker, A. J. M. (1987). Emotional Reactions to Ethnic Minorities. *European Journal of Social Psychology, 17*, 305–325.

Dion, K. H. (1973). Cohesiveness As a Determinant of Ingroup-Outgroup Bias. *Journal of Personality and Social Psychology, 28*, 163–171.

Dollard, J. (1957). *Caste and Class in a Southern Town.* Garden City, NY: Doubleday.

Donohue, J. J., III, and Heckman, J. (1994). Continuous Versus Episodic Change: The Impact of Civil Right Policy on the Economic Status of Blacks. In P. Burstein (ed.), *Equal Employment Opportunity.* New York: Aldine De Gruyter.

Dovidio, J. F., and Gaertner, S. L. (1986). Prejudice, Discrimination and Racism: Historical Trends and Contemporary Approaches. In J. F. Dovidio and S. L. Gaertner (eds.), *Prejudice, Discrimination and Racism.* Orlando, FL: Academic Press.

Dowdall, G. W. (1974). White Gains From Black Subordination in 1960 and 1970. *Social Problems, 22*, 162–183.

Drake, St. C. and Cayton, H. R. (1962). *Black Metropolis,* revised ed. New York: Harper and Row.

Driver, E. D. (1969). Self-Conceptions in India and the United States. *The Sociological Quarterly, 10*, 341–354.

D'Souza, D. (1995). *The End of Racism: Principles for a Multiracial Society.* New York: Free Press.

Dusek, J. B., and Joseph, G. (1983). The Bases of Teacher Expectancies: A Meta-Analysis. *Journal of Educational Psychology, 75*, 327–346.

Dyer, J., Vedlitz, A., and Worchel, S. (1989). Social Distance Among Racial and Ethnic Groups in Texas: Some Demographic Correlations. *Social Science Quarterly, 70*, 607–616.

Eagly, A. H., and Chaiken, S. (1993). *The Psychology of Attitudes.* Fort Worth, TX: Harcourt Brace Jovanovich.

Edley, C., Jr. (1996). *Not All Black and White: Affirmitive Action, Race, and American Values.* New York: Hill and Wang.

Edwards, J. (1995). *When Race Counts: The Morality of Racial Preference in Britain and the United States.* London: Rutledge.

Eisenstein, J., and Flemming, R. B. (1988). *The Tenor of Justice: Criminal Courts and the Guilty Plea Process.* Urbana: University of Illinois Press.

Eitzen, D. S. (1985). *In Conflict and Order: Understanding Society.* Boston: Allyn and Bacon.

Eitzen, D. S., and Zinn, M. B. (1994). *In Conflict and Order: Understanding Society.* Needham Heights, MA: Allyn and Bacon.

Elliott, D. S., and Ageton, S. S. (1980). Reconciling Race and Class Differences in Self-Reported and Official Estimates of Delinquency. *American Sociological Review, 45*, 95–110.

Ellwood, D. T. (1988). *Poor Support: Poverty in the American Family.* New York: Basic Books.

Enloe, C. H. (1973). *Ethnic Conflict and Political Development.* Boston: Little, Brown.

Epstein, J. (1985). After the Bus Arrives. *Journal of Social Issues, 41*, 23–44.

Eshel, S., and Peres, Y. (1973). *The Integration of a Minority Group: A Causal Model.* Tel Aviv, Israel: Tel Aviv University.

Esman, M. J. (1972). Malaysia: Communal Coexistence and Mutual Deterrence. In E. Q. Campbell, (ed.), *Racial Tensions and National Identity.* Nashville,TN: Vanderbilt University Press.

———. (1994). *Ethnic Politics.* Ithaca, NY: Cornell University Press.

Espiritu, Y. L. (1992). *Asian American Panethnicity: Bridging Institutions and Identities.* Philadelphia: Temple University Press.

Ezorsky, G. (1993). *Racism and Justice: The Case for Affirmative Action.* Ithaca, NY: Cornell University Press.

Farkas, G., and Vicknair, K. (1996). Appropriate Tests of Racial Wage Discrimination Require Controls for Cognitive Skill: Comment on Cancio, Evans, and Maume. *American Sociological Review, 61*, 557–560.

Farley, J. E. (1987). Excessive Black and Hispanic Unemployment in U.S. Metropolitan Areas: The

Roles of Racial Inequality, Segregation, and Discrimination in Male Joblessness. *American Journal of Economics and Sociology, 46*, 129–150.

——. (1995). *Majority-Minority Relations*, 3d ed. Englewood Cliffs, NJ: Prentice-Hall.

Farley, R. (1984). *Blacks and Whites: Narrowing the Gap?* Cambridge, MA: Harvard University Press.

——. (1993). The Common Destiny of Blacks and Whites: Observations About the Social and Economic Status of the Races. In H. Hill and J. E. Jones, Jr. (eds.), *Race in America: The Struggle for Equality*. Madison: University of Wisconsin Press.

Farley, R., Schuman, H., Bianchi, D.C., and Hatchett, S. (1978). Chocolate City, Vanilla Suburbs: Will the Trend Toward Racially Separate Communities Continue? *Social Science Research, 7*, 319–344.

Farley, R., Hatchett, S., and Schuman, H. (1979). A Note on Changes in Black Racial Attitudes in Detroit, 1968–1976. *Social Indicators Research, 6*, 439–443.

Farley, R., and Frey, W. H. (1994). Changes in the Segregation of Whites from Blacks During the 1980s: Small Steps Toward a More Integrated Society. *American Sociological Review, 59*, 23–45.

Farley, R., Steeh, C., Krysan, M., Jackson, T., and Reeves, K. (1994). Stereotypes and Segregation: Neighborhoods in the Detroit Area. *American Journal of Sociology, 100*, 750–780.

Farnen, R. F. (ed.) (1994). *Nationalism, Ethnicity, and Identity: Cross National and Comparative Perspectives*. New Brunswick: NJ: Transaction Publishers.

Feagin, J. R. (1980). School Desegregation: A Political-Economic Perspective. In W. G. Stephan and J. R. Feagin (eds.), *School Desegregation: Past, Present, and Future*. New York: Plenum Press.

Feagin, J. R., and Vera, H. (1995). *White Racism: The Basics*. New York: Routledge.

Fernandez, R. R., and Guskin, J. T. (1981). Hispanic Students and School Desegregation. In W. D. Hawley (ed.), *Effective School Desegregation*. Beverly Hills, CA: Sage.

Firefighters Local Union No. 1784 v. Stotts et al. (1987). *United States Reports*, vol. 467: 561–621.

Fisher, R. (1994). *Beyond Machiavelli: Tools for Coping With Conflict*. Cambridge, MA: Harvard University Press.

Foner, P. (1947). *History of the Labor Movement in the United States*. New York: International Publishers.

Ford, W. S. (1972). *Interracial Public Housing in Border City*. Lexington, MA: Lexington Books.

Fordham, S. (1990). Racelessness as a Factor in Black Students' School Success: Pragmatic Strategy or Pyrrhic Victory? In N. M. Hidalgo, C. McDowell, and E. Siddle (eds.), *Facing Racism in Education*. Cambridge: Harvard Educational Review Reprint Series.

Fordham, S., and Ogbu, J. U. (1986). Black Students School Success: Coping with the Burden of 'Acting White.' *Urban Review, 18*, 10–11, 13–23.

Forehand, G. A., Ragosta, A., and Rock, D. A. (1976). *Conditions and Processes of Effective School Desegregation*. Princeton, NJ: Educational Testing Service.

Franklin, J. H. (1969). *From Slavery to Freedom: A History of Negro Americans*, 3d ed. New York: Vintage Books.

——. (1994). *From Slavery to Freedom: A History of African Americans*. New York: Knopf.

Friedman, T. L. (1995). *From Berut to Jerusalem*. New York: Doubleday.

Frijda, N. H., Kuipers, P., and Ter Schure, E. (1989). Relations Among Emotion, Appraisal, and Emotional Action Readiness. *Journal of Personality and Social Psychology, 57*, 212–228.

Fromkin, H. L., and Sherwood, J. J. (eds.). (1974). *Integrating the Organization: A Social Psychological Analysis*. New York: Free Press.

Fuchs, L. H. (1990). *The American Kaleidoscope: Race, Ethnicity, and the Civic Culture*. Hanover, NH: Wesleyan University Press.

——. (1993). An Agenda for Tomorrow: Immigration Policy and Ethnic Policies. *Annals of the American Academy of Political and Social Science, 530*, 171–186.

Fuerst, J. S. (1981). Report Card: Chicago's All-Black Schools. *Public Interest, 64*, 79–91.

Fyfe, J. J. (1981). Observations on Police Deadly Force. *Crime and Delinquency, 27*, 376–389.

Gaertner, S. L., and Dovidio, J. F. (1986). The Aversive Form of Racism. In J. F. Dovidio and S. L. Gaertner (eds.), *Prejudice, Discrimination and Racism*. Orlando, FL: Academic Press.

Galster, G. C., and Keeney, W. M. (1988). Race, Residence, Discrimination, and Economic Opportunity: Modeling the Nexus of Urban Racial Phenomena. *Urban Affairs Quarterly, 24*, 87–117.

Gamoran, A. (1992). The Variable Effects of High School Tracking. *American Sociological Review, 57*, 812–829.

Garth, A. (1974). *The Invisible China: The Overseas Chinese and the Politics of Southeast Asia*. New York: MacMillan.

Garza, R. T., and Santos, S. J. (1991). Ingroup/Outgroup Balance and Interdependent Inter-Ethnic Behavior. *Journal of Experimental Social Psychology, 27*, 124–137.

Geffner, R. A. (1978). The Effects of Interdependent Learning on Self-Esteem, Inter-Ethnic Relations, and Intra-Ethnic Attitudes of Elementary School Children. Unpublished Doctoral Thesis. Santa Cruz: University of California.

Geller, W. A., and Scott, M. (1992). *Deadly Force: What We Know*. Washington, DC: Police Executive Research Forum.

Geller, W., and Karales, K. J. (1981). Shootings of and by Chicago Police: Uncommon Crises. Part I: Shootings by Chicago Police. *Journal of Criminal Law and Criminology, 72*, 1813–1866.

Gellner, E. (1983). *Nations and Nationalism*. Oxford, England: Blackwell.

Gerard, H. B., and Miller, N. (1975). *School Desegregation: A Long-range Study*. New York: Plenum.

Gibson, M., and Ogbu, J. U. (1991). *Minority Students and Schooling*. New York: Garland Press.

Giles, M. W. (1977). Percent Black and Racial Hostility: An Old Assumption Reexamined. *Social Science Quarterly, 58*, 412–417.

Giles, M. W., Cataldo, E. F., and Gatlin, D. S. (1975). White Flight and Percent Black: The Tipping Point Reexamined. *Social Science Quarterly, 56*, 85–92.

Gill, R. T., Glazer, N., and Thernstrom, S. A. (1992). *Our Changing Population*. Englewood Cliffs, NJ: Prentice Hall.

Glazer, N. (1958). The American Jews and the Attainment of Middle-Class Rank: Some Trends and Explanations. In M. Sklare (ed.), *The Jews: Social Patterns of an American Group*. Glencoe, IL: The Free Press.

———. (1993). Is Assimilation Dead? *Annals of American Academy of Political and Social Science, 530*, 122–136.

Glazer, N., and Moynihan, D. P. (1970). *Beyond the Melting Pot: The Negroes, Puerto Ricans, Jews, Italians, and Irish of New York City*. Cambridge, MA: Massachusetts Institute of Technology Press.

Gleason, P. (1980). American Identity and Americanization. *Harvard Encyclopedia of Ethnic Groups*. Cambridge, MA: Harvard University Press.

Glenn, N. D. (1963). Occupational Benefits to Whites from the Subordination of Negroes. *American Sociological Review, 28*, 443–448.

Gollnick, D. M., and Chinn, P. C. (1990). *Multicultural Education in a Pluralistic Society*, 3d ed. New York: Merrill.

Gonzales, A. (1979). Classroom Cooperation and Ethnic Balance. Paper presented at annual meeting of American Psychological Assn., New York.

Gonzales, R., and LaVelle, M. (1988). *The Hispanic Catholic in the U.S.* New York: Northeastern Pastoral Center.

Good, T. L. (1981). Teacher Expectations and Student Perceptions: A Decade of Research. *Educational Leadership, 38*, 415–421.

Goodlad, J. I. (1984). *A Place Called School: Prospects for the Future*. New York: McGraw-Hill.

Goodwin, C. (1979). *The Oak Park Strategy*. Chicago: University of Chicago Press.

Goodwin, D. K. (1994). *No Ordinary Time: Franklin and Eleanor Roosevelt: The Home Front in World War II*. New York: Simon and Schuster.

Gordon, M. M. (1964). *Assimilation in American Life*. New York: Oxford University Press.

Grant, G., and Ogawa, D. M. (1993). Living Proof: Is Hawaii the Answer? *Annals of the American Academy of Political and Social Science, 530*, 137–154.

Grant, P. R. (1991). Ethnocentrism Between Groups of Unequal Power Under Threat in Intergroup Competition. *Journal of Social Psychology, 131*, 21–28.

Gray, H. (1995a). Television, Black Americans, and the American Dream. In G. Dines and J. M. Humez (eds.), *Gender, Race, and Class in Media*. Thousand Oaks, CA: Sage.

——. (1995b). *Watching Race: Television and the Struggle for "Blackness"*. Minneapolis: University of Minnesota Press.

Grebler, L., Moore, J. W., and Guzman, R. C. (1970). *The Mexican-American People: The Nation's Second Largest Minority*. New York: Free Press.

Grofman, B., Handley, L., and Niemi, R. G. (1992). *Minority Representation and the Quest for Voting Equality*. Cambridge, England: Cambridge University Press.

Guinier, L. (1992). Voting Rights and Democratic Theory. In B. Grofman and C. Davidson (eds.), *Controversies in Minority Voting*. Washington, DC: Brookings Institution.

Gurr, T. R. (1993). *Minorities at Risk: A Global View of Ethnopolitical Conflicts*. Washington, DC: United States Institute of Peace Press.

Gurr, T. R., and Harff, B. (1994). *Ethnic Conflict in World Politics*. Boulder, CO: Westview.

Gutmann, A. (ed.). (1994). *Multiculturalism: Examining the Politics of Recognition*. Princeton, NJ: Princeton University Press.

Haas, M. (1992). *Institutional Racism: The Case of Hawaii*. Westport, CT: Praeger.

Hacker, A. (1992). *Two Nations: Black and White, Separate, Hostile, Unequal*. New York: Scribners.

Hallinan, M. T. (1992). The Organization of Students for Instruction in the Middle School. *Sociology of Education, 65*, 114–127.

Hallinan, M. T., and Smith, S. S. (1985). The Effects of Classroom Racial Composition on Students' Interracial Friendliness. *Social Psychology Quarterly, 48*, 3–16.

Hamilton, D. L., Carpenter, S., and Bishop, G. D. (1984). Desegregation of Suburban Neighborhoods. In N. Miller and M. B. Brewer (eds.), *Groups in Contact*. Orlando, FL: Academic Press.

Harding, J., and Hogrefe, R. (1952). Attitudes of White Department Store Employees Toward Negro Co-Workers. *Journal of Social Issues, 8*, 18–28.

Harris, R. (1972). *Prejudice and Tolerance in Ulster*. Manchester, England: Manchester University Press.

Harvey, D. G., and Slatin, G. T. (1975). The Relationship Between Child's SES and Teacher Expectations. *Social Forces, 54*, 140–159.

Hassan, M. K. (1987). Parental Behavior, Authoritarianism, and Prejudice. *Manas, 34*, 41–50.

Hechter, M. (1975). *Internal Colonialism: The Celtic Fringe in British National Development 1536–1966*. London: Routledge & Kegan Paul.

——. (1987). *Principles of Group Solidarity*. Berkeley: University of California Press.

Heineman, B. W., Jr., and Associates. (1987). *Work and Welfare: The Case for New Directions in National Policy*. Washington, DC: Center for National Policy.

Hellman-Rajanayagam, D. (1994). *The Tamil Tigers: Armed Struggle for Identity*. Stuttgart: F. Steiner.

Helper, R. (1986). Success and Resistance Factors in the Maintenance of Racially Mixed Neighborhoods. In J. M. Goering, (ed.), *Housing Desegregation and Federal Policy*. Chapel Hill: University of North Carolina Press.

Hewstone, M., and Brown, R. (1986). Contact Is Not Enough: An Intergroup Perspective on the Contact Hypothesis. In M. Hewstone and R. Brown (eds.), *Contact and Conflict in Intergroup Encounters*. Oxford, England: Blackwell.

Higham, J. (1955). *Strangers in the Land: Patterns of American Nativism: 1860*. New Brunswick, NJ: Rutgers University Press.

Hill, R. C., and Negrey, C. (1985). Deindustrialization and Racial Minorities in the Great Lakes Region, U.S.A. In D. S. Eitzen and M. B. Zinn (eds.), *The Reshaping of America*. Englewood Cliffs, NJ: Prentice-Hall.

Hilliard, A. G. (1988). Conceptual Confusion and the Persistence of Group Oppression Through Education. *Equality and Excellence, 1*, 36–43.

Hinkle, S., and Brown, R. J. (1990). Intergroup Comparisons and Social Identity: Some Links and Lacunae. In D. Abrams and M. Hogg (eds.), *Social Identity Theory: Constructive and Critical Advances*. New York: Springer-Verlag.

Hiro, D. (1993). *Fire and Embers: A History of the Lebanese Civil War*. New York: St. Martin's Press.

Hogg, M. A., and Abrams, D. (1988). *Social Identifications: A Social Psychology of Intergroup Relations and Group Processes*. London: Rutledge.

Holli, M. G., and Green, P. H. (1984). *The Making of the Mayor: Chicago, 1983*. Grand Rapids, MI: William B. Eerdmans.

Hollinger, D. A. (1995). *Postethnic America: Beyond Multiculturalism*. New York: Basic Books.

Hollister, J. E., and Boivin, M. J. (1987). Ethnocentrism Among Free Methodist Leaders and Students. *Journal of Psychology and Theology, 15*, 57–67.

Homans, G. C. (1974). *Social Behavior: Its Elementary Forms*, revised edition. New York: Harcourt Brace Jovanovich.

Hope, R. O. (1979). *Racial Strife in the U.S. Military*. New York: Praeger.

Horowitz, D. L. (1975). Ethnic Identity. In N. Glazer and D.P. Moynihan (eds.), *Ethnicity: Theory and Experience*. Cambridge, MA: Harvard University Press.

———. (1985). *Ethnic Groups in Conflict*. Berkeley: University of California Press.

Horowitz, M., and Rabbie, J. M. (1989). Stereotypes of Groups, Group Members, and Individuals in Categories: A Differential Analysis. In D. Bar-Tal et al. (eds.), *Stereotyping and Prejudice: Changing Conceptions*. New York: Springer-Verlag.

Horton, J. (1995). *The Politics of Diversity: Immigration, Resistence and Change in Monterey Park, California*. Philadelphia, PA: Temple University Press.

Howard-Pitney, D. (1990). *The Afro-American Jeremiad: Appeals for Justice in America*. Philadelphia, PA: Temple University Press.

Huckfeldt, R., and Kohfeld, C. W. (1989). *Race and the Decline of Class in American Politics*. Urbana: University of Illinois Press.

Hudson, M. C. (1968). *The Precarious Republic: Political Modernization in Lebanon*. New York: Random House.

Hughes, R. (1993). *The Culture of Complaint: The Fraying of America*. New York: Oxford University Press.

Hunt, C. L., and Walker, L. (1974). *Ethnic Dynamics: Patterns of Intergroup Relations in Various Societies*. Homewood, IL: Dorsey Press.

Hunter, A. (1974). *Symbolic Communities: The Persistence and Change of Chicago's Local Communities*. Chicago: University of Chicago Press.

Hutnik, N. (1986). Patterns of Ethnic Minority Identification and Modes of Social Adaptation. *Ethnic and Racial Studies, 9*, 150–167.

Irish, D. P. (1952). Reactions of Caucasian Residents to Japanese-American Neighbors. *Journal of Social Issues, 8*, 10–17.

Irvine, J. J. (1990). *Black Students and School Failure: Policies, Practices, and Prescriptions*. Westport, CT: Greenwood.

Jackman, M. R., and Muha, M. J. (1984). Education and Intergroup Attitudes: Moral Enlightenment, Superficial Democratic Commitment, or Ideological Refinement? *American Sociological Review, 49*, 751–769.

Jackson, J. W. (1993). Contact Theory of Intergroup Hostility. *International Journal of Group Tensions, 23*, 43–65.

Jackson, S. (ed.) (1992). *Diversity in the Workplace: Human Resource Initiatives*. New York: Guilford Press.

Jackson, S., and Ruderman, M. N. (eds.) (1995). *Diversity in Work Teams: Research Paradigms for a Changing Workplace*. Washington, DC: American Psychological Association.

James, H. (1907). *The American Scene.* New York: Harper.

Jankowski, M. S. (1995). *The Rising Significance of Status in U.S. Race Relations.* In M. P. Smith and Feagin, J. R. (eds.), *The Bubbling Cauldron: Race, Ethnicity, and the Urban Crisis.* Minneapolis: University of Minnesota Press.

Jaret, C. (1995). *Contemporary Racial and Ethnic Relations.* New York: Harper Collins.

Jaynes, G., and Williams, R. M., Jr. (eds.) (1989). *Common Destiny: Blacks and American Society.* Washington, DC: National Academy Press.

Jenness, V., and Grattet, T. R. (1993). The Criminalization of Hate: The Social Context of Hate Crimes Legislation in the U.S. Paper presented at Annual Meetings of American Sociological Assn., Miami.

Johnson, K. A. (1987). Black and White in Boston. *Columbia Journalism Review.* (May–June):50–52.

Johnson, J. H., and Oliver, M. L. (1990). Economic Restructuring and Black Male Joblessness in U.S. Metropolitan Areas. Los Angeles: UCLA Center for the Study of Urban Poverty.

Johnson, D. W., Johnson, R. T., and Maruyama, G. (1984). Goal Interdependence and Interpersonal Attraction in Heterogeneous Classrooms: A Metaanalysis. In N. Miller and M. B. Brewer (eds.), *Groups in Contact.* New York: Academic Press.

Jones, M. A. (1992). *American Immigration.* Chicago: University of Chicago Press.

Jonsson, M. (1966). Teacher and Pupil Attitudes Toward Busing, Integration, and Related Issues in Berkeley Elementary Schools. Berkeley, CA: Berkeley Unified School District.

Kallen, H. (1924). *Culture and Democracy in the United States.* New York: Arno Press.

Kamenetsky, J., Burgess, G. G., and Rowen, T. (1956). The Relative Effectiveness for Four Attitude Assessment Techniques in Predicting Criterion. *Educational and Psychological Measurement, 16,* 187–194.

Kanter, R. M. (1972). *Commitment and Community: Communes and Utopias in Sociological Perspective.* Cambridge, MA: Harvard University Press.

Kasarda, J. D. (1989). Urban Industrial Transition and the Underclass. *Annals of the American Academy of Political and Social Science, 501,* 26–47.

Katz, D. (1965). Nationalism and Strategies of International Conflict Resolution. In H. C. Kelman (ed.), *International Behavior: A Social-Psychological Analysis.* New York: Holt, Rinehart, and Winston.

Katz, I. (1968). Factors Influencing Negro Performance in the Desegregated Schools. In M. Deutsch, I. Katz, and A. Jensen (eds.), *Social Class, Race, and Psychological Development.* New York: Holt, Rinehart, and Winston.

Katz, I., Wackenhut, J., and Hass, R. G. (1986). Racial Ambivalence, Value Duality, and Behavior. In J. F. Dovidio and S. L. Gaertner (eds.), *Prejudice, Discrimination and Racism.* Orlando, FL: Academic Press.

Katz, I., and Hass, R. G. (1988). Racial Ambivalence and American Value Conflict: Correlation and Priming Studies of Dual Cognitive Structures. *Journal of Personality and Social Psychology, 55,* 893–905.

Kaufman, P., McMillen, M. M., and Bradby, D. (1992). *Dropout Rates in the United States, 1991.* Washington, DC: National Center for Education Statistics.

Keefe, S. E., and Padilla, A. M. (1987). *Chicano Ethnicity.* Albuquerque: University of New Mexico Press.

Kellas, J. G. (1991). *The Politics of Nationalism and Ethnicity.* New York: St. Martin's Press.

Kelley, H. H., and Thibaut, J. W. (1978). *Interpersonal Relations: A Theory of Interdependence.* New York: John Wiley and Sons.

Kelly, C. (1988). Intergroup Differentiation in a Political Context. *British Journal of Social Psychology, 27,* 319–332.

Kelman, H. C. (1975). Interracial Exchanges. In W. C. Coplin and C. Kegley (eds.), *Analyzing International Relations.* New York: Praeger.

Kemal, A. A., and Maruyama, G. (1990). Cross-Cultural Contact and Attitudes of Qatari Students in the United States. *International Journal of Intercultural Relations, 14,* 123–134.

311

Kennedy, M. M., Jung, R. K., and Orland, M. E. (1986). *Poverty, Achievement, and the Distribution of Compensatory Education Services*. Washington, DC: U.S. Department of Education.

Kilson, M., and Cottingham, C. (1991). Thinking About Race Relations: How Far Are We Still From Integration? *Dissent, 38*, 520–530.

Kirschenman, J. (1990). Tales from the Survivors: Business Relocation in a Restructured Urban Economy. Paper presented at Joint Annual Meetings of North Central Sociological Association and Southern Sociological Society. March, Louisville, KY.

Kirschenman, J., and Neckerman, K. M. (1991). "We'd Love to Hire Them, But . . ." The Meaning of Race for Employers. In C. Jencks and P. E. Peterson (eds.), *The Urban Underclass*. Washington, DC: Brookings Institution.

Kitano, H. H. L., and Daniels, R. (1988). *Asian Americans*. Englewood Cliffs, NJ: Prentice-Hall.

———. (1988). *Asian Americans: Emerging Minorities*. Englewood Cliffs, NJ: Prentice Hall.

Kitano, H. H. L., and Young, W. (1982). Chinese Interracial Marriage. In G. A. Cretser and J. J. Leon (eds.), *Intermarriage in the United States*. New York: Haworth Press.

Kluegel, J. R., and Bobo, L. (1993). Opposition to Race-Targeting. *American Sociological Review, 58*, 443–464.

Kluegel, J. R., and Smith, E. R. (1986). *Beliefs About Inequality: American's View of What Is and What Ought To Be*. Hawthorne, NY: Aldine de Gruyter.

Knapp, M. S., and Shields, P. M. (1990). Reconceiving Academic Instruction for Children of Poverty. *Phi Delta Kappan* (June) 753–758.

Kolchin, P. (1994). *American Slavery, 1619–1877*. New York: Hill and Wang.

Koning, H. (1993). *The Conquest of America: How the Indian Nations Lost Their Continent*. New York: Monthly Review Press.

Kosmin, B. A., and Lachman, S. P. (1993). *One Nation Under God: Religion in Contemporary American Society*. New York: Harmony Books.

Kovel, J. (1970). *White Racism: A Psychohistory*. New York: Pantheon.

Kozol, J. (1991). *Savage Inequalities: Children in America's Schools*. New York: Crown.

Kramer, J. H., and Ulmer, J. T. (1996). Sentencing Disparity and Guidelines Departures. *Justice Quarterly, 13*, 81–105.

Kriesberg, L. (1982). *Social Conflicts*, 2d ed. Englewood Cliffs, NJ: Prentice-Hall.

Kuhn, M. H., and McPartland, T. (1954). An Empirical Investigation of Self-Attitudes. *American Sociological Review, 19*, 68–76.

Kymlicka, W. (1995). *Multicultural Citizenship: A Liberal Theory of Minority Rights*. Oxford, England: Clarendon Press.

Lafayette Journal and Courier. (1995). Tear Gas Breaks Up Indy Crowd, July 27, A–4.

LaFree, G. D. (1980). The Effect of Sexual Stratification By Race on Official Reactions to Rape. *American Sociological Review, 45*, 842–854.

Lakin, M., Lomranz, J., and Lieberman, M. A. (1969). *Arab and Jew in Israel*. Washington, DC: NTL-Institute for Applied Behavioral Science.

Lambert, W. E., and Tucker, G. R. (1972). *Bilingual Education of Children: The St. Lambert Experiment*. Rowley, MA: Newbury House.

Lamm, R., and Imhoff, G. (1985). *The Immigration Time Bomb*. New York: Truman Talley.

Landis, D., Hope, R. O., and Day, H. R. (1984). Training for Desegregation in the Military. In N. Miller and M. B. Brewer (eds.), *Groups in Contact*. Orlando, FL: Academic Press.

LaPiere, R. T. (1934). Attitudes vs. Actions. *Social Forces, 13*, 230–237.

Lawler, E. J. (1992). Affective Attachments to Nested Groups: A Choice-Process Theory. *American Sociological Review, 57*, 327–339.

Leacock, E. B. (1969). *Teaching and Learning in City Schools*. New York: Basic Books.

Lemmons, J. S. (1977). Black Stereotypes as Reflected in the Popular Culture, 1880–1920. *American Quarterly, 29*, 102–116.

312

Lenski, G. (1966). *Power and Privilege: A Theory of Social Stratification*. New York: McGraw-Hill.

LeVine, G. N., and Rhodes, C. (1981). *The Japanese-American Community: A Three-Generation Study*. New York: Praeger.

LeVine, L. W. (1996). *The Opening of the American Mind*. Boston: Beacon Press.

LeVine, R. A., and Campbell, D. T. (1972). *Ethnocentrism: Theories of Conflict, Ethnic Attitudes, and Group Behavior*. New York: John Wiley and Sons.

Lewis, A. C. (1991). Washington News: Bilingual Education. *Education Digest* (January), 63–64.

Lewis, R., and St. John, N. (1974). The Contribution of Cross-Racial Friendship to Minority Group Achievement in Desegregated Classrooms. *Sociometry, 37*, 79–91.

Lichter, J. H., and Johnson, D. W. (1969). Changes in Attitudes Toward Negroes of White Elementary School Students After Use of Multiethnic Readers. *Journal of Educational Psychology, 60*, 148–152.

Liebkind, K. (1992). Ethnic Identity—Challenging the Boundaries of Social Psychology. Surrey, England: Surrey University Press.

Lijphart, A. (1977). *Democracy in Plural Societies: A Comparative Exploration*. New Haven, CT: Yale University Press.

Lincoln, C. E., and Mamiya, L. H. (1990). *The Black Church in the African-American Experience*. Durham, NC: Duke University Press.

Lind, M. (1995). *The Next American Nation: The New Nationalism and the Fourth American Revolution*. New York: Free Press.

Lindsey, K. (1995). Race, Sexuality, and Class in Soapland. In G. Dines and J. M. Humez (eds.), *Gender, Race, and Class in Media*. Thousand Oaks, CA: Sage.

Liyama, P., and Kitano, H. H. L. (1982). Asian-Americans and the Media. In G. L. Berry and C. Mitchell-Kerman (eds.), *Television and the Socialization of the Minority Child*. New York: Academic Press.

Loden, M., and Rosener, J. B. (1991). *Workforce America! Managing Employee Diversity as a Vital Resource*. Homewood, IL: Business One Irwin.

Loewen, J. W. (1971). *The Mississippi Chinese: Between Black and White*. Cambridge, MA: Harvard University Press.

Longabaugh, R. (1966). The Structure of Interpersonal Behavior. *Sociometry, 29*, 441–460.

Longshore, D. (1982). School Racial Composition and Blacks' Attitudes Toward Desegregation. *Social Science Quarterly, 63*, 674–687.

Lundberg, S. J. (1994). Equality and Efficiency: Antidiscrimination Policies in the Labor Market. In P. Burstein (ed.), *Equal Employment Opportunity*. New York: Aldine De Gruyter.

Lustick, I. (1994). *Conflict with the Arabs in Israeli Politics and Society*. New York: Garland.

Lyman, S. (1974). *Chinese Americans*. New York: Random House.

MacDonald, H. (1993). The Diversity Industry. *The New Republic,* July 5, 23.

MacDonald, J. F. (1983). *Black and White TV: Afro-Americans in Television Since 1948*. Chicago: Nelson Hall.

MacKenzie, B. K. (1948). The Importance of Contact in Determining Attitudes Toward Negroes. *Journal of Abnormal and Social Psychology, 43*, 417–441.

Mackie, D. M., and Hamilton, D. L. (eds.). (1993). *Affect, Cognition, and Stereotyping*. New York: Academic Press.

Mahard, R. E., and Crain, R. L. (1980). The Influence of High School Racial Composition on the Academic Achievement and College Attendance of Hispanics. Paper presented at annual meeting of American Sociological Society, New York.

Mannheimer, D., and Williams, R. M., Jr. (1949). A Note on Negro Troops in Combat. In S. A. Stouffer, E. A. Suchman, L. C. DeVinney, S. A. Star, and R. M. Williams, Jr. (eds.), *The American Soldier, 1*. Princeton, NJ: Princeton University Press.

Marable, M. (1984). *Race, Reform, and Rebellion: The Second Reconstruction in Black America, (1945–1982)*. Jackson: University Press of Mississippi.

Marger, M. N. (1994). *Race and Ethnic Relations: American and Global Perspectives*, 3d ed. Belmont, CA: Wadsworth.

Marshall, H., and Jiobu, R. (1975). Residential Segregation in United States Cities: A Causal Analysis. *Social Forces, 53*, 449–460.

Martin, P. (1992). *Farm Labor in California: Past, Present, and Future*. Davis, CA: Report for the Farm Worker Services Coordination Council.

Martinez, T. M. (1972). Advertising and Racism. In E. Simon (ed.), *Pain and Promise: The Chicano Today*. New York: Mentor Books.

Martire, G., and Clark, R. (1982). *Antisemitism in the United States*. New York: Praeger.

Maruyama, G., and Miller, N. (1979). Reexamination of Normative Influence Processes in Desegregated Classrooms. *American Educational Research Journal*, 16, 173–283.

Marx, K. (1964). *Selected Works in Sociology and Social Philosophy*. New York: McGraw Hill.

Massey, D. S. (1985). Ethnic Residential Segregation: A Theoretical Analysis and Empirical Review. *Sociology and Social Research, 69*, 315–350.

Massey, D. S., and Denton, N. A. (1988). Suburbanization and Segregation in U.S. Metropolitan Areas. *American Journal of Sociology, 94*, 592–626.

——. (1993). *American Apartheid: Segregation and the Making of the Underclass*. Cambridge, MA: Harvard University Press.

Matthews, D. R., and Prothro, J. W. (1963). Social and Economic Factors and Negro Voter Registration in the South. *American Politician Science Review, 57*, 24–44.

McAll, C. (1990). *Class, Ethnicity and Social Inequality*. Montreal: McGill-Queens University Press.

McConahay, J. B. (1981). Reducing Racial Prejudice in Desegregated Schools. In W. P. Hawley (ed.), *Effective School Desegregation*. Beverly Hills, CA: Sage.

McGarry, J., and O'Leary, B. (1993). The Macro-regulation of Ethnic Conflict. In J. McGarry and B. O'Leary (eds.), *The Politics of Ethnic Conflict Resolution*. London: Routledge.

McGuire, W. J., and Padawer-Singer, A. (1976). Trait Salience in the Spontaneous Self-Concept. *Journal of Personality and Social Psychology, 33*, 743–754.

McGuire, W. J., McGuire, C. V., Child, P., and Fujioka, T. (1978). Salience of Ethnicity in the Spontaneous Self-Concept as a Function of One's Ethnic Distinctiveness in the Social Environment. *Journal of Personality and Social Psychology, 36*, 511–520.

McKay, S., and Pittam, J. (1993). Determinants of Anglo-Australian Stereotypes of the Vietnamese in Australia. *Australia Journal of Psychology, 45*, 17–23.

McLemore, S. D. (1994). *Racial and Ethnic Relations in America*, 4th ed. Boston: Allyn and Bacon.

McPartland, J., and York, R. (1967). Further Analyses of Equality of Educational Opportunity Survey. In U.S. Commission on Civil Rights. *Racial Isolation in the Public Schools*, vol 2. Washington, DC: U.S. Government Printing Office.

Means, B., and Knapp, M. S. (1991). Cognitive Approaches to Teaching Advanced Skills to Educationally Disadvantaged Students. *Phi Delta Kappan* (December), 282–289.

Medina, M., Saldate, M., and Mishra, S. P. (1985). The Sustaining Effects of Bilingual Education: A Follow-up Study. *Journal of Instructional Psychology, 12*, 132–139.

Meer, B., and Freedman, E. (1966). The Impact of Negro Neighbors on White Home Owners. *Social Forces, 45*, 11–19.

Melson, R. (1992). *Revolution and Genocide: On the Origins of the Armenian Genocide and the Holocaust*. Chicago: University of Chicago Press.

Meo, L. M. T. (1965). *Lebanon: Improbable Nation*. Bloomington: Indiana University Press.

Messick, D. M., and Mackie, D. M. (1989). Intergroup Relations. *Annual Review of Psychology, 40*, 45–81.

Messner, S. F., and South, S. J. (1986). Economic Deprivation, Opportunity Structure, and Robbery Victimization. *Social Forces, 64*, 975–991.

Meyer, M. M., and Fienberg, S. E. (1992). *Bilingual Education Strategies*. Washington, DC: National Academy Press.

Michener, H. A., DeLamater, J. D., and Schwartz, S. H. (1990). *Social Psychology*, 2d ed. San Diego: Harcourt Brace Jovanovich.

Miller, L. S. (1995). *An American Imperative: Accelerating Minority Educational Advancement*. New Haven, CT: Yale University Press.

Miller, L. S., Fredisdorf, M., and Humphrey, D. C. (1992). *Student Mobility and School Reform*. New York: Council for Aid to Education.

Miller, N. (1980). Making School Desegregation Work. In W. G. Stephan and J. R. Feagin (eds.), *School Desegregation: Past, Present, and Future*. New York: Plenum Press.

Miller, N., and Brewer, M. B. (1986). Categorization Effects on Ingroup and Outgroup Perception. In J. F. Dovidio and S. L. Gaertner (eds.), *Prejudice, Discrimination, and Racism*. Orlando, FL: Academic Press.

Minard, R. D. (1952). Race Relations in the Pocahontas Coal Field. *Journal of Social Issues, 8*, 29–44.

Miracle, A. W. (1981). Factors Affecting Interracial Cooperation: A Case Study of a High School Football Team. *Human Organization, 40*, 150–154.

Moerman, M. (1968). Being Lue: Uses and Abuses of Ethnic Identification. In J. Helm (ed.), *Essays on the Problems of Tribe*. Seattle: University of Wahington Press.

Molotch, H. (1972). *Managed Integration*. Berkeley: University of California Press.

Monahan, T. P. (1976). An Overview of Statistics on Interracial Marriage in the United States, with Data on its Extent from 1967–1970. *Journal of Marriage and the Family, 38*, 223–231.

Moore, J. W., and Pachon, H. (1970). *Hispanics in the United States*. Englewood Cliffs, NJ: Prentice-Hall.

Morgenthau, T. (1993). America: Still a Melting Pot? *Newsweek*, August 9, 16–23.

Morrill, R. (1965). The Negro Ghetto: Problems and Alternatives. *Geographical Review, 55*, 339–361.

Morris, D. C. (1970). White Racial Orientations Toward Negroes in an Urban Context. Doctoral Dissertation, Ohio State University. Ann Arbor, MI: University Microfilms, No. 70–14, 078.

Morrison, G. A., Jr. (1972). An Analysis of Academic Achievement Trends for Anglo-American, Mexican-American and Negro-American Students in a Desegregated School Environment. Unpublished doctoral Dissertation, University of Houston.

Morrison, E. W., and Herlihy, J. M. (1992). Becoming the Best Place to Work: Managing Diversity at American Express Travel Related Services. In S. E. Jackson and Associates (eds.), *Diversity in the Workplace*. New York: Guilford.

Murdock, S. H. (1995). *An America Challenged: Population Change and the Future of the United States*. Boulder, CO: Westview.

Murguia, E. (1982). *Chicano Intermarriage: A Theoretical and Empirical Study*. San Antonio, TX: Trinity University Press.

Myerson, M., and Banfield, E. C. (1955). *Politics, Planning and Public Interest*. Glencoe, IL: Free Press.

Nagel, J. (1982). The Political Mobilization of Native Americans. *Social Science Journal,19*, 37–45.

——. (1995). Resource Competition Theories. *American Behavioral Scientist, 38*, 442–488.

Nair, K. (1962). *Blossoms in the Dust: The Human Factor in Indian Development*. New York: Praeger.

Nash, G. B. (1974). *Red, White, and Black*. Englewood Cliffs, NJ: Prentice-Hall.

National Center for Education Statistics (1991). *Digest of Education Statistics 1991*. Washington DC: U.S. Government Printing Office.

——. (1994). *Digest of Education Statistics, 1994*. Washington DC: U.S. Government Printing Office.

——. (1995). *Digest of Education Statistics, 1995*. Washington, DC: U.S. Government Printing Office.

Neimeyer, R. A., and Mitchell, K. S. (1988). Similarity and Attraction: A Longitudinal Study. *Journal of Social and Personal Relationships, 5*, 131–148.

Newcomb, T. M. (1961). *The Acquaintance Process*. New York: Holt, Rinehart, and Winston.

Newman, L., and Buka, S. L. (1990). *Every Child A Learner: Reducing Risks of Learning Impairment During Pregnancy and Infancy*. Denver, CO: Education Commission of the States.

Newman, P. (1964). *British Guiana: Problems of Cohesion in an Immigrant Society*. London: Oxford University Press.

Newman, W. M. (1973). *American Pluralism*. New York: Harper and Row.

New York Times (1995a). Review of Recent Supreme Court Decisions. July 2, IV, 1:1.

——. (1995b). Farrakhan's Attack on Jews Unleashes Backlash. October 16, A-10.

——. (1996). High Court Voids Race-Based Plans for Redistricting. June 14, A-1.

Nielson, F. (1985). Toward a Theory of Ethnic Solidarity in Modern Societies. *American Sociological Review*, 50, 133–149.

Nixon, H. L., II. (1979). *The Small Group*. Englewood Cliffs, NJ: Prentice-Hall.

Noel, D. L. (1968). A Theory of the Origin of Ethnic Stratification. *Social Problems, 16*, 157–172.

Northwood, L. K., and Barth, E. A. T. (1965). *Urban Desegregation: Negro Pioneers and Their White Neighbors*. Seattle: University of Washington Press.

Nostrand, R. L. (1992). *The Hispanic Homeland*. Norman: University of Oklahoma Press.

Novak, M. (1971). *The Rise of the Unmeltable Ethnics*. New York: Macmillan.

Oakes, J. (1985). *Keeping Track: How Schools Structure Inequality*. New Haven, CT: Yale University Press.

Oakes, J., Gamoran, A., and Page, R. N. (1992). Curriculum Differentiation: Opportunities, Outcomes, and Meanings. In P. W. Jackson (ed.), *Handbook of Research on Curriculum*. Washington DC: American Educational Research Association.

Oakes, P., Halsam, S. A., and Turner, J. C. (1994). *Stereotyping and Social Reality*. Oxford, England: Blackwell.

O'Brien, R. M. (1987). The Interracial Nature of Violent Crime. *American Journal of Sociology, 92*, 817–835.

Ogbu, J. U. (1990). Overcoming Racial Barriers to Equal Access. In J. I. Goodlad and P. Keating (eds.), *Access to Knowledge: An Agenda for Our Nation's Schools*. New York: The College Board.

Ogbu, J. U., and Matute-Bianchi, M. E. (1986). Understanding Sociocultural Factors: Knowledge, Identity, and School Achievement. In *Beyond Language: Social and Cultural Factors in Schooling Language Minority Students*. Sacramento: California State Department of Education.

Olzak, S. (1983). Contemporary Ethnic Mobilization. *Annual Review of Sociology, 9*, 355–374.

——. (1992). *The Dynamics of Ethnic Competition and Conflict*. Stanford, CA: Stanford University Press.

Omi, M., and Winant, H. (1994). *Racial Formation in the United States*, 2d ed. New York: Routledge.

Orfield, G. (1981). *Toward a Strategy For Urban Integration*. New York: Ford Foundation.

Osborne, R., and Cormack, R. (1991). *Discrimination and Public Policy in Northern Ireland*. Oxford, England: Clarendon Press.

Osgood, C. E., Suci, G., and Tannenbaum, P. (1957). *The Measurement of Meaning*. Urbana: University of Illinois Press.

Padilla, F. M. (1985). *Latino Ethnic Consciousness: The Case of Mexican Americans and Puerto Ricans in Chicago*. Notre Dame, IN: Notre Dame University Press.

Parisi, N., Gottfredson, R., Hindelang, M. J., and Flanigan, T. J. (eds.) (1979). *Sourcebook of Criminal Justice Statistics—1978*. Washington, DC: U.S. Government Printing Office.

Parker, J. H. (1968). The Interaction of Negroes and Whites in an Integrated Church Setting. *Social Forces, 46*, 359–366.

Parman, D. (1994). *Indians and the American West in the Twentieth Century*. Bloomington: Indiana University Press.

Parsons, T. (1951). *The Social System*. New York: Free Press.

Patchen, M. (1970). *Participation, Achievement, and Involvement on the Job*. Englewood Cliffs, NJ: Prentice-Hall.

———. (1975). *The Relation of Inter-racial Contact and Other Factors to Outcomes in the Public High Schools of Indianapolis*. W. Lafayette, IN: Department of Sociology and Anthropology.

———. (1982). *Black-White Contact in Schools: Its Social and Academic Effects*. West Lafayette, IN: Purdue University Press.

———. (1983). Student's Own Racial Attitudes and Those of Peers of Both Races, as Related to Inter-racial Behaviors. *Sociology and Social Research, 68*, 59–77.

———. (1995). Attitudes and Behavior Toward Ethnic Out-groups: How Are They Linked? *International Journal of Group Tensions, 25*, 169–188.

Patterson, O. (1977). *Ethnic Chauvinism: The Reactionary Impulse*. New York: Stein and Day.

Pescosolido, B. A., Grauerholz, E., and Milkie, M. A. (1997). Culture and Conflict: The Portrayal of Blacks in U.S. Children's Picture Books Through the Mid- and Late-Twentieth Century. *American Sociological Review, 62*, forthcoming.

Peterson, R. D., and Hagan, J. (1984). Changing Conceptions of Race: Towards an Account of Anomalous Findings of Sentencing Research. *American Sociological Review, 49*, 56–70.

Peterson, W. (1971). *Japanese Americans*. New York: Random House.

Pettigrew, T. F. (1958). Personality and Sociocultural Factors in Intergroup Attitudes: A Cross-National Comparison. *Journal of Conflict Resolution, 2*, 29–42.

———. (1969). The Negro and Education: Problems and Proposals. In I. Katz and P. Gurin (eds.), *Race and the Social Sciences*. New York: Basic Books.

———. (1971). *Racially Separate or Together?* New York: McGraw-Hill.

———. (1973). Attitudes on Race and Housing. In A. H. Hawley and V. P. Rock (eds.), *Segregation in Residential Areas*. Washington, DC: National Academy of Sciences.

———. (1985). New Black-White Patterns: How Best to Conceptualize Them? *Annual Review of Sociology, 11*, 329–346.

———. (1986). The Intergroup Contact Hypothesis Reconsidered. In M. Hewstone and R. Brown (eds.), *Contact and Conflict in Intergroup Encounters*. Oxford, England: Blackwell.

———. (1996). The Affective Component of Prejudice: Empirical Support for the New View. In S. A. Tuch and J. K. Martin (eds.), *Racial Attitudes in the 1990's: Continuity and Change*. Westport, CT: Praeger.

Pettigrew, T. F., and Cramer, M. R. (1959). The Demography of Desegregation. *Journal of Social Issues, 15*, 61–71.

Pettit, A. G. (1980). *Images of the Mexican American in Fiction and Film*. College Station, TX: Texas A&M Press.

Phinney, J. S. (1990). Ethnic Identity in Adolescents and Adults: Review of Research. *Psychological Bulletin, 108*, 499–514.

Piliavin, I. M., Rodin, J., and Piliavin, J. A. (1969). Good Samaritanism: An Underground Phenomenon. *Journal of Personality and Social Psychology, 13*, 289–299.

Pisko, V. W., and Stern, J. D. (eds.) (1985). *The Condition of Education, 1985 Edition. Statistical Report*. National Center for Education Statistics. Washington, DC: U.S. Government Printing Office.

Piven, F. F., and Cloward, R. A. (1971). *Regulating the Poor: The Functions of Public Welfare*. New York: Pantheon Books.

Policy Information Center. (1989). *What Americans Study*. Princeton, N.J.: Educational Testing Service.

Porter, J. (1965). *The Vertical Mosaic: An Analysis of Class and Power in Canada*. Toronto, Canada: University of Toronto Press.

Prager, J., Longshore, D., and Seeman, M. (eds.) (1986). *School Desegregation Research: New Directions in Situational Analysis*. New York: Plenum Press.

Premdas, R. R. (1993). Balance and Ethnic Conflict in Fiji. In J. McGarry and B. O'Leary (eds.), *The Politics of Ethnic Conflict Regulation*. London: Routledge.

Probert, B. (1978). *Beyond Orange and Green: The Political Economy of the Northern Ireland Crisis*. London: Zed Press.

Pruitt, D. G., and Rubin, J. Z. (1986). *Social Conflict: Escalation, Stalemate, and Settlement*. New York: Random House.

Quillian, L. (1995). Prejudice as a Response to Perceived Group Threat: Population Composition and Anti-Immigrant and Racial Prejudice in Europe. *American Sociological Review, 60,* 586–611.

Quinley, H. E., and Glock, C. Y. (1979). *Anti-Semitism in America.* New York: The Free Press.

Radelet, M. L. (1981) Racial Characteristics and the Imposition of the Death Penalty. *American Sociological Review, 46,* 918–927.

Raley, R. K. (1996). Cohabitation, Marriageable Men, and Racial Differences in Marriage. *American Sociological Review, 61,* 973–983.

Randolph, G., Landis, D., and Tzeng, O. C. S. (1977). The Effects of Time and Practice Upon Culture Assimilator Training. *International Journal of Intercultural Relations, 1,* 105–119.

Reed, M. (1991). *Black Pioneers in White Denominations.* Boston: Skinner House.

Reich, M. (1981). *Racial Inequality: A Political-Economic Analysis.* Princeton, NJ: Princeton University Press.

Reich, R. B. (1991). *The Work of Nations.* New York: Alfred A. Knopf.

Reiss, A. J., Jr. (1971). *The Police and the Public.* New Haven, CT: Yale University Press.

Rhodes, J. (1995). Television's Realist Portrayal of African American Women and the Case of *L.A. Law.* In G. Dines and J. M. Humez (eds.), *Gender, Race, and Class in Media.* Thousand Oaks, CA: Sage.

Richmond, A. H. (1986). Racial Conflict in Great Britain. *Contemporary Sociology, 15,* 184–187.

Ridgeway, C. L. (1983). *The Dynamics of Small Groups.* New York: St. Martin's Press.

Riordan, C. (1978). Equal Status Interracial Contact: A Review and Revision of the Concept. *International Journal of Intercultural Relations, 2,* 161–185.

Rist, R. C. (1970). Student Social Class and Teacher Expectations: The Self-Fulfilling Prophecy in Ghetto Education. *Harvard Educational Review, 40,* 411–451.

Roebuck, J. B., and Murty, K. S. (1993). *Historically Black Colleges and Universities: Their Place in American Higher Education.* Westport, CT: Praeger.

Roediger, D. R. (1991). *The Wages of Whiteness: Race and the Making of the American Working Class.* London: Verso.

Rogers, M., Hennigan, K., Bowman, C., and Miller, N. (1984). Intergroup Acceptance in Classrooms and Playground Settings. In N. Miller and M. B. Brewer (eds.), *Groups in Contact.* New York: Academic Press.

Rokeach, M. (1960). *The Open and Closed Mind.* New York: Basic Books.

Rokeach, M., and Mezei, L. (1966). Race and Shared Belief as Factors in Social Choice. *Science, 151,* 167–172.

Rokkan, S., and Urwin, D. W. (1983). *Economy, Territory, Identity: Politics of West European Peripheries.* London: Sage.

Roof, W. C., and McKinney, W. (1987). *American Mainline Religion: Its Changing Shape and Future.* New Brunswck, NJ: Rutgers University Press.

Rooney-Rebek, P., and Jason, L. (1986). Prevention of Prejudice in Elementary School Students. *Journal of Primary Prevention, 7,* 63–73.

Roosens, E. (1989). *Creating Ethnicity: The Process of Ethnogenesis.* Newbury Park, CA: Sage.

Roosevelt, T. (1923–26). *The Works of Theodore Roosevelt.* 4 vol. New York: C. Scribner's Sons.

Rose, D. L. (1994). Twenty-Five Years Later: Where Do We Stand on Equal Opportunity Law Enforcement? In P. Burstein (ed.), *Equal Employment Opportunity.* New York: Aldine DeGruyter.

Rose, P. I. (1990). *They and We: Racial and Ethnic Relations in the United States,* 4th ed. New York: McGraw-Hill.

Roseman, I. J. (1984). Cognitive Determinants of Emotion: A Structural Theory. In P. Shaver (ed.), *Review of Personality and Social Psychology,* vol. 5. Beverly Hills: Sage.

Rosenbaum, J. E., Popkin, S. J., Kaufman, J. E., and Rusin, J. (1991). Social Integration of Low-Income Black Adults in Middle-Class White Suburbs. *Social Problems, 38,* 448–461.

Rosenberg, M. (1979). *Conceiving the Self.* New York: Basic Books.

Roskies, D. K., and Roskies, D. G. (1975). *The Shtetl Book.* New York: Anti-Defamation League of B'nai B'rith.

Rossell, C. H. (1983). Desegregation Plans, Racial Isolation, White Flight and Community Response. In C. H. Rossell and W. D. Hawley (eds.), *The Consequences of School Desegregation*. Philadelphia, PA: Temple University Press.

———. (1990). *The Carrot or the Stick for School Desegregation Policy: Magnet Schools or Forced Busing*. Philadelphia, PA: Temple University Press.

Rothbart, M. (1993). Intergroup Perception and Social Conflict. In S. Worchel and J. A. Simpson (eds.), *Conflict Between People and Groups*. Chicago: Nelson-Hall.

Rothbart, M., and John, O. P. (1985). Social Categorization and Behavioral Episodes: A Cognitive Analysis of the Effects of Intergroup Contact. *Journal of Social Issues, 41,* 81–104.

Rotheram-Borus, M. J. (1993). Biculturalism Among Adolescents. In M. E. Bernal and G. P. Knight (eds.), *Ethnic Identity: Formation and Transmission Among Hispanics and Other Minorities*. Albany: State University of New York Press.

Royal Commission on Bilingualism and Biculturalism. (1969). *The Work World*. Ottawa, Canada: Queen's Printer, report. vol. 3.

Rudwick, E. M. (1964). *Race Riot at East St. Louis, 1917*. Carbondale: Southern Illinois University Press.

St. John, N. H. (1975). *School Desegregation: Outcomes for Children*. New York: Wiley.

Saltman, J. (1990). *A Fragile Movement: The Struggle for Neighborhood Stabilization*. New York: Greenwood Press.

Sampson, R. J. (1984). Group Size, Heterogeneity, and Intergroup Conflict. *Social Forces, 62,* 618–639.

Sandefur, G. D., and McKinnell, T. (1986). American Indian Intermarriage. *Social Science Research, 15,* 347–371.

Sawyer, J. (1967). Dimensions of Nations: Size, Wealth, and Politics. *American Journal of Sociology, 73,* 145–172.

Schelbert, L. (1970). *New Glarus, 1845–1970. The Making of a Swiss American Town*. Glarus, WI: Komm. Tschudi.

Scherer, K. R. (1988). Cognitive Antecedents of Emotion. In V. Hamilton, G. H. Bower, and J. Frijda (eds.), *Cognitive Perspectives on Emotion and Motivation*. Dordrecht, Netherlands: Kluwer.

Schermerhorn, R. A. (1978). *Comparative Ethnic Relations: A Framework for Theory and Research*. Chicago: University of Chicago Press.

Schlesinger, A. M., Jr. (1992). *The Disuniting of America*. New York: W.W. Norton and Company.

Schmid, C. L. (1981). *Conflict and Consensus in Switzerland*. Berkeley: University of California Press.

Schneider, B., and Lee, Y. (1990). A Model for Academic Success: The School and Home Environment of East Asian Students. *Anthropology and Education Quarterly, 21,* 358–375.

Schneider, M., and Phelan, T. (1990). Blacks and Jobs: Never the Twain Shall Meet? *Urban Affairs Quarterly, 26,* 299–312.

Schoen, R., and Cohen, L. E. (1980). Ethnic Endogamy Among Mexican American Grooms: A Reanalysis of Generational and Occupational Effects. *American Journal of Sociology, 86,* 359–366.

Schofield, J. W. (1982). *Black and White in School: Trust, Tension, or Tolerance*. New York: Praeger.

Schofield, J. W., and Sager, H. A. (1979). The Social Context of Learning in an Interracial School. In R. Rist (ed.), *Inside Desegregated Schools*. San Francisco, CA: Academic Press.

Schuman, H. (1991). Changing Racial Norms in America. *Michigan Quarterly Review* (summer), 460–477.

Schuman, H., and Johnson, M. P. (1976). Attitudes and Behavior. In A. Inkeles, J. Coleman, and N. Smelser (eds.), *Annual Review of Sociology, 2*.

Schuman, H., Steeh, C., and Bobo, L. (1985). *Racial Attitudes in America: Trends and Interpretations*. Cambridge, MA: Harvard University Press.

Schuman, H., and Bobo, L. (1988). Survey-Based Experiments on White Racial Attitudes Toward Residential Integration. *American Journal of Sociology, 94,* 273–299.

319

Schwemm, R. G. (1990). *Housing Discrimination: Law and Litigation.* New York: Clark Boardman.

Sears, D. O. (1988). Symbolic Racism. In P. A. Katz and D. A. Taylor (eds.), *Eliminating Racism: Profiles in Controversy.* New York: Plenum.

Shafir, G. (1995). *Immigrants and Nationalists.* Albany: State University of New York Press.

Shanklin, E. (1994). *Anthropology and Race.* Belmont, CA: Wadsworth.

Sharan, S. (1980). Cooperative Learning in Small Groups: Recent Methods and Effects on Achievement, Attitudes, and Ethnic Relations. *Review of Educational Research, 50,* 241–271.

Sheley, J. F. (1995). *Criminology: A Contemporary Handbook,* 2d ed. Belmont, CA.: Wadsworth.

Sherif, M., Harvey, O. J., White, B. J., Hood, W. R., and Sherif, C. W. (1961). *Intergroup Conflict and Cooperation: The Robber's Cave Experiment.* Norman: University of Oklahoma.

Sherif, M., and Hovland, C. I. (1961). *Social Judgment: Assimilation and Contrast Effects in Communication and Attitude Change.* New Haven, CT: Yale University Press.

Sherwood, J. J., Barron, J. W., and Fitch, H. G. (1969). Cognitive Dissonance: Theory and Research. In R. V. Wagner and J. J. Sherwood (eds.), *The Study of Attitude Change.* Belmont, CA: Brooks/Cole.

Shibutani, T., and Kwan, K. M. (1965). *Ethnic Stratification: A Comparative Approach.* New York: Macmillan.

Shulman, S. (1989). A Critique of the Declining Discrimination Hypothesis. In S. Shulman and W. Darity Jr., (eds.) *The Question of Discrimination: Racial Inequality in the U.S. Labor Market.* Middletown, CT: Wesleyan University Press.

Sigelman, L., and Welch, S. (1993). The Contact Hypothesis Revisited: Black-White Interaction and Positive Racial Attitudes. *Social Forces, 71,* 781–795.

Siguan, M., and Mackey, W. F. (1987). *Education and Bilingualism.* London: Kogan Page.

Simmons, O. G. (1971). The Mutual Images and Expectations of Anglo-Americans and Mexican-Americans. In N. N. Wagner and M. J. Haug (eds.), *Chicanos: Social and Psychological Perspectives.* St. Louis, MO: Mosby.

Simon, J. L. (1990). Immigrants Help the U.S. Economy. In D. L. Bender and B. Leone (eds.), *Immigration: Opposing Viewpoints.* San Diego, CA: Greenhaven Press.

Simpson, G. E., and Yinger, J. M. (1985). *Racial and Cultural Minorities: An Analysis of Prejudice and Discrimination,* 5th ed. New York: Plenum.

Singh, V. (1977). Some Theoretical and Methodological Problems in the Study of Ethnic Identity: A Cross-Cultural Perspective. *New York Academy of Sciences: Annals, 285,* 32–42.

Skardel, D. B. (1974). *The Divided Heart: Scandinavian Immigrant Experience Through Literary Sources.* Lincoln: University of Nebraska Press.

Slavin, R. E. (1983). *Cooperative Learning.* New York: Longman.

———. (1985). Cooperative Learning: Applying Contact Theory in Desegregated Schools. *Journal of Social Issues, 41,* 45–62.

Slavin, R. E., and Madden, N. A. (1979). Social Practices that Improve Race Relations. *American Educational Research Journal, 16,* 169–180.

Sleeter, C., and Grant, C. (1988). *Making Choices for Multicultural Education.* New York: Merrill.

Smith, A. (1979). *Nationalism In the Twentieth Century.* New York: New York University Press.

Smith, A. D. (1987). *The Ethnic Origin of Nations.* Oxford, England: Basil Blackwell.

Smith, A. W. (1981). Racial Tolerance as a Function of Group Position. *American Sociological Review, 46,* 558–573.

Smith, D. A., and Visher, C. A. (1981). Street-Level Justice: Situational Determinants of Police Arrest Decisions. *Social Problems, 29,* 167–177.

Smith, E. (1993). Social Identity and Social Emotions: Toward New Conceptualizations of Prejudice. In D. M. Mackie and D. L. Hamilton (eds.), *Affect, Cognition, and Stereotyping.* New York: Academic Press.

Smith, T. W. (1990). Ethnic Images. GSS Topical Report No. 19. National Opinion Research Center, IL: University of Chicago.

Sniderman, P. (1975). *Personality and Democratic Politics*. Berkeley: University of California Press.

Sniderman, P., and Piazza, T. (1993). *The Scar of Race*. Cambridge, MA: Harvard University Press.

Snipp, C. M. (1989). *American Indians: The First of This Land*. New York: Russell Sage Foundation.

Snyder, M. (1981). On the Self-Perpetuating Nature of Social Stereotypes. In D. L. Hamilton (ed.), *Cognitive Processes in Stereotyping and Intergroup Behavior*. Hillsdale, NJ: Erlbaum.

South African Institute of Race Relations (1992). *Race Relations Survey, 1991–1992*. Johannesburg: South Africa Institute of Race Relations.

South, S. J., and Felson, R. B. (1990). The Racial Patterning of Rape. *Social Forces, 69*, 71–93.

Sowell, T. (1974). Black Excellence: The Case of Dunbar High School. *The Public Interest, 35,* 3–21.

——. (1990). *Preferential Policies: An International Perspective*. Boulder, CO: William Morrow.

——. (1993). Effrontery and Gall, Inc. *Forbes, September 27*, 52.

Spady, W. (1976). The Impact of School Resources on Students. In W. Sewell, R. Hauser, and D. Featherman (eds.), *Schooling and Achievement in American Society*. New York: Academic Press.

Spicer, E. H. (1980). American Indians, Federal Policy Toward. In S. Thernstrom, A. Orlov, and O. Handlin (eds.), *Harvard Encyclopedia of American Ethnic Groups*. Cambridge, MA: Belknap Press.

Spilerman, S. (1970). The Causes of Racial Disturbances: A Comparison of Alternative Explanations. *American Sociological Review, 35*, 627–649.

Spohn, C.S. (1995). Courts, Sentences, and Prisons. *Daedalus, 124*, 119–143.

Stark, R. (1964). Through a Stained Glass Darkly: Reciprocal Protestant-Catholic Images in America. *Sociological Analysis, 25*, 159–166.

Steele, C. (1992). Race and the Schooling of Black Americans. *Atlantic Monthly, April*, 68–78.

Stein, A. A. (1976). Conflict and Cohesion: A Review of the Literature. *Journal of Conflict Resolution, 20*, 143–165.

Steinberg, L., Dornbusch, S. M., and Brown, B. B. (1992). Ethnic Differences in Adolescent Achievement: An Ecological Perspective. *American Psychologist, 47*, 728.

Steinfels, P. (1992). Debating Intermarriage and Jewish Survival. *New York Times, October 18*, 1, 16.

Stephan, W. G. (1978). School Desegregation: An Evaluation of Predictions Made in Brown vs. Board of Education. *Psychological Bulletin, 85*, 217–238.

——. (1987). The Contact Hypothesis in Intergroup Relations. In C. Hendrick (ed.), *Group Processes and Intergroup Relations*. Beverly Hills, CA: Sage.

——. (1991). School Desegregation: Short-Term and Long-Term Effects. In H. J. Knopke, R. J. Norrell, and R. W. Rogers (eds.), *Opening Doors: Perspectives on Race Relations in Contemporary America*. Tuscaloosa: University of Alabama Press.

Stephan, W. G., and Stephan, C. W. (1984). The Role of Ignorance in Intergroup Relations. In N. Miller and M.B. Brewer (eds.), *Groups in Contact*. Orlando, FL: Academic Press.

Stoesz, D., and Karger, H. J. (1992). *Reconstructing The American Welfare State*. Lanham, MD: Rowman and Littlefield.

Stroebe, W., and Insko, C.A. (1989). Stereotypes, Prejudice and Discrimination: Changing Conceptions in Theory and Research. In D. Bar-Tal, C. Graumann, A. Kruglanski, and W. Stroebe (eds), *Stereotyping and Prejudice: Changing Conceptions*. New York: Springer-Verlag.

Struch, H., and Schwartz, S. H. (1989). Intergroup Aggression: Its Predictors and Distinctions from Ingroup Bias. *Journal of Personality and Social Psychology, 56*, 364–373.

Stryker, S. (1981). Symbolic Interactionism: Themes and Variations. In M. Rosenberg and R. H. Turner (eds.), *Social Psychology: Sociological Perspectives*. New York: Basic Books.

Stryker, S., and Serpe, R. T. (1994). Identity Salience and Psychological Centrality. *Social Psychology Quarterly, 57*, 16–35.

Sumner, W. G. (1906). *Folkways*. Lexington, MA: Ginn.

Szymanski, A. (1976). Racial Discrimination and White Gain. *American Sociological Review, 41*, 403–414.

Tajfel, H. (1981). *Human Groups and Social Categories: Studies in Social Psychology*. Cambridge, MA: Cambridge University Press.

———. (1982). Social Psychology of Intergroup Relations. *Annual Review of Psychology, 33*, 1–39.

Tajfel, H., and Turner, J. C. (1986). The Social Identity Theory of Intergroup Behavior. In S. Worchel and W. G. Austin (eds.), *Psychology of Intergroup Relations*, 2d. ed. Chicago: Nelson-Hall.

Takagi, D. Y. (1992). *The Retreat from Race: Asian-American Admissions and Racial Politics*. New Brunswick, NJ: Rutgers University Press.

Takaki, R. (1993). Multiculturalism: Battleground or Meeting Ground. *Annals of the American Academy of Political and Social Science, 530*, 109–121.

———. (1994). The Myth of the "Model Minority." In R. C. Monk (ed.), *Taking Sides: Clashing Views on Controversial Issues in Race and Ethnicity*. Guilford, CT: Dushkin.

Taub, R. P., Taylor, D. G., and Durham, J. D. (1984). *Paths of Neighborhood Change: Race and Crime in Urban America*. Chicago: University of Chicago Press.

Taylor, B. R. (1991). *Affirmative Action at Work: Law, Politics, and Ethics*. Pittsburgh, PA: University of Pittsburgh Press.

Taylor, G. (1979). Housing, Neighborhoods, and Race Relations. *Annals of American Academy of Political and Social Science, 441*, 26–46.

Taylor, P. S. (1934). *An American-Mexican Frontier, Nuecas County, Texas*. Chapel Hill: University of North Carolina Press.

Thernstrom, A. M. (1987). *Whose Vote Counts: Affirmative Action and Minority Voting*. Cambridge, MA: Harvard University Press.

Thiederman, S. (1991). *Bridging Cultural Barriers for Corporate Success*. New York: Lexington Books.

Thomas, W. (1976). *Bail Reform in America*. Berkeley: University of California Press.

Thompson, L. M., and Prior, A. (1982). *South African Politics*. New Haven, CT: Yale University Press.

Ting-Toomey, S. (1981). Ethnic Identity and Close Friendship in Chinese-American College Students. *International Journal of Intercultural Relations, 5*, 383–406.

Tinker, J. N. (1982). Intermarriage and Assimilation in a Plural Society: Japanese Americans in the United States. In G. A. Cretser and J. J. Leon (eds.), *Intermarriage in the United States*. New York: Haworth Press.

Tobin, G. A. (1987). *Divided Neighborhoods: Changing Patterns of Residential Segregation*. Newbury Park, CA: Sage.

Tomaskovic-Devey, D., and Roscigno, V. J. (1996). Racial Economic Subordination and White Gain in the U.S. South. *American Sociological Review, 61*, 565–589.

Tonry, M. (1993). Sentencing Commissions and their Guidelines. *Crime and Justice: A Review of Research, 17*, 137–195.

Trejo, A. (ed.) (1979). *The Chicanos: As We See Ourselves*. Tucson: University of Arizona Press.

Trew, K. (1986). Catholic-Protestant Contact in Northern Ireland. In M. Hewstone and R. Brown. (eds.), *Contact and Conflict in Intergroup Encounters*. Oxford, England: Basil Blackwell.

Triandis, H. C. (1975). Culture Training, Cognitive Complexity, and Interpersonal Attitudes. In R. Brislin, S. Bochner, and W. Lonner (eds.), *Cross-Cultural Perspectives on Learning*. Beverly Hills, CA: Sage.

Tucker, M. B., and Mitchell-Kernan, C. (1990). New Trends in Black Interracial Marriage: The Social Structure Context. *Journal of Marriage and the Family, 52*, 209–218.

Turner, B. M., and Wilson, W. J. (1976). Dimensions of Racial Ideology: A Study of Urban Black Attitudes. *Journal of Social Issues, 32*. 139–152.

Turner, B. M., Lovell, R. D., Young, J. C., and Denny, W. F. (1986). Race and Peremptory Challenges During Voire Dire: Do Prosecution and Defense Agree? *Journal of Criminal Justice, 14*, 61–69.

Turner, B. M., Hogg, M. A., Oakes, P. J., Reicher, S. D., and Wetherell, M. (1987). *Rediscovering the Social Group: A Self-Categorizaiton Theory*. Oxford, England: Blackwell.

Unneever, J. D., and Hembroff, L. A. (1988). The Prediction of Racial/Ethnic Sentencing Disparities: An Expectation States Approach. *Journal of Research in Crime and Delinquency, 25*, 53–82.

Useem, E. L. (1991). Tracking Students Out of Mathematics. *Education Digest, May*, 54–58.

U.S. Bureau of the Census (1991). *The Hispanic Population in the United States, March 1991*. Washington, DC: U.S. Government Printing Office.

———. (1992a). Marital Status and Living Arrangements: March 1992. Washington, DC: U.S. Government Printing Office.

———. (1992b). *The Asian and Pacific Islander Population in the United States, March 1991 and 1990*. Washington, DC: U.S. Government Printing Office.

———. (1993a). *The Hispanic Population in the United States, March 1992*. Washington, DC: U.S. Government Printing Office.

———. (1993b). *Statistical Abstract of the United States, 1993*. Washington, DC: U.S. Government Printing Office.

———. (1994a). *Statistical Abstract of The United States, 1994*. Washington, DC: U.S. Government Printing Office.

———. (1994b). *The Hispanic Population in the United States: March 1993*. Washington, DC: U.S. Government Printing Office.

———. (1995). *The Black Population in the United States, March 1994 and 1993*. Washington, DC: U.S. Government Printing Office.

U.S. Commission on Civil Rights (1967). *Racial Isolation in the Public Schools*. Washington, DC: U.S. Government Printing Office.

———. (1976). *Fulfilling the Letter and Spirit of the Law: Desegregation of the Nation's Schools*. Washington, DC: U.S. Government Printing Office.

U.S. Department of Justice (1990). Sourcebook of Criminal Justice Statistics, 1990. Washington, DC: U.S. Government Printing Office.

———. (1994). Crime in the United States, 1993: Uniform Crime Reports. Washington, DC: U.S. Government Printing Office.

U.S. National Advisory Commission on Civil Disorders (1968). Report of the National Advisory Commission on Civil Disorders. New York: Bantam Books.

Utley, R. M. (1984). *The Indian Frontier of the American West, 1846–1890*. Albuquerque: University of New Mexico Press.

Valdez, A. (1983). Recent Increases in Intermarriage by Mexican-American Males. *Social Science Quarterly, 64*, 136–144.

Valenzuela De La Garza, J., and Medina, M. (1985). Academic Achievement as Influenced by Bilingual Instruction for Spanish-Dominant Mexican-American Children. *Hispanic Journal of Behavior Science, 7*, 247–259.

van den Berghe, P. L. (1978). *Race and Racism: A Comparative Perspective*, 2d ed. New York: Wiley.

Van Fossen, B. E. (1968). Variables Related to Resistance to Desegregation in the South. *Social Forces, 47*, 39–74.

Vega, M. M., and Greene, C. Y. (1993). *Voices from the Battlefront: Achieving Cultural Equity*. Trenton, NJ: Africa World Press.

Wagner, U., and Machleit, U. (1989). Contact and Prejudice Between Germans and Turks: A Correlation Study. *Human Relations, 42*, 561–574.

Walker, S. (1993). *Taming the System: The Control of Discretion in Criminal Justice, 1950–1990*. New York: Oxford University Press.

Wall Street Journal. (1996). "How Did Everyone Vote?" November 7; A–17.

Walzer, M. (1992). *What It Means To Be An American*. New York: Marsilio.

Ward, G. C. (1994). *Shadow Ball: The History of the Negro Leagues*. New York: Knopf.

Warner, L. G., and DeFleur, M. L. (1969). Attitude as an Interactional Concept: Social Constraint and Social Distance as Intervening Variables Between Attitudes and Action. *American Sociological Review, 34*, 153–169.

Waters, M. C. (1990). *Ethnic Options: Choosing Identities in America*. Berkeley: University of California Press.

Weber, G. (1971). *Inner City Children Can Be Taught to Read: Four Successful Schools*. Washington, DC: Center for Basic Education.

Weber, R., and Crocker, J. (1983). Cognitive Processes in the Revision of Stereotypic Beliefs. *Journal of Personality and Social Psychology, 45*, 961–977.

Weigel, R., and Howes, P. W. (1985). Conceptions of Racial Prejudice: Symbolic Racism Reconsidered. *Journal of Social Issues, 41*, 117–138.

Weigel, R., Wiser, P. L., and Cook, S. W. (1975). The Impact of Cooperative Learning Experiments in Cross-Ethnic Relations and Attitudes. *Journal of Social Issues, 31*, 219–244.

Weinberg, M. (1977). *Minority Students: A Research Appraisal*. Washington, DC: National Institute of Education.

Weiner, M. (1978). *Sons of the Soil: Migration and Ethnic Conflict in India*. Princeton: Princeton University Press.

Weingrod, A. (1965). *Israel: Group Relations in a New Society*. New York: Praeger.

Weinstein, R. S., Marshall, H. H., Sharp, L., and Botkin, M. (1987). Pygmalion and the Student: Age and Classroom Differences in Children's Awareness of Teacher Expectations. *Child Development, 58*, 1090–1091.

Wellisch, J. B., Marcus, A., MacQueen, A., and Duck, G. (1976). *An In-depth Study of Emergency School Aid Act (ESAA) Schools: 1974–1975*. Washington, DC: System Development Corporation.

Wertheimer, M. (1950). The Laws of Organization of Perceptual Forms. In W. D. Ellis (ed.), *A Sourcebook of Gestalt Psychology*. London: Routledge and Kegan Paul.

West, C. (1993). *Race Matters*. Boston: Beacon Press.

Weyr, T. (1988). *Hispanic U.S.A.: Breaking the Melting Pot*. New York: Harper and Row.

Wheeler, W. (1986). *An Analysis of Social Change in a Swedish-Immigrant Community: The Case of Lindsborg, Kansas*. New York: AMS Press.

Wicker, A. W. (1969). Attitudes vs. Action: The Relationship of Verbal and Overt Behavioral Responses to Attitude Objects. *Journal of Social Issues, 25*, 41–78.

Wiggins, J. A., Wiggens, B. B. and VanderZanden, J. (1994). *Social Psychology*, 5th ed. New York: McGraw-Hill.

Wilkerson, I. (1991). Interracial Marriage Rises, Acceptance Lags. *The New York Times*, December 2, Al.

Williams, N. (1990). *The Mexican-American Family: Tradition and Change*. Dix Hills, NY: General Hall.

Williams, R. M., Jr. (1964). *Strangers Next Door*. Englewood Cliffs, NJ: Prentice-Hall.

Willig, A. (1987). Meta-Analysis of Selected Studies in the Effectiveness of Bilingual Education. *Review of Educational Research, 57*, 351–362.

Wilner, D. M., Walkley, R. P., and Cook, S. W. (1955). *Human Relations in Interracial Housing*. Minneapolis: University of Minnesota Press.

Wilson, C. C., II, and Gutierrez, F. (1995). *Race, Multiculturalism, and the Media: From Mass to Class Communication*, 2d ed. Thousand Oaks, CA:Sage.

Wilson, F. D. (1985). The Impact of School Desegregation Programs on White Public School Enrollment, 1968–1976. *Sociology of Education, 58*, 137–153.

Wilson, P. J. (1967). *A Malay Village and Malaysia: Social Values and Rural Development*. New Haven, CT: HRAF Press.

Wilson, W. J. (1973). *Power, Racism, and Privilege*. New York: Free Press.

——. (1978). *The Declining Significance of Race: Blacks and Changing American Institutions*. Chicago: University of Chicago Press.

——. (1987). *The Truly Disadvantaged: The Inner City, the Underclass, and Public Policy*. Chicago: University of Chicago Press.

Winter, G. (1962). *The Suburban Captivity of the Churches*. Garden City, NY: Doubleday.

Woodmansee, J. J., and Cook, S. W. (1967). *Dimensions of Verbal Racial Attitudes: Their Identification and Measurement*. Journal of Personality and Social Psychology, 7, 240–250.

Woodward, C. V. (1957). The Strange Career of Jim Crow. New York: Oxford University Press.

Worchel, W. (1979). Cooperation and the Reduction of Intergroup Conflict: Some Determining Factors. In W. G. Austin and S. Worchel (eds.), *The Social Psychology of Intergroup Relations*. Monterey, CA: Brooks/Cole.

Works, E. (1961). The Prejudice-Interaction Hypothesis from the Point of View of the Negro Minority Group. *American Journal of Sociology, 67*, 47–52.

Wortman, P. M., and Bryant, F. B. (1985). School Desegregation and Black Achievement: An Integrative Review. *Sociological Methods and Research, 13*, 289–324.

Wright, G. C. (1977). Contextual Models of Electoral Behavior: The Southern Wallace Vote. *American Political Science Review, 71*, 497–508.

Yarrow, M. R., Campbell, J. D., and Yarrow, L. J. (1958). Interpersonal Change: Process and Theory. *Journal of Social Issues, 14*, 60–62.

Yinger, J. M. (1985). Assimilation in the United States: The Mexican Americans. In W. Conner (ed.), *Mexican Americans in Comparative Perspective*. Washington DC: Urban Institute Press.

Zak, I. (1973). Dimensions of Jewish-American Identity. *Psychological Reports, 33*, 891–900.

Zander, A., Stotland, E., and Wolfe, D. (1960). Unity of Group, Identification with Group and Self-Esteem of Members. *Journal of Personality, 28*, 463–478.

Zanna, M. P., and Rempel, J. K. (1988). Attitudes: A New Look at an Old Concept. In D. Bar-Tal and A. W. Kruglanski (eds.), *The Social Psychology of Knowledge*. Cambridge: Cambridge University Press.

Ziegler, D. W. (1993). *War, Peace, and International Politics*, 6th ed. New York: HarperCollins.

Zirkel, P. A. (1971). Self-Concept and the Disadvantage of Ethnic Group Membership and Mixture. *Review of Educational Research, 41*, 211–225.

Zuel, C. R., and Humphrey, C. R. (1971). The Integration of Black Residents in Suburban Neighborhoods. *Social Problems, 18*, 462–474.

Index

327